HANS REICHENBACH

SELECTED WRITINGS: 1909–1953

VOLUME TWO

VIENNA CIRCLE COLLECTION

VOLUME 4

EDITOR: ROBERT S. COHEN

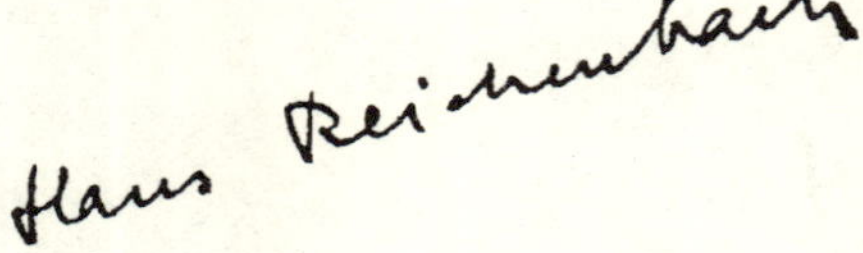

HANS REICHENBACH AT U.C.L.A.

HANS REICHENBACH

SELECTED WRITINGS
1909-1953

VOLUME TWO

Principal Translations by

ELIZABETH HUGHES SCHNEEWIND

Edited by

MARIA REICHENBACH *and* ROBERT S. COHEN

D. REIDEL PUBLISHING COMPANY
DORDRECHT : HOLLAND / BOSTON : U.S.A.
LONDON : ENGLAND

Library of Congress Cataloging in Publication Data

CIP

Reichenbach, Hans, 1891–1953.
Selected writings, 1909–1953.

(Vienna circle collection; v. 4)
"Bibliography of writings of Hans Reichenbach": v. 1, p.
Includes index.
1. Philosophy–Addresses, essays, lectures. 2. Science–Philosophy–Addresses, essays, lectures. 3. Physics–Philosophy–Addresses, essays, lectures. 4. Social problems–Addresses, essays, lectures. I. Reichenbach, Maria. II. Cohen, Robert Sonné. III. Series.
B29. R423 1978 191 78-18446
ISBN 90-277-0291-8 (vol. 1) ISBN 90-277-0292-6 (vol. 1) pbk.
ISBN 90-277-0909-2 (vol. 2) ISBN 90-277-0910-6 (vol. 2)
ISBN 90-277-0892-4 (set) ISBN 90-277-0893-2 (set)

Principal translations from the German by Elizabeth Hughes Schneewind, further translations by Laurent Beauregard (Article 50), and Maria Reichenbach (Articles 44, 45, 55, and 56)
Translations edited by M.R. and R.S.C.

Published by D. Reidel Publishing Company,
P.O. Box 17, Dordrecht, Holland

Sold and distributed in the U.S.A., Canada, and Mexico
by D. Reidel Publishing Company, Inc.
Lincoln Building, 160 Old Derby Street, Hingham,
Mass. 02043, U.S.A.

Printed in The Netherlands

TABLE OF CONTENTS

CONTENTS OF VOLUME ONE

PUBLISHER'S NOTE

The publishers of this volume wish to thank the following publishers for permission to include translations or reprints of the respective articles.

Full details can be found in the Bibliography of Writings of Hans Reichenbach at the end of the book.

Part V

art. 44, 45	
" 55	– Routledge & Kegan Paul, London
" 56	
" 47	– Bayerische Akademie der Wissenschaften, Munich
" 48	– Springer-Verlag, Berlin-Heidelberg-New York
" 49	– *Zeitschrift für Naturforschung* (Tübingen)
" 50	– Gauthier-Villars, Paris

Part VI

art. 52a, b, 53	– Springer-Verlag, Berlin-Heidelberg-New York
" 54, 57 58	– Felix Meiner, Leipzig

PART V

PHILOSOPHY OF PHYSICS

44. THE PRESENT STATE OF THE DISCUSSION ON RELATIVITY*

A Critical Investigation

[1922f]

[A few pages have been omitted dealing with Herman Weyl's generalization of Riemannian space, and with some minor criticisms of the theory of relativity which are of no historical significance. – M.R.]

INTRODUCTION

In the years during which the theory of relativity has been the subject of philosophical discussion, a considerable number of publications concerning it have appeared. It would therefore seem appropriate to summarize the philosophical viewpoints expressed and the conclusions arrived at in these investigations. Of course, one cannot hope to distil one final judgment or one common interpretation from such a variety of works. The same reasons which make it impossible to unify the various philosophical systems also make it impossible to unify the various treatments of a specific problem in the philosophy of science, for these treatments have usually been mere projections of their author's philosophical position. It is obvious that each philosophical school selects from the theory of relativity exactly those philosophical issues that appear interesting from its particular vantage point, and either accepts, rejects, or 'interprets' the physical theory depending on whether or not the theory agrees with the doctrines of the school. Such an attitude is not fruitful for an understanding of the philosophical content of a new physical discovery. Moreover, philosophical analysis meets with difficulties in this case because a much greater knowledge of physics is required than has commonly been necessary for studies in the philosophy of science. The adequacy of any philosophical criticism of the theory of relativity therefore depends on the author's grasp of the *physical* content of the theory, and on his understanding of the physical significance of particular assertions. This accounts for the amazing variety of opinions, the violent controversies about the importance, consistency, and scientific applicability of Einstein's discoveries. The kaleidoscopic chaos of opinions is heightened, in turn, by the participation of the general public in this controversy.

Under these circumstances, any attempt to impartially review the present state of the discussion would be a rather useless enterprise because no coherent

*From *Modern Philosophy of Science: Selected Essays,* ed. and tr. by Maria Reichenbach, Routledge & Kegan Paul, London, 1959, pp. 1–45. Copyright © Maria Reichenbach 1959 except in U.S. Copyright © in U.S. by Maria Reichenbach 1959.

picture can be constructed from the different formulations. Everything is still quite unsystematic; dogmatic understanding is found next to clear insight, and the voice of the amateur has not yet been stilled. It almost seems as if Einstein's great physical discovery might split philosophy even further; convictions about the most elementary functions of knowledge become irreconcilable at a time when their adherents should humbly learn from this new development in science. Let us, therefore, try to critically examine these writings, separating the important from the unimportant contributions, correcting the mistakes, and exposing the fundamental philosophical problems arising from Einstein's discovery.

We must admit that such a presentation itself constitutes a point of view, but that is a fate which no productive work can escape. However, this need not prevent us from making new attempts at arriving at *objective* statements by means of analysis; there is no other way. The following critical presentation may therefore be regarded not only as a review of, but also as a participation in the discussion. Later developments will show to what extent our criticism is justified. I am convinced that such participation is useful, even necessary, because the number of obvious misunderstandings is so large, and it is futile for some philosophers to attack the theory so violently simply because they do not fully understand its physical content. I rather believe that it will be possible to bring about agreement on many questions in a relatively short time. It is the aim of the present essay to contribute to such a clarification.

In our presentation, we shall have to discuss the physical significance of the theory of relativity, and even specific physical questions, but this will always be done from the viewpoint of philosophical analysis, and merely physical problems will be excluded. Therefore, the purely factual assertions of the theory will not be criticized; that may be left to physics, and for the present there is no other course but to accept these assertions, subject to possible later modifications. It will be one of our problems, however, to discover what empirical assertions the theory does, in fact, contain. This question has not yet been clarified sufficiently, and even the physicist usually loses his bearings when he is expected to give a clear answer. Here is the source of most of the philosophical misunderstandings. In this respect, we hope that the following presentation will clarify the discussion. We shall closely adhere to the physical theory.

A presentation of the theory itself is not necessary. The literature in this field is extensive enough, and there are some excellent expositions of the theory. In addition to the original publications by Einstein and the outstanding works by Weyl (69),* v. Laue (31), Kopff (24), and Pauli (38), which are accessible only to the expert, there are popular presentations by Einstein (10), Bloch (2), Freundlich (17), Schlick (56), Born (4),[1] Thirring (65), and the short and well-organized article by Sommerfeld (64).

The philosophical works treated below do not exhaust the literature – that is impossible – but I presume that I have not missed any of the more important works published in recent years. Unfortunately, I had no access to the philosophical literature of foreign countries. A bibliography of the publications mentioned in the text will be found at the end of this essay.

1. CONCEPTIONS INFLUENCED BY VAIHINGER

It is true that the theory of relativity sometimes makes use of fictions. This is the reason why Vaihinger's philosophy of 'As If' seemed especially suited to a clarification of the philosophical problems of Einstein's theory. Vaihinger himself did not participate in these discussions, but his adherents did. At a meeting that took place in Halle in 1920, their attitude was negative throughout; Kraus, who has become the mouthpiece of this school in regard to relativity, is especially critical of the theory.

What are fictions? According to Vaihinger, they are "mental structures" (68, p. 12) "which do not directly correspond to reality" (68, p. 17). In a certain sense, all conceptual constructions can be called fictions because all of them are mental structures that may never be construed as 'mirrors' of reality; they are determined by the nature of our thought processes and are nothing but tools for the understanding of reality. However, the latter view corresponds neither to Vaihinger's terminology nor to that of the recent fictionist school. Vaihinger expressly (and rightly) distinguishes fiction from hypothesis: "Whereas every hypothesis seeks to be an adequate expression of some reality still unknown and to mirror this objective reality, the fiction is advanced with the consciousness that it is an inadequate subjective and pictorial manner of conception, whose coincidence with reality is, from the start, excluded and which cannot, therefore, be afterwards verified, as we hope to be able to verify an hypothesis" (68, p. 268). Thus there are two classes of mental structures used in acquiring knowledge; one of them has a

* [The numbers in the text refer to the bibliography at the end of this essay. In the case of the author's works, we also cite Reichenbach's bibliography number in square brackets. – Ed.]

certain relation to reality ('description' is a better term in this context than the word 'mirror' used by Vaihinger), the other one has no relation to reality, and is only a conceptual tool. The question arises: which assertions of the theory of relativity are mere fictions?

Obviously, it is a fiction when Einstein speaks of an observer who sets his watch at the arrival of a light signal. It is impossible to perform this operation because the error which would result from the faulty sensory reaction of the observer would be much larger than the correction in the setting of the watches required by relativity theory. Although popular presentations of the theory make frequent use of such fictions, their purely pedagogical and heuristic purpose is apparent. Such fictions can easily be eliminated, but this fact is so obvious to the expert that, instead of avoiding them, he often uses them successfully.

For the same reason, such fictions are not the concern of the philosophy of 'As If' since no philosophical analysis is needed to discover them. The fictionist school maintains, rather, that the theory of relativity contains different, more fundamental fictions whose fictitious character, far from being obvious, is not known even to the physicists, including Einstein himself.

We shall turn to a discussion of Kraus' views. Basic to the theory of relativity is Einstein's theory of the measurements of spatial and temporal intervals; Kraus calls this theory a fiction. He declares: "One can interpret the Lorentz contraction only as a mathematical consequence resulting from certain fictitious measuring operations; more precisely: one obtains the value of the Lorentz contraction when one figures out what the results of certain measuring operations would be. Actually, such a measurement has never been performed. Thus the calculation of measuring results under certain fictitious conditions is confused with actual measurement" (26, p. 359). These words contain Kraus' main ideas, and it will be worth while to examine them critically. We admit that a direct measurement according to Einstein's methods has never been performed; but the *direct* measurement is certainly one of those trivial fictions which we excluded from consideration on the grounds of being obvious. Kraus overlooks the possibility of arriving at statements about the behaviour of measuring rods and clocks without making *direct* measurements with these objects. The Michelson experiment shows that there is a certain connection between rigid bodies and the velocity of light; from such experiments we can infer how rigid rods would behave if measurements were performed. Kraus is mistaken when he infers that, according to the theory of relativity, only those bodies accidentally used as measuring instruments would undergo the Einsteinian changes (26, p. 358). On the contrary, for the theory of relativity every rigid rod is a measuring rod and the theory has empirical evidence for its assertions. Of course, these assertions contain *hypotheses* since our inferences are only indirect, but that does not make them *fictions*.

For Einstein, clocks, like measuring rods, are physical things, not fictions, and he has suggested experiments designed to test the retardation of clocks empirically.[2] Einstein's assertions might be *false*, but this is an empirical question. Kraus does not notice that *his* statements about measuring rods and clocks are also empirical, and therefore may be false. Let two clocks be synchronized and placed next to each other; let them then be transported at different velocities to a distant point. According to Einstein, the faster travelling clock will be retarded compared to the other. This is a factual statement that may be true or false. Direct experimental confirmation of the statement has not been possible so far, but that does not render it a *fiction*, only an *hypothesis*. How does Kraus know that Einstein's retardation of clocks does *not* occur? The assertion that it does not occur is also only an hypothesis.

It turns out that Kraus himself confuses fiction and hypothesis, that he is even less certain about this distinction than the physicists. It is true that Einstein's theory contains statements that cannot be confirmed empirically; we shall discuss them in Section 4. But the statements mentioned by Kraus are empirical. He calls Einstein's law of the constancy of light a fiction; yet this assertion is essentially a factual statement. Kraus writes: "The statement that the relative velocity of light, unaffected by the motion of the light source, remains the same with respect to every system which is in rectilinear uniform motion is false. It does not violate our thinking habits, but violates *a priori* necessary judgments" (26, pp. 363–4). Yet this statement asserts only the following simple fact: If I use the same method and the same measuring rods and clocks to measure the velocity of light in different systems, the result is always 300,000 km/sec. This is a statement about the relation of rigid rods and clocks to the propagation of light, and it is to be tested in experimental physics. Physics has not yet fully confirmed this statement but considers it highly probable. *A priori*, we can know as little about this relation as we can about the ability of light to penetrate matter. Kraus does not see that his polemics are directed against a simple and intuitively plausible factual assertion. As so often in the history of philosophy, we are faced with the strange case that a philosopher wants to prescribe on the basis of 'self-evident' knowledge what phenomena the physicists can observe. When Kraus calls Einstein's light principle a *fiction*, he maintains that a measurement with rigid rods and clocks in a moving system will *not* yield 300,000 km/sec as the velocity of light. On the basis of an apriorist philosophy, he wants to assert something about the behaviour of physical things; he wants to *deduce* physics from philosophy. In the century since Kant's death, philosophy should have outgrown this point of view. Regardless of whether or not the light principle is factually true and regardless of whether future experiments will confirm this tentative principle, Kraus' criticism must be firmly rejected; it is impermissible for philosophers to disregard the limits of knowledge.

How little Kraus understands the physical significance of Einstein's light principle is manifested by his statement that, according to this principle, $c + v = c$ (26, p. 368). This equation is indeed a contradiction. The addition of velocities constitutes a *physical* process and it cannot be demonstrated, in any way, that this addition is represented by the above equation (cf. 49 [1921f]). Given a system in which the measured velocity of light is c, and given a second system moving with the velocity v, what value is obtained for the velocity of a light ray in the first system if it is measured by clocks and rigid rods which have been transported to the moving system? This is the problem of the addition of velocities; the answer depends on the behaviour of clocks and measuring rods. It is indeed a problem of *combining* velocities, but the operation can be called 'addition' only in a generalized sense. *Algebraic* addition is only one way of combining quantities, and what mathematical operation applies to the physical combination of velocities is an empirical question. The equation should rather be written in the following way:

$$c\,(+)\,v = c$$

where the plus sign in parentheses stands for addition in the generalized sense.[3] The special case of the addition of velocities in relativity theory has been formulated mathematically (i.e. it can be reduced to algebraic addition). Einstein achieved this formulation by means of his addition theorem

$$\frac{u + v}{1 + \dfrac{u \cdot v}{c^2}} = w$$

This formula is the interpretation of the plus sign in parentheses for the addition of velocities in relativity theory. As can easily be seen, $w = c$ if $u = c$, i.e. it yields the physically required result.

In a more recent work, Kraus reproaches Einstein for confusing the *means* of description with the *object* of description. He identifies the 'classical principle of the equivalence of co-ordinates' with the Newtonian principle of relativity according to which the acceleration of a body is the same in all uniformly moving inertial systems. For Einstein, the special principle of relativity is the assertion that, relative to these systems, the same laws of nature hold. Kraus continues with the following criticism: "With these statements, Einstein confuses himself and his readers. The statement does not express Einstein's special principle of relativity, but merely the classical principle of the equivalence of co-ordinates, which has always been maintained on the tacit, but obvious, assumption of the *invariance of the units of measurement*. Einstein fails to mention the *fundamental change in this assumption* which he himself introduced" (27, p. 472). In the first place, this criticism by Kraus contains a factual error. It is not true that, according to Newton's principle of relativity, the same laws of nature hold in all inertial systems; only the same laws of mechanical acceleration do. The propagation of light, for instance, is different for all these systems. Einstein's assumption, which also applies to the propagation of light, is therefore *different* from that of Newton. It is not Einstein who is wrong, but Kraus. Secondly, Kraus' criticism is not clear on logical grounds. He speaks of the invariance of the units of measurement without defining this concept. He overlooks the fact that there are no other means of comparing measuring units than by physical processes. What does it mean to say that a meter in one system has the same length as a meter in another system? No physical application of this statement is possible except by defining 'one meter' by reference to a physical thing. The simplest definition is to say that the same stick should be called a meter no matter where it is located. Other definitions, for instance by reference to light waves, do not differ in principle. Every statement about the value of a velocity is, therefore, a statement about the relation of *two* physical things, the moving object and the object defining the unit. It is Einstein's great merit to have deliberately reduced all physical measuring values to such *correlated values of physical things* (cf. 48 [1920a]). His light principle is also a statement about such a relation. It is a matter of taste whether one wants to call the transported unit *changed*. It is impossible to measure this change objectively because the result of such a measurement depends upon a previously given definition of congruence. It seems to be expedient to say that the same measuring rod always has the same length.

Now we can understand Kraus' objection that Einstein has confused the means and the objects of description. Kraus believes that Einstein's light principle is a statement about changes in measuring rods, not a statement about light. But now Kraus' mistake becomes obvious; there are no empirical statements about measuring rods alone or about light alone, but only about the *relation* between light and measuring rods; Einstein's light principle is such a statement.

In addition to attacking the light principle, Kraus objects to Einstein's definition of simultaneity. He contends that it violates the logical law of non-contradiction (26, p. 336). This erroneous and short-sighted conception has been corrected repeatedly, by Linke (34, p. 413), by Thirring (66, 67), and by me (50 [1921g]). I shall therefore only point out that, in the theory of relativity, simultaneity (incidentally, only simultaneity at *different* places, not at the *same* place) is a relative concept, like right and left, and thus no principle of logic is violated. The only problem is whether one may properly regard simultaneity as a relative concept, and this is not a logical, but an epistemological problem. Elsewhere (28), Kraus regards it as *self-evident* that simultaneity is *not* a relative concept. But the problem cannot be solved by such a declaration; obviously, there are many people to whom it is *not* evident, to whom, on the contrary, the relativity of simultaneity is evident. Kant introduced a new requirement for synthetic *a priori* principles, he showed that such principles cannot simply be taken for granted but must be demonstrated to be *conditions of experience*. Only then can they be regarded as objective truths. This requirement enables us to replace assumptions based on untutored experience by scientifically founded assertions. Kraus nowhere attempts to demonstrate that absolute simultaneity is a condition of experience. Of course, such an attempt would be in vain since the theory of relativity has shown that experience is possible on the basis of the relativity of simultaneity. The assumption of absolute simultaneity is thus based on *primitive* evidence, and this assumption can easily be shown to be unsatisfactory.

Although Kraus believes that Einstein's simultaneity is impossible, he has no better objections against it than those we have mentioned. It is very strange that nevertheless he retains relative simultaneity as a useful fiction. If Einstein's conception of simultaneity were *false*, the physicist would certainly not accept it even as a fiction. Indeed, we cannot require that simultaneity be defined *correctly* because it is not an *hypothesis*. But it is not a *fiction* either; there is a third possibility. We shall pursue this problem in Section 4.[4]

Lipsius (35) has also maintained the fictional character of Einstein's theory in a lecture at the 'As If' Congress. His statements, however, contain so many misunderstandings of the physical theory that we can dispense with a detailed criticism. He contends that the theory of relativity contradicts Maxwell's theory since it denies the existence of a material ether. Apparently, he does not notice that this remark is only *historically*, not *logically* correct; logically, a wave theory of light without a material elastic medium is quite tenable. He denies, furthermore, that "it is epistemologically feasible to assign the same logical status to the time co-ordinate as to the three spatial co-ordinates" (35, p. 441). He does not realize that, in fact, physics does *not* treat them alike; the negative sign of the time parameter in ds^2 (in the 'indefinite' metric) *distinguishes* the time co-ordinate from the spatial co-ordinates. It is this circumstance which prevents the reversibility of causal chains[5] which Lipsius mistakenly imputes to the theory of relativity. Finally, Lipsius believes erroneously that "the relativism destroys the unity of the world of experience" (35, p. 441); see pp. 17 and 29. Lipsius' statements about the fictitious character of the theory of relativity cannot have great importance if they contain so many misunderstandings.

Another attempt to demonstrate the fictitious character of Einstein's theory was made by L. Höpfner (22, 23). Compared to Kraus' writings, Höpfner's presentations have the advantage of being rigorous formulations, and it is commendable that he singles out certain particular assertions as examples of fictions; in this way, one can at least decide whether they are fictions. It turns out, however, that all of Höpfner's contentions except one are false, and that one exception is trivial.

Höpfner lists six fictions in Einstein's theory; he tries to account for them by formulating statements about them using the words 'as if'. He succeeds in constructing such formulations, but he does not notice that in doing so, he alters the physical significance of the statements. Thus he calls Einstein's light principle a fiction: "The velocity of light should be treated in all calculations and equations as if it were an absolute constant magnitude" (22, p. 473). Einstein has never made such an assertion, and the theory of relativity certainly could not be based upon this statement. The velocity of light is not an absolute magnitude, but, as we explained above, a ratio; this ratio is *in fact* the same everywhere. If this were not the case, the theory of relativity would be false because all formulas of the theory are based on the assumption that this ratio holds objectively. If the theory were incorrect, the optical experiments would not be in accord with it. Höpfner contends further that, if Einstein's theory were correct, the positions of the hands on the clocks would causally influence the light signal; this statement, too, he calls a fiction. It certainly is a fiction, and a deliberately false one, but it is an obvious misunderstanding to believe that Einstein's theory needs such naïve assumptions. Höpfner's other fictions are of a similar kind. The only correct statement which one could make is that Einstein occasionally employs the fiction of observers and clocks existing everywhere. But this is a trivial fiction, of which the author of the theory of relativity is undoubtedly aware.

The extent to which Höpfner misunderstands the physical problem which was the source of the theory of relativity is shown by his statement, appearing elsewhere (23), about Newton's absolute space. He calls this a "scientifically justified fiction" (23, p. 482), but does not see that he would have to *defend* his assertion against Einstein. One should certainly search for a *cause* of the occurrence of inertial forces, i.e. one should interpret these forces as resulting from physical phenomena for which there is independent evidence; this is the reason why Newton quite consistently ascribed reality to absolute space. Since it was not possible, however, to find some other evidence for absolute space, this space remained a *fiction*, and Einstein is perfectly justified in calling it a "merely fictitious cause". In contrast, Einstein adduces the masses of the fixed stars as the cause of inertia, i.e. he adduces a cause for which there is independent evidence. Höpfner incorrectly compares the introduction of the fixed stars with the introduction of the zero point on the thermometer. The zero point is not introduced as the *cause* of heat, but as a *reference point of comparison* for the measurement of temperature; it corresponds to the co-ordinate system to which measurements of velocities refer. Yet Newton's absolute space is *more* than a reference point for measurement; this follows from the fact that the orientation of his absolute space is not arbitrary, but can be inferred from the observed motions of the celestial bodies. In contrast, both the co-ordinate system from which measurements are made and the zero point of the thermometer are arbitrary.

Whereas in the other cases mentioned the philosophy of 'As If' erroneously blames physics by contending that the hypotheses of physics are actually fictions, in this case it will not permit physics to remove a fiction and replace it by an hypothesis. It seems 'as if' this philosophy wants to introduce fictions into physics at any price. This method would be quite inappropriate to philosophy – unless the words 'as if' allow us to hope for something better. It would be advisable for the philosophy of 'As If' to study the significance of the physical theory more attentively if it intends to participate in the philosophical analysis of the theory of relativity. One must have advanced beyond the point of misunderstanding physics in order to be able to criticize the most profound physical theory.

2. CONCEPTIONS INFLUENCED BY MACH

It is well known that Mach expressed one of the essential ideas of the general theory of relativity as long as forty years ago. He recognized that motion can be defined only in relation to *bodies*, and that it is meaningless to speak of motion 'relative to space'. In his criticism of Newton's principles, he says about motion: "In reality, therefore, we are simply cognizant of a relation of a body *K* to *A, B, C* . . . If now we suddenly neglect *A, B, C* . . . and attempt to speak of the deportment of the body *K* in absolute space, we implicate ourselves in a twofold error. In the first place, we cannot know how *K* would act in the absence of *A, B, C* . . .; and in the second place, every means would be wanting of forming a judgment of the behaviour of *K* and of putting to the test what we had predicted – which latter therefore would be bereft of all scientific significance" (36, p. 282). These words contain a clear criticism of the doctrine of absolute motion, but, beyond that, they point the way to a possible development of a precise theory of motion. The idea that motion, as a spatial process, is *recognizable* only in relation to other bodies antedates Mach's writings. Even earlier, Leibniz had expressed such an idea; a passage, quoted in Lange's historically interesting work, *Die geschichtliche Entwicklung des Bewegungsbegriffes* (30), shows Leibniz to be convinced of kinematic relativity.[6] Whenever motion is characterized as a *change of spatial distances*, it is relative; this idea is implicit in the concept of kinematic motion. What distinguishes Mach's view so greatly from previous conceptions is his awareness that *dynamic* relativity must be maintained in addition to kinematic relativity. Motion can be recognized by the occurrence of *forces* – this is the significance of Newton's equation 'force = mass × acceleration' – and Mach asserts that the occurrence of secondary forces of motion, the so-called forces of inertia, likewise depends on the existence of other bodies. This idea was unknown before Mach. He infers it from the fact that every interpretation of motion is, in principle, reversible, so that it is always possible to regard the inertial forces as an effect of other bodies. This idea is best illustrated by Mach's treatment of the problem of rotation. "But if we take our stand on the basis of facts, we shall find we have knowledge only of *relative* spaces and motions. *Relatively*, not considering the unknown and neglected medium of space, the motions of the universe are the same whether we adopt the Ptolemaic or the Copernican mode of view. Both views are, indeed, equally *correct*; only the latter is more simple and more *practical*. The universe is not *twice* given, with an earth at rest and an earth in motion, but only *once*, with its *relative* motions alone determinable. It is, accordingly, not permitted us to say how things would be if the earth did not rotate. We may interpret the one case that is given us in different ways. If, however, we so interpret it that we come into conflict with experience, our interpretation is simply wrong. The principles of mechanics can, indeed, be so conceived, that even for relative rotations centrifugal forces arise" (36, p. 284). It is admirable how convincingly the idea of dynamic relativity is expressed in this passage, and nobody has acknowledged this ingenious insight more than Einstein. In his obituary for Ernst Mach (11), he quotes the relevant passages from Mach's *The Science of Mechanics*.

In spite of our admiration for Mach's ideas, we must not forget that Mach errs when he accepts general relativity on *a priori* grounds. Schlick was the first to point out this mistake (58, pp. 166–8).[7] The principle of dynamic relativity is *empirical*, since observable consequences concerning the behaviour of rotating masses (for instance, fly wheels) can be inferred from it. It is not possible to *logically* derive dynamic relativity from kinematic relativity. Schlick shows this clearly by translating the idea of relativity into the language of Mach's positivism: "Where optical experiences do not teach us anything, kinesthetic ones may not teach us anything either". It is strangely ironic that Mach inadvertently became the victim of apriorism.

Admirers of Mach's analysis of space experienced a great disappointment when *The Principles of Physical Optics* was published posthumously last year [1921]. In the preface of this book, the seventy-four-year-old author declares himself to be a convinced opponent of the theory of relativity. "I gather from the publications which have reached me, and especially from my correspondence, that I am gradually becoming regarded as the forerunner of relativity. I am able even now to picture approximately what new expositions and interpretations many of the ideas expressed in my book on mechanics will receive in the future from the point of view of relativity.

"It was to be expected that philosophers and physicists should carry on a crusade against me, for, as I have repeatedly observed, I was merely an unprejudiced rambler, endowed with original ideas, in varied fields of knowledge. I must, however, as assuredly disclaim to be a forerunner of the relativists as I withhold from the atomistic belief of the present day" (37, Preface).

Although we must accept Mach's declaration, it cannot disprove the close connection between his criticism of the problem of motion and the general theory of relativity, nor the fact that Einstein's theory carries out Mach's program. We must remember that the passage quoted from Mach's book stems from the year 1912 – that this passage was written before the final version of the *general* theory of relativity appeared, even though the principle of equivalence and the theory of the deviation of light had already been published. Mach's opposition to the special theory seems much easier to understand because this electrodynamic theory has no relation to the passage in Mach's *Mechanics*. We might ask, therefore, whether the ultimate general theory would not have reconciled the old 'rambler endowed with original ideas'. This remains doubtful, however, since the special theory is an element of the general one; we have to accept the fact that the originator of the idea of relativity did not recognize it in his old age when he met it in a new form.

Mach's reversal of his own position has led his philosophical heirs to adopt different attitudes towards the theory of relativity. Petzold, following the younger Mach, has become a convinced relativist; Dingler, however, appealing to the remarks by the older Mach, declares himself to be in strong opposition to Einstein's theory. F. Adler occupies an intermediate position.

Petzold sees Mach's theory of motion in close relation to Mach's phenomenalistic epistemology. If only observable phenomena are real, only relative motion exists, since our senses perceive only the relative motion of bodies. "Those who have become convinced that the shapes in visual space are inseparable from the colours, and who, in addition, understand the perfect epistemological equality of the sense of touch and the sense of sight, know that physical space, the space of experience, shows us only relative motion, and that there exists no observation, and no means – whether optical, or electromagnetic, or mechanical – that would permit us to *perceive* absolute motion" (43, p. 7). For Petzold, Einstein's main contribution consists in the fact that the theory of relativity regards only *coincidences* as realities accessible to experience. We know that when Einstein established the general theory, he regarded the fact that, in principle, only coincidences of physical things are observable as an essential premise; the general theory, therefore, considers coincidences as the only invariants, and relativizes merely the *metrical relations* between the coincidences. Petzold views this conception as the only way of describing nature *objectively* and of eliminating metaphysics. Metaphysics, not sensation, is the deceptive element in knowledge according to Petzold. "Today the doctrine of Heraclitus and Parmenides that the senses deceive us would be ridiculous in the face of experimental science . . . what we observe and judge are always complexes of sense impressions, or of memories of them: Einstein calls them coincidences of perceptions . . . Physical phenomena are determined when these coincidences can be uniquely co-ordinated in pairs. This point is of utmost importance. Whoever states more about nature than that, as far as we know it is always possible to uniquely correlate coincidences to each other, goes beyond the limits of experience. In principle, this was Hume's assertion as well as Mach's doctrine, and the same idea is at the foundation of Kirchhoff's judgment about the vagueness which he found inherent in the concepts of forces and their effects. Einstein accepted this principle completely, and it is here, above all, that we have to seek one of the sources of the far-reaching generalization of the idea of relativity which he has achieved in his general theory" (42, p. 64). Classical physics and epistemology committed the error of regarding mechanical motion as the explanation of all phenomena. This was a prejudice because mechanics is not closer to the senses than optics and acoustics. Mechanics originated from the sense of touch, while optics was born of the sense of sight and, in principle, it does not matter which kind of sense perception physics takes as its starting point (39). All those objections to the theory of relativity which charge it with being inconceivable arise merely because one is still preoccupied with a mechanical conception of the world; progress in science consists in overcoming mechanistic materialism, and the theory of relativity is the last great element in this development. It is interesting to see how Petzold pre-empts the theory of relativity for a specific philosophical development – positivism. He discusses these ideas in his work *Die Stellung der Relativitätstheorie in der geistigen Entwicklung der Menschheit* (42). In the following section, we shall encounter a similar claim by Neo-Kantianism. It cannot be denied that there is a certain justification for Petzold's view; instinctively, he emphasizes the fundamentally philosophical character of the theory of relativity. The insistence upon observability is truly Machian, and Einstein himself acknowledges his close relationship to Mach. For example, Einstein quotes (11) the passage in which Mach says that a statement has no scientific significance if there exists no possibility of testing it. Furthermore, Petzold seems to agree with Einstein's conception of causality (44). Petzold believes that the task of causal explanation is completed when a unique functional co-ordination of all events has been established; in such a pure 'description', metaphysical concepts like 'causal forces transcending phenomena' are meaningless. Thus, for Petzold, the question about a force causing the Lorentz contraction becomes meaningless;[8] this phenomenon is sufficiently explained as a function of relative motion. The question whether the Lorentz contraction is 'real' or 'apparent' is likewise a vacuous one according to Petzold; for him, 'real' means 'observable', and thus something can be real for one observer which is not real for another.

This is the point at which Petzold's positivism deviates from the theory of relativity. Einstein does not assert the relativity of *truth* (cf. 45), and only a one-sided positivistic interpretation can read this assertion into the theory; indeed, Petzold seems to be of the opinion that the theory of relativity becomes comprehensible only through positivism. In the new adition of his book *Das Weltproblem* (41), Petzold discusses this question in detail[9] and answers a remark by Cassirer (41, p. 208; cf. also the quotation from Cassirer on p. 28 in this essay). Cassirer had contended that the facts are determined objectively, not by the measurement of a *single* system, but by the measurements of *all* systems, or by the discovery of the transformation formulas which indicate the relation of the systems. Only this invariant relation has objective significance; the statements of a single system no more provide an exhaustive description of the world than a blueprint does of a building. Petzold, however, claims that the measurements of a single system are sufficient because the measurements of other systems can be derived from these by means of the transformation formulas, and he asserts that the theory of relativity is a "confirmation of Protagoras' *homo mensura* statement".

There is no doubt that Cassirer has the correct interpretation in this controversy. Petzold forgets that the knowledge of the transformation formulas provides that element which points beyond the measurements of a *single* system. The transformation formulas are not empty definitions, but empirical discoveries; they express the causal relationships between the observations in different systems. If we are merely given the measurements in *one* system we are not thereby given the transformation formulas and the measurements in other systems. It is a matter of indifference whether one characterizes the objective state of affairs, as Cassirer does, by the measurements of the totality of systems, or, as Petzold does, by the measurements of *one* system *and* the added transformation formulas; both versions say that the world is *not exhaustively* characterized by the way it appears to one observer. *The fact that there is a functional relation between the measurements of the different observers expresses a property of reality.* The law of the constancy of the velocity of light asserts more than that the measurement of this velocity by rigid rods and clocks yields the value 300,000 km/sec in *one* system; the connection between light and rigid rods and clocks is determined objectively only if we add that we obtain this value for measurements made with moving rods and clocks as well. For a complete characterization of the velocity of light, the measurements of *all* systems is required. It is possible to find the value of the velocity of light in moving systems by means of the Lorentz transformation, but conversely, the Lorentz transformation can be derived from the value of the velocity of light in the different systems. These statements represent merely two different mathematical descriptions of the same state of affairs. Petzold writes that the method of the theory of relativity "is in accordance with Protagoras' statement which can be formulated as follows: the world 'is' for every system as it 'appears' to be in that system, or it 'is' for every reference point as it 'appears' to be from that point, i.e. as it is discovered to be from that point, and all the different reference points are compatible with each other" (41, p. 208). The reference points are not only compatible, but their relation is governed by a law; this relation is accessible to knowledge and makes our discoveries independent of the limitations of an accidental reference point.

Let us emphasize, at this point, that one should not think that it is necessary for an observer at rest in a given system to describe phenomena by means of measurements in that system. The terminology of the usual presentations of the theory of relativity is often misleading. A moving observer is not compelled to use just those methods of measurement which are the simplest for him. He can, for instance, define simultaneity in such a way that it corresponds to the simultaneity of an observer 'at rest'; then the law of the constancy of the velocity of light does *not* hold for him, but he can nevertheless achieve a unique determination of events. Einstein's definition of simultaneity is not 'truer' than any other definition; cf. below p. 39. There is no necessary description of the world for a given observer; a description is determined only after certain metrical presuppositions for measurement have been established by *definition*. The transformation formulas which express the relations between measurements in all possible systems eliminate the arbitrariness of these assumptions. Differences in the measurements obtained in the different systems have nothing to do with the *subjectivity of the observer*, but result from the indeterminateness of the *concept of measurement*; measuring is *comparing*, and a measurement is definite only after the object of comparison has been designated. The subjectivity of perception is of a different sort. Even 'the world of one system' is a world which goes beyond immediate perception and therefore contains arbitrary elements; it is not possible to eliminate these elements by reducing them to the immediate perceptions of *one* observer, but only by determining the causal relations between the perceptions which result from a *variation* of the arbitrary elements. The causal relations are the same for all observers.

The danger of interpreting the theory of relativity in terms of a preconceived philosophical position manifests itself in Petzold's rejection of certain consequences necessarily connected with the theory. We are referring to

Einstein's assumptions concerning the limiting character of the velocity of light and the finiteness of space. Petzold is aware of his opposition to Einstein. He believes that Einstein goes beyond the "range of our sense organs", and that Einstein's doctrine constitutes a "retrogression to the rationalistic mistake of Kant" for whom "the things had to behave according to reason" (44, p. 473). But Petzold's interpretation is erroneous, and the above assumptions of the theory of relativity are compatible even with Petzold's positivism. These theorems of Einstein's are not, in principle, different from the other doctrines of the theory of relativity; they represent observable phenomena. Petzold objects to the statement that velocities faster than light are impossible on the grounds that this statement is negative; yet every positive statement can be transformed into a negative one. The statement Petzold objects to is identical with the positive one that any motion of a physical thing requires a finite amount of kinetic energy; for velocities faster than light, the kinetic energy becomes infinite. The statement is logically of the same type as, for instance, the principle of energy; this principle asserts that there does not exist a natural process for which the energy of a closed system increases. Similar considerations hold for the finiteness of space; if space were finite, then light rays would return to their point of origin after a finite time without being reflected. The fact that this assertion has not yet been sufficiently confirmed experimentally and that it will always have to be inferred inductively does not deprive it of its empirical character. (The spherical shape of the other side of the moon can also be inferred only inductively.[10])

Petzold's final judgment of the theory of relativity constitutes a misunderstanding of its logical character. The theory is definitely *objective*. He writes: "The quest for the absolute was expressed in the general theory of relativity in a quite unnecessary manner. The equations which described natural relations, and which remained unaffected by the arbitrary transformations of the theory, were considered to be absolute invariants, absolute laws of nature, if not explicitly, then tacitly, in terms of the underlying concepts" (41, p. 214). The word 'absolute' is, of course, open to misinterpretations; undoubedly, the invariant equations *make the laws independent of the conditions of measurement*. If these conditions are known for one particular measurement, then that measurement is connected with all possible measurements by virtue of the invariant equations; in this sense, and only in this sense, can the measurement be regarded as absolute. It is only in this sense that Einstein's light principle – contrary to Petzold's conception (41, p. 215) – is an absolute truth; its specific form, the constancy of the velocity of light, is dependent on certain conditions of measurement which, however, can be precisely formulated.

Mach's view concerning the theory of relativity is of special interest to Petzold. He published a rather long analysis of it in the eighth edition of Mach's *The Science of Mechanics* (40). He agrees with Einstein that Mach was very close to finding the theory of relativity, even though he emphasizes that Mach's epistemological orientation hindered his productive mathematical and physical work. "Einstein's theory is not simply a continuation of Mach's. The latter directs his attention so much upon the system of the fixed stars that Einstein had to emancipate himself from this idea before he could express the essential points of Fizeau's and Michelson's experiments in the most economical form . . . But the physicists are right when they do not see any contradiction between Mach's views and the implications of the theory of relativity for theoretical physics" (40, pp. 501–2). It is fortunate that Petzold who is such a competent interpreter of Mach's ideas, stresses this fact so vigorously.

Dingler has as much right to call himself one of Mach's pupils as Petzold does; Mach assigned Petzold the task of bringing out a new edition of his *Mechanics*, while, in the preface of this work, he delegated to Dingler the responsibility for the further development of his philosophy of space and time. According to the statement from *The Principles of Physical Optics* quoted above (p. 13), one can hardly doubt that the seventy-four-year-old philosopher would have agreed with Dingler's rejection of the theory of relativity. When Petzold tries to show that Dingler has abandoned Mach's conception (he sees in Dingler's theories a false interpretation of Mach's principle of economy (40, p. 503)), Dingler has the right to cite Mach's late statement in his defense. It is tragic that Dingler can avail himself of this justification. No one who has read Mach's excellent critical analysis of classical mechanics without preconceived notions could doubt that the young Mach, the gifted critic of Newton, would have become a convinced supporter of Einstein's theory. It frequently happens that the author of a revolutionary idea shuts himself off from those who attempt to develop his idea further. It seems that Mach exhausted his revolutionary strivings in this one achievement, and became insensitive to the imaginative daring of others; apparently the liberating effect of a fruitful idea consists, for its creator, more in the intensity with which he has produced it than in its content. How much more pronounced must this attitude be when age and illness prevent one from sharing the enthusiasm of other thinkers. "I myself – seventy-four years old and struck down with a grave malady – shall not cause any more revolutions"

(36, Preface). Neither could he any longer participate in a revolution, yet no one who sees in Einstein's theory the ingenious completion of ideas which Mach only divined will have any less esteem for Mach because of these circumstances. It is vey regrettable, however, that Dingler plays off the seventy-four-year-old Mach against the young author of *The Science of Mechanics*.

Dingler's criticism of the theory of relativity shows so little insight into Einstein's ideas that his evaluation of Mach's attitude towards Einstein cannot be taken at face value. Dingler has taken issue with the theory of relativity in numerous publications (cf. 6, 7); one is surprised at the dogmatism that they reveal. Dingler believes he has found a completely new method for the philosophy of science. The method consists in describing the world of natural phenomena in terms of certain logical principles; contradictions of these principles cannot occur since all experiences are reinterpreted until they conform to these principles. Dingler's idea is not essentially new; Kant also held that experience never contradicts the principles of reason because experience can be gained only by their help. But Dingler's assertion is completely *false*. It cannot be proved that the totality of the principles will never *contradict* experience since these principles may constitute an *over*-determination for the interpretation of experiences. For Kant, consistency was guaranteed, to a certain extent, by the fact that the principles were derived from reason and reason was thought to be orderly; however, this necessary presupposition of *a priori* philosophy has turned out to be unjustifiable, as I have shown in another publication (47, Section IV [1920f]). There is absolutely no reason why Dingler's principles should always permit a consistent interpretation of experience because their origin remains obscure.

Dingler contends that the theory of relativity is false because it employs the notion of non-Euclidean space, and it is one of Dingler's principles that space is Euclidean. Euclidean space is supposed to be *simpler* than non-Euclidean space. Dingler overlooks the fact that 'simple' is a relative term, and that the issue is whether the total corresponding system of physics becomes simpler. According to Dingler, a rigid body is necessarily Euclidean because it is defined in terms of Euclidean geometry. He will not admit that there exist other definitions of *rigid body*. It is supposedly impossible to demonstrate that space is non-Euclidean because such a demonstration presupposes Euclidean space. I have answered this objection (and several other misunderstandings) in another publication (52 [1921c]). It can very well be demonstrated that space is *not* Euclidean even if it is at first assumed to be Euclidean; if by assuming $A = B$ one can, with the help of other statements, derive $A \neq B$, this is a contradiction, and therefore '$A \neq B$' must be true. If one wants to prove positively that space has a certain non-Euclidean structure, one can use the *method of successive approximation*, accepting Euclidean geometry as an approximation for small dimensions.

Dingler's objections did not seriously hamper the theory of relativity, and one cannot say that his elaboration of Mach's ideas has been fruitful. F. Adler's objections (1) offer greater difficulties. Adler does not start with epistemological objections, and that is his strength. He just wants to investigate whether the observational data used by Einstein – Michelson's and Fizeau's experiments – necessarily lead to Einstein's theory. He restricts himself to the special theory of relativity. Not only does Adler arrive at the conclusion that Einstein's theory is not a *necessary* consequence of the experiments – no scientist would ever claim it was a necessary consequence since no theory can be deductively derived from observations – but he also asserts that Einstein's theory is *false*. This is not the place to discuss these questions in detail since they lead deep into special problems of physics; elsewhere (54 [1922d]) I have given a refutation of Adler's main thesis. Another of his claims, which I criticized in a letter to Mr. Adler, was subsequently given up by him. It is, however, of philosophical interest that, ultimately, epistemological motives induce Adler to reject the theory of relativity. He believes that it is impossible to obtain the value c for the velocity of light by using a moving clock; he calls such a result a "miracle of nature" (1, p. 176), and demands that the physical theory abandon this idea. He believes, however, that there is good reason to say that clocks transported along different paths will again be synchronized when they are brought together (1, p. 208). He does not offer any adequate justification for this statement. One can just as well see a miracle of nature in the undisturbed transportability of the clocks; what mysterious relation regulates the periods of the clocks in such a way that they remain the *same*? It makes no sense to impose rules upon nature that stem from human reason. Either assumption is possible; which of them corresponds to reality only experience can tell. Moreover, I was able to show (54) that the two assumptions do not exclude each other, but are logically compatible; the retardation of clocks when transported follows only if certain other assumptions of Einstein's theory are accepted.

Adler's writings show no definite relationship to Mach's but one may perhaps sense a hint of Mach's spirit in the respect and the admiration which this representative of Mach's doctrines expresses for Einstein.

3. THE NEO-KANTIAN CONCEPTION

It is strange that schools as divergent as Mach's positivism on the one hand, and Neo-Kantianism on the other, can interpret the theory of relativity in conformity with their own epistemology, and that each is to a certain extent justified in doing so. Whereas the right of positivism to interpret the theory of relativity is based on the relativistic restriction to observables, that of Kantianism is based on the interpretation which the theory of relativity gives to the unobservables, space and time. The foremost problem of Kantian epistemology is to investigate how statements can be made about these two unobservable forms of appearance, but it was Kant's sin of omission simply to take over the *content* of such statements without criticism. Since this mistake had inadvertently slipped into Neo-Kantian philosophy, Neo-Kantianism had all the more reason to focus upon the physical theory that turned with unequalled success against the *content* of the classical space-time theory. There arose the problem of whether the analysis of the content of space-time statements would lead to a confirmation or a refutation of Kant's philosophy of the forms of phenomena.

It is possible to uphold Kant's philosophy in the face of the theory of relativity by proving that the objects to which the theory of relativity refers are *different* from the objects referred to in the Transcendental Aesthetics. It would be futile for the philosopher to doubt the correctness of the physical theory, but Kant's doctrine of pure intuition enables us to restrict the statements of epistemology to an isolated domain existing independently of experience. This point of view is represented by the radical Kantians; I will mention Sellien, Ilse Schneider, and Lenore Ripke-Kühn.[11] The radical Kantians do not want to admit that Einstein's theory refers to the content of the intuition of space and time; they maintain that the theory concerns only the *measurement* of space and time magnitudes, not space and time proper.

Sellien proceeds dogmatically in his justification. "Since it is in the nature of geometry to refer to the 'pure' intuition of space, experience cannot influence geometry at all. Conversely, experience becomes possible only through geometry. Under these conditions, the theory of relativity is deprived of its right to assert that the 'true' geometry is non-Euclidean. It may at most say: the laws of nature can conveniently be expressed in a very general form when we base them on non-Euclidean metrical relations" (63, p. 48). About time he writes in a similar way: "First of all, it is a matter of fact that Einstein's theory cannot concern pure time as a form of intuition". If pure intuition is *defined* in such a way that it cannot be affected by experience, the statement is true; but one must not believe that one is still on Kantian ground with such an empty definition. For Kant, pure intuition is not divorced from empirical intuition; pure intuition is the form of, and hence determines empirical intuition. It is the profound significance of his philosophy that he does not take his *a priori* principles for granted, but derives them from the possibility of experience. Ilse Schneider has also overlooked *this* meaning of the transcendental philosophy. She charges that my criticism of the Kantian *a priori* does not "correspond to the meaning of the transcendental philosophy emphasized by Kant" (61, p. 73); but if one searches for this meaning in her writings, one finds the claim that the "general laws and the concepts of the *a priori* are immutable" (61, p. 17). I will admit that this is an assertion of the transcendental philosophy, but this assertion does not exhaust its significance; this is precisely the reason why I have objected to the transcendental philosophy. Kant does not want to say merely that the general *a priori* laws are logically correct – this would be trivial – but rather that empirical knowledge cannot dispense with them. The theory of relativity, in turn, is an empirical theory which does *not* make use of *a priori* laws of space and time; the theory shows that it is possible to attain empirical knowledge by means of conditions of experience that are *different* from the Kantian ones. If, in spite of this fact, one wants to defend Kant's philosophy of space and time, one would have to show that the validity of Kant's forms of pure intuition must be presupposed for any application of non-Euclidean geometry and relativistic time. The Neo-Kantians have never attempted to give such a demonstration. They merely dogmatically assert that an empirical theory *cannot* affect pure intuition. They nowhere attempt to relate the empty and untouchable *a priori* to the observable world, to empirical knowledge. Such a demonstration would be impossible. Undoubtedly, one can arrive at a perfectly adequate understanding of the relativity of simultaneity without presupposing absolute simultaneity.[12] We cannot prevent anyone from asserting that absolute simultaneity exists, although it is unknowable; but one must not believe that this absolute time is a condition of experience. Kant himself would have rejected such a 'chimera' because he held that the forms of intuition were not only products of reason, but also presuppositions of knowledge. Similar considerations apply to Euclidean space; one must not assert that it is necessary to presuppose Euclidean space in order to understand non-Euclidean space. One might, however, attempt to demonstrate that it is necessary to presuppose Euclidean space by referring to the fact that Riemannian space contains Euclidean space as a differential element;[13]

however, Kant's Euclidean space is employed for statements about finite dimensions, and as such it is *not* a presupposition of Riemannian space. The theory of relativity has achieved a scientific system *without* presupposing the Kantian forms of intuition, and thus Euclidean space and absolute time no longer have the unique position which Kant ascribed to them.

It cannot be maintained that Kant understood the forms of pure intuition to be more general structures, and that he did not restrict himself to the specific forms we have mentioned. For Kant, it was a matter of course that pure intuition is identical with Euclidean space and absolute time. When he distinguished the determination of empirical time as an empirical problem, he did not mean that physics should be permitted to construct any kind of definition of uniformity or of simultaneity; he meant that the task of determining what physical mechanism corresponds best to absolute uniform time, and of indicating what empirical procedure measures absolute simultaneity in the most precise manner is an empirical one. It did not occur to him that any given time interval can be *defined* with equal justification as a measure of uniformity. He may have known that *kinematically* one is not forced to choose a particular definition, but he believed too firmly in the Newton-Euler theory not to regard the law of inertia as a criterion supplying an approximate *dynamic* determination of absolute time. Certainly, one must not interpret Kant so naïvely as to believe that he took the psychological experience of time as a basis for determining absolute uniform time; he saw that an approximate determination of absolute time could only be achieved by appealing to physical laws, i.e. by means of systematic knowledge. But Kant could not know that the totality of physical laws do *not* provide such a determination, and that the laws retain their structure when the time metric is arbitrarily changed since this was not known until Einstein developed the theory of relativity. The same considerations apply to the space of pure intuition which Kant took to be Euclidean space. It is futile to show by means of quotations from *Die metaphysischen Anfangsgründe der Naturwissenschaft* – as Ilse Schneider does (61, p. 69) – that Kant had on occasion considered the possibility of constructing multi-dimensional geometries. He took such a geometry to be a *conceptual* system, "a geometry constructible by a finite mind" (61, p. 69). It would contradict the *Critique of Pure Reason* were he to consider these geometries as possible forms of *pure intuition*. The *Critique* begins by asking for the source of the apodictic certainty with which we assert the axioms of geometry. Kant's answer is that it is pure intuition which forces us to accept these axioms. What would be the significance of this answer if the same pure intuition were to force us to deny the truth of these axioms? Certainty in the choice of the axioms would be ruled out, and synthetic judgments *a priori* about space would be impossible. Kant's pure intuition is *not* compatible with the space-time doctrine of the theory of relativity. Some Neo-Kantians try to blur this incompatibility by citing appropriately chosen quotations. Surely, one would perform a better service to Kant if, in the face of modern physics, one were to abandon the content of his assertions and, following the great plan of his system, search for the conditions of experience on new paths instead of clinging dogmatically to his specific statements. Nothing can be defended any longer by reverence for Kant's every word when a new physical science knocks at the door of philosophy.

It is Cassirer's great achievement to have awakened Neo-Kantianism from its 'dogmatic slumber', while its other adherents carefully tried to shield it from any disturbance by the theory of relativity. It is no accident that this role was left to Cassirer. Those who have always taken the development of the natural sciences to be a development towards greater conceptual clarity will see in the theory of relativity the ultimate and most profound advance in this direction, and consequently will not hesitate to uncover those ideas in Kant's philosophy which were conditioned by his time and which have been superseded in the evolutionary process. In Kant's critique of knowledge, one must distinguish the method of formulating questions, the 'transcendental method', from the specific answers that Kant gives to particular questions; it is possible to reject the particular answers without abandoning the critical method itself. I see Cassirer's merit in the fact that he proceeded in this manner, and did not evade modern physics like the other Kantians. His work (5) is the masterful presentation of a historian to whom systematic analysis gave breadth of vision, and whose superior competence lacks any dogmatism. His every sentence evinces a command of critical analysis that is bent, not on a *preservation of Kant's doctrines*, but on a *continuation* of Kant's *methods*. The transcendental method searches for the presuppositions of knowledge; if the system of knowledge has changed since Kant, then Kant's presuppositions of knowledge must be corrected. There is no doubt that the contradiction between Kant and Einstein can be resolved in this way.

Cassirer therefore abandons the idea of interpreting pure intuition in Kant's sense. He separates the metric from intuition, and takes pure intuition to be the general law of coexistence that even Riemannian geometry retains. According to Cassirer, the metrical axioms are no longer dictated by pure intuition. Riemann began by searching for the most general type of three-dimensional manifold; in this manifold, Euclidean 'plane space' turns out to be a special case resulting from a certain form of the metric. If one takes as the space of pure intuition this general Riemannian structure which has certain continuity and order properties but which leaves the choice of the metric open, all contradictions to the theory of relativity disappear. Cassirer is aware of his deviation from Kant even though he manages to soften his rejection of the older theory by a very wide interpretation. He writes: "The most general meaning of this term [pure intuition], a meaning to which Kant does not always adhere since he inadvertently gives the term more specific meanings and applications, is simply that of the order of coexistence and of succession. Nothing is presupposed concerning the particular metrical relations holding in either of these orders" (5, p. 85). The narrow requirement of Euclidean space and uniform time is replaced by the general rule of the theory of relativity that all metrical determinations are to be equivalent and uniquely correlated to one another. This rule determines a definite manifold, but of a much more general type than that determined by Kant's laws of pure intuition. "It is true that here we have gone beyond Kant; he established his analogies of experience essentially in accordance with Newton's three fundamental laws" (5, p. 82). The acceptance of the theory of relativity therefore requires a modification of Kant's doctrine of pure intuition.[14]

One must not forget that the modification required by the theory of relativity is not only consistent with Kantian philosophy, but also in a sense serves to complete it. It was Kant's great contribution to have pointed out that space and time have no physical reality, that they are merely structural laws of knowledge. One may say that the conception of the ideality of the forms of intuition finds its mathematical expression in the principle of general relativity.[15] Elsewhere I have voiced my astonishment (47, p. 8 [1920f]) that the principle of relativity had not been asserted long before Einstein by Kantian-oriented philosophy. Cassirer is correct when he writes: "If Einstein sees the essential feature of the theory of relativity in the fact that it deprives space and time of the 'last vestiges of physical thinghood', he opens the way for a definite application of the philosophy of critical idealism to, and for its development within empirical science" (5, p. 79). In this context, Kant's philosophy is more compatible with Einstein's theory than with Newton's, and it is surprising that Kant himself did not realize that his views were inherently incompatible with those of Newton. Cassirer may rightly consider himself Kant's successor when he denies that pure intuition determines a metric and claims that non-Euclidean geometry is better suited than Euclidean to the things of experience. The theory of relativity merely asserts the greater adequacy of non-Euclidean geometry;

a Euclidean physics is also possible, but the Euclidean character disappears when the congruence of two segments is defined as coincidence with the same rigid measuring rod. It is due to this idea of the 'naturalness' of non-Euclidean geometry that Cassirer writes: "The actual advantage of Euclidean geometry seems, at first glance, to consist in its concreteness and intuitive plausibility compared to which all 'pseudo-geometries' become empty logical 'possibilities'. These possibilities exist only in theory, not in practice; they seem like an empty play with concepts which can be neglected when we deal with experience and 'nature', i.e. the synthetic unity of empirical science. This view is strangely and paradoxically reversed when we look back upon our earlier considerations. It turns out that pure Euclidean space deviates more from the fundamental requirements of empirical physical science than do the non-Euclidean manifolds. It is precisely because Euclidean space is the logically simplest form of space that it fails to do justice to the inherent complexity of the physical structure of the world" (5, p. 113).

The theory of relativity is a confirmation of Kantian and Neo-Kantian doctrines in a further sense; it lends support to Kant's analysis of the *concept of an object*. In contrast to naïve realism, Kant holds that a physical object is not a directly given thing but is *defined* by physical laws during the process of acquiring knowledge. Thus, Cassirer speaks of magnitudes rather than objects, and he regards it as the primary characteristic of scientific development that concepts of objects are continuously eliminated in favour of concepts of magnitude. The meanings of 'temperature', 'atom', and 'mass' are not given intuitively, but are determined only by the totality of physical laws. Consequently, 'truth' for natural science does not mean correspondence with a thing – that would be an impossible requirement – but rather internal consistency of the conceptual system (Schlick calls it "uniqueness of co-ordination".) In this way, mechanistic materialism is overcome, and Cassirer shows, in a consummate presentation, that the theory of relativity must be understood in the light of this development. This critical formulation of the concept of truth ultimately clarifies the significance of the principle of relativity. The theory of relativity does not entail the subjectivity of truth; to believe that it does would be to completely misunderstand Einstein's frequent references to 'measuring results for one observer'. In addition to 'measurements in one system', there are transformation formulas for all other systems, and both these components are required to determine the invariant state of affairs, a state which can be described in the various languages of different systems, but which holds objectively. Cassirer is correct when he objects to Petzold: "In this respect, the physical principle of relativity has little more than a name in common with the 'relativistic positivism' to which it has been compared. If one sees in it a revival of the philosophy of the ancient Sophists, a confirmation of Protagoras' statement that man is 'the measure of all things', one misunderstands its essential contribution. The physical theory of relativity does not maintain that what is true is what appears to be true to the individual; on the contrary, it warns us not to regard statements holding only in an individual system as true in the scientific sense, i.e. as expressing comprehensive and ultimate empirical laws. Laws are neither discovered nor confirmed by observations and measurements made in an individual system, not even by those made in any given number of such systems, but only by the mutual co-ordination of results obtainable in *all possible* systems" (5, p. 56; cf. also above p. 16).

It is remarkable how Cassirer's theory, derived from his historical criticism, coincides with the concept of objectivity in the theory of relativity.

I should like to conclude the discussion of Cassirer's Kantianism with a criticism of his brilliant and excellent work, a criticism I have been intending to make for a long time. With the elimination of the metric from pure intuition, certain axioms of the theory of space and time, the metrical axioms are deprived of their character as synthetic judgments *a priori*. If both a judgment and its denial are acceptable to reason – and Riemannian geometry, for instance, accepts the denial of the Euclidean axiom of parallels – then that judgment can no longer be regarded as synthetic *a priori*. Yet it was the belief that there are synthetic *a priori* judgments which Kant took as his starting point, and he considered the principal achievement of his transcendental method to be its demonstration of the eternal applicability of such judgments to science. If some of these judgments have now lost their privileged status, more has been shattered than these judgments alone: the *certainty of the transcendental method* has been undermined, and there is no guarantee that the hitherto unaffected axioms will hold for ever. If physics should proceed, under the influence of quantum theory, to conceive of space as a discrete manifold (a matter that is undecidable at the moment), Cassirer's concept of pure intuition would require a further extension. Under such conditions, a continuous, metric-free space would no longer be an adequate framework for empirical reality. It seems to me that one should renounce the apodictic certainty of all statements about the form of knowledge. I have shown elsewhere that epistemology does not thereby become impossible; rather, as a method of scientific analysis, it is concerned with discovering what principles of knowledge hold at a given time.

One should not contend that, since the logical equivalence of all geometries leaves the internal consistency of Euclidean geometry unaffected, the validity of synthetic *a priori* judgments remains unchallenged; such an appeal to conventionalism is denied to Kantianism. The internal consistency of geometry is *analytic*. Conventionalism would admit all *logically consistent* conceptual systems as possible structural forms of empirical knowledge, but the significance of synthetic *a priori* principles consists in the fact that they constitute a specific choice among the *logical* possibilities. Similarly, the law of causality is not *logically* necessary since uncaused events are logically possible; according to Kant, it is a synthetic *a priori* principle which excludes uncaused events. If 'synthetic *a priori*' meant nothing but 'internally consistent', Kantians would have to admit that, at some future time, we might gain knowledge of uncaused events. But on such an interpretation, synthetic *a priori* principles would ultimately degenerate into empty formulas that would impose no limits upon experience. This is the reason why eliminating the metric from pure intuition leads to a denial of synthetic judgments *a priori*.

Cassirer resolved the contradiction between Kant's epistemology and the theory of relativity by extending the concept of pure intuition. I agree that, in this way. Kant's philosophy is rendered consistent with present-day physics, that this consistency is achieved with the minimum number of changes in Kant's philosophy, and that there are even certain doctrines in Kantianism (e.g. the ideality of the forms of intuition) which point to such a reconciliation. Nevertheless, I maintain that such an approach is tantamount to a denial of synthetic *a priori* principles, and that there is no other remedy but to renounce the apodictic character of epistemological statements.[16]

4. THE RELATIVISTIC CONCEPTION

Under this heading, I wish to present a position which adheres very closely to the physical content of the theory of relativity and which finds support in Einstein's writings. Its aim is not to incorporate the theory into some philosophical system, but rather to formulate the philosophical consequences of the theory independently of any point of view, and to assimilate them as a permanent part of philosophical knowledge.

It would be erroneous to say that Einstein's contribution consists only in the establishment of a *physical* theory; he has always been aware of the fact that his theory is based upon a *philosophical* discovery. The starting point of the special theory of relativity – the contradiction between two optical experiments – was a problem of *interpretation*, not a problem of discovery. The two optical experiments[17] were contradictory only because a unifying principle was lacking; the *physical* discovery was completed with the performance of the experiments, whereas the *logical* discovery of its explanation was still missing. Every physical theory is, of course, a *logical* achievement because it establishes the theoretical connections between observed facts. But in this case all known logical methods seemed to fail. Lorentz had worked out his theory within the framework of traditional methods, but this theory itself had led to a new riddle, the contraction of rigid rods. I do not wish to suggest that the contraction in Lorentz' theory contradicts the principle of causality (it is certainly a functional explanation); a paradox consists in the fact that, given the contraction, all effects originating from the ether are said to be *quantitatively of such a kind that motion relative to the ether cannot be observed*. Physics cannot regard such effects as accidental; it must look for an explanation of this *consistent unobservability*. The significance of Einstein's solution consists in the fact that it explains this unobservability by the principle of relativity, in this case by giving up the view that a material ether can serve to determine a state of motion.

According to Ehrenfest (9, p. 19), Einstein's theory combines the following assertions:

I. Light sources send us light signals as independent phenomena through empty space.

II. The measured velocities of light rays from a source moving towards us, and from a source at rest would be observed to be the same.

III. We declare that we are satisfied with the combination of these two statements.

Ehrenfest's view goes right to the heart of the theory; but the 'combination of these two statements' becomes comprehensible only when Einstein's definition of simultaneity is assumed. The measurement of the velocity of light presupposes the definition of simultaneity. If this definition is not the same for observers moving with different velocities, then (II) does not contradict (I), i.e. the constancy of the velocity of light (II) does not contradict the principle of relativity (I), since (I) entails giving up the material ether. The contradiction between the optical experiments could no longer be solved within the framework of traditional epistemological concepts, and a philosophical analysis of the concepts of space and time was needed in order to construct relativistic physics.

Einstein himself is aware of the philosophical character of his analysis. Witness his remarks concerning the related contribution by Mach, remarks which are contained in his obituary for Ernst Mach: "Concepts that have proved to be useful for a systematic description of things easily acquire such an authority for us that we forget their earthly origin and accept them as immutable entities. In this way they are stamped 'logically necessary presuppositions', '*a priori* given', etc. The path of scientific progress often is obstructed by such errors for a long time. It is therefore not just an idle game when we are trained to analyse familiar concepts and to show what the conditions of their justification and their usefulness are, and how in particular they have grown out of the given phenomena of experience. In this way, their all too great authority is broken. They are eliminated if they are not legitimate, corrected if their co-ordination to the given phenomena is not accurate, replaced if a new system can be devised that is preferable for certain reasons. Such analyses usually seem unnecessary, artificial, sometimes even ridiculous to the special scientist who is more interested in particular problems. The situation changes, however, when one of the traditionally used concepts is to be replaced by a more rigorous one because the development of the respective science requires it. Then those who have not been careful with their own concepts protest vigorously, and complain of a revolutionary threat to their most sacred possessions. This lament is re-enforced by the voices of those philosophers who think they cannot dispense with the concept to be replaced because they had claimed it for their jewel-case of the 'absolute', of the '*a priori*' or, in short, because for some reason they had proclaimed the eternal truth of that concept. The reader will have guessed that I am hinting, in particular, at certain concepts of the theory of space and time, as well as of mechanics, which have been modified by the theory of relativity. No one will deny that the epistemologists have prepared the way for this development; for my own part, I know at least that I was directly and indirectly greatly stimulated by Hume and Mach" (11).

The significance of the theory of relativity consists in its analysis of certain concepts of science, or, in Kantian terminology, of the conditions of experience; the attempts of some philosophers to relegate the achievement of the theory of relativity to physics show only that these philosophers are no match for the physicist Einstein in philosophical competence. Einstein's discovery is not a philosophical triviality. Einstein once wrote the following in a letter to me: "The value of the theory of relativity for philosophy seems to me to be that it has revealed the dubiousness of certain concepts that were also regarded as small coin even in philosophy. Concepts are empty when they cease to be closely connected with experience. They are like upstarts who are ashamed of their origin and want to deny it" – I must, with all my admiration for this confident rejection of any dogmatic rationalism, object to the modesty of the first sentence; unfortunately philosophy had not yet recognized the doubtfulness of those concepts, although this should have been its task. It is a serious failing of post-Kantian philosophy to have left the development of the epistemological space-time problem to the mathematicians and physicists. The line of development extends from Bolyai, Lobatschefskij, Riemann, Helmholtz and Mach to Poincaré, Hilbert, Einstein, and Weyl; it is high time that philosophy adjust itself to this development and begin its work where the mathematicians left off instead of guarding itself with a *noli me tangere*.

There is a short essay by Einstein, entitled 'Geometry and Experience' (14), which formulates the relation between geometry and physics with classical brevity. To the question about the apodictic certainty of the geometrical axioms, Einstein answers: "As far as the laws of mathematics refer to reality, they are not certain; and as far as they are certain, they do not refer to reality". Subsequently, he distinguishes the purely logical 'axiomatic geometry' from 'practical geometry', i.e. geometry applied to physical things; the relation between the two is expressed by the sentence: "Solid bodies are related, with respect to their possible dispositions, as are bodies in Euclidean geometry of three dimensions" (14, p. 32). It is an empirical question whether this statement is true; according to Einstein, it holds only approximately for small dimensions. It is not possible to define *solid body* without referring to physical laws. These laws can also be chosen in such a way that the *rigid body* is defined by Euclidean geometry; in this case, the quoted sentence becomes empty, but the choice of physical laws is restricted. This is the reason why, according to Einstein's formulation, only the sum $G + P$ of geometry and physics is testable by experience.

Such philosophical considerations are not by-products of Einstein's works in physics, but rather the logical foundation which makes his work possible. He writes: "I attach special importance to the view of geometry which I have just set forth, because without it I should have been unable to formulate the theory of relativity ... The decisive step in the transition to general co-variant equations would certainly not have been taken if the above interpretation had not served as a stepping-stone" (14, p. 33). Those who know Einstein's clear and comprehensive way of thinking from personal conversations will understand the significance of this remark. Einstein is not a formal mathematician concerned with developing purely mathematical theories;

rather, he thinks analytically, i.e. he is concerned with clarifying the meaning of concepts. Mathematics is, for him, only a means of expressing an intuitive process – a process which operates from unconscious sources and for which the formal language is merely the framework. It is a rare gift of fortune to find philosophical intuition and a talent for mathematics and physics combined in one mind; only a mind combining these traits could create the theory of relativity.

The essay 'Ether and the Theory of Relativity' (13) is further evidence of Einstein's competence as a philosopher. In this essay, he gives several precise formulations of the old problem of the existence of ether, a problem that had remained unsolved because it had been stated in equivocal terms. One can say that the space between material complexes is filled with ether, since space has the physical property of affecting the shape of the measuring instruments; in this way one can define a practical geometry. But ether, so understood, does not have the property of a substance whose individual particles can be traced in time; the concept of motion is not applicable to it. In addition to ether or 'metrical field', there exists matter or the 'electromagnetic field', which comprises light rays (i.e. electric waves) as well as ordinary matter. The two fields interact, and the relation between them is expressed by the equations of gravitation. Thus, the problem of the ether has been solved by renouncing a concept which is not applicable to reality, the concept of *substance*. Philosophy is confronted with the fact that physics creates new categories which cannot be found in traditional dictionaries.

The first philosopher to accept the theory of relativity whole-heartedly was Schlick. In this regard, he occupies a leading position among philosophers, and his conception of the theory of relativity, which is related to Poincaré's conventionalism, is shared by Einstein. Schlick opposes Kant and represents philosophical empiricism, but he also defines his position in contrast to Mach's positivism.

For Schlick, the problem of intuition is the foremost philosophical problem posed by the theory of relativity. The space of physics is not identical with tactile or visual space, but "is conceived as being independent of our perceptions, though not as being independent of physical objects since it is real only in relation to them" (56, p. 75). In the last clause, Schlick refers to Einstein's idea that the structure of space is determined by the masses of bodies, even at those places which are not themselves filled with matter, the metrical field of the $g_{\mu\nu}$ would not exist if there were no masses. Here Schlick opposes Kant's doctrine of the ideality of space (cf. also 57). Only with regard to the arbitrariness of the co-ordinates is space an ideal structure; its metric, however, expresses an objective property of reality.[18] This conception does not contradict conventionalism. Schlick must not be interpreted as saying that a certain metric has been prescribed; a metric emerges only after the physical laws have been established (the *P* of Einstein's formula). One can also change the *metric*, provided one changes the laws of physics correspondingly. But the relation between these modifications expresses an *invariant fact*. In this case, as in the case of all other problems treated in the theory of relativity, the laws of the universe cannot be given in a single structural formula. One cannot say that real space is non-Euclidean. Only the invariant relation of changeable structural formulas determines a property of reality. This is the language which the theory of relativity has developed for the description of nature.

Yet precisely because the space of physics describes a real state, it is *unintuitive*; the spatial metric determines a state just as temperature or electric intensity do, and it possesses reality in the same sense as these parameters. "Physical objects are therefore unintuitive; physical space is not given in perception, but is a *theoretical construct*. One may not ascribe to physical objects the intuitive spatiality which we encounter in our visual or tactile sensations, but only an unintuitive order which we call objective space and describe theoretically by a manifold of numbers (co-ordinates). Intuitive space is similar to the sensible qualities, colours, sounds, etc. Physics does not know colour, but only frequencies of the oscillations of electrons, not heat qualities, but only kinetic energy of molecules" (56, p. 76).

Thus Schlick gives up Kant's concept of pure intuition. Since space is not intuitive, "intuition cannot, in spite of the conviction of many adherents of Kant's philosophy, tell us anything about whether space is Euclidean or not" (56, p. 77). The logical construction of physical space is opposed to the intuitive spaces of the different sense organs; Kant's pure intuition occupies an intermediate position between the two, and has no epistemological significance. Schlick formulates his conception most clearly in his brilliantly written answer to Cassirer. He writes: "When, in other publications, I contrasted psychological space (and time) as purely intuitive with physical space as a purely logical construction, I was well aware that 'intuition' is defined by Kant in a completely different way. A number of critics have misunderstood this point. Cassirer calls Kant's pure intuition a certain 'method of objectification'; (5, p. 123, 124 n.) it certainly is that, but this does not exhaust its significance. Kant wanted to eliminate all psychological aspects, but I shall never become convinced that he succeeded. Such an elimination is not possible without applying the only method that is capable of separating the purely logical aspect of geometry from the psychological-intuitive one: the method of implicit definition which has been developed in modern mathematics. (Cf. 60, p. 30ff.) Without this method, we cannot even express the idea of a logical concept nor understand it independently of all psychological aspects. Kant's space of pure intuition necessarily contains such aspects; without the content provided by these aspects, the concept of space would be 'empty' for Kant" (57, p. 108). These statements may be regarded as an interpretation supplementing Schlick's *Erkenntnislehre* (60, p. 301).

There remains the peculiar psychological hold which the Euclidean metric has over us. Schlick deals with this problem, in his comments on Helmholtz' writings, in the following way: "Our Euclidean intuition fights with the logical analysis which recognizes that Euclidean intuition is not quite adequate for reality. Yet it holds sway over us: thus we call the light rays passing close to the sun *curved*, and represent them as such in our drawings, even though they are the straightest lines of our space. However, the actual situation does not correspond to the assumptions of the foregoing remarks (Helmholtz had constructed an extreme example) since the deviations from Euclidean geometry are so small that they are not revealed in ordinary perception. We agree with Helmholtz that, if these deviations were visible, our intuition would be perfectly adjusted to a non-Euclidean metric because it would have developed under the influence of non-Euclidean experiences; in this case. there would be no discrepancy between intuition and the theoretical description of nature" (59, p. 174, n. 83).

To this remark, I should like to add that I personally regard Schlick's presentation as essentially correct, but I have the impression that the investigation of this question is not yet closed. The intimate connection between intuition and conceptualization is a problem that will raise difficult questions for psychology.

Schlick's analysis of the problem of intuition, which was developed on the basis of the theory of relativity, is a heartening sign of how fruitful the co-operation between philosophy and physics can be. Schlick proves himself to be an outstanding student of physical theory, and his candid judgment is never impaired by his philosophical 'viewpoint'. 'Viewpoints' are most dangerous when physical discoveries are to be evaluated philosophically. The representative of a certain philosophical *Weltanschauung* will always be inclined to interpret the physical sciences in accordance with his own predilections, and he may easily overlook new philosophical discoveries. The philosophers of the 'As If' provide a striking example; they noticed that in some disciplines, such as jurisprudence, fictions play an important part, and therefore they believed that the theory of relativity must also be a fiction. Schlick never shows such prejudices. He exhibits, in all his writings, the same sure understanding of physics, and a readiness to profit from the exact sciences for the purpose of philosophical analysis. This is the reason why Schlick's expositions find more and more recognition among physicists. The physicist feels relieved when a philosophical method reminds him of his own objective way of thinking. I have heard physicists express their amazement at the lack of a body of generally accepted philosophical truths; every philosophical school wants to start all over again, and attacks the result of its predecessors. There is profound insight in this naïve remark. A 'point of view' belongs at the *end* of philosophy, not at the *beginning*, and there are enough specific problems left before one can even think of giving final answers to the most general questions.

By calling his philosophical point of view 'empiricism', Schlick invites an identification of his views with that naïve position which neglects the problems of conceptualization. Formerly I believed, on the basis of his opposition to Kant, that Schlick overlooked the constitutive significance of the categories in the concept of object, and in an earlier publication (47, p. 110 [1920f]), I objected to Schlick's views on these grounds. However, in the course of our correspondence, it turned out that this objection was based on a misunderstanding and therefore I should like to take it back. Recently Schlick has made his views more precise (57, pp. 98 and 111). Accordingly, it seems to me that Schlick's empiricism must not be regarded as a 'system' comparable to the traditional systems of philosophy; rather, it must be regarded as a method based on the belief that reality is given in experience, a method which takes as its task the *analysis of the process of experience* in the widest sense, without adhering to any specific interpretation. I am quite ready to accept this method because my so-called "analytic method" (47, p. 71 [1920f]) represents exactly this objective approach.

In this connection, I should like to mention two of my own studies. The first (47 [1920f]) is devoted to an analysis of Kant's philosophy. Kant had the idea that reason prescribes a certain system of principles by means of which our knowledge of the physical world is established. According to Kant, this system can never be falsified by experience because experience is possible only by means of these principles; therefore these principles can be called synthetic judgments *a priori*.[19] But Kant uses an assumption for which he gives no evidence (and for which he can give no evidence); he assumes that the system of principles is not *overdetermined* (hypothesis of the arbitrariness of co-ordination). It is possible for a contradiction to arise between the totality of experience and a system of principles if that system provides more rules for experience than experience will admit. The ultimate instrument in the judgment of empirical truths, *perception*, is *independent* of reason; although perception always permits different *interpretations*, the combination of the interpretations is no longer arbitrary. It is the significance of the theory of relativity to have discovered the limits of arbitrariness. According to the theory of relativity, the choice of a geometry is arbitrary; but it is *no longer* arbitrary once congruence has been defined by means of rigid bodies. The combination of the principles of Euclidean geometry and this definition of congruence is excluded by experience. Thus we can show that there *are* overdetermined combinations; it is not possible to prove that the system offered by reason is not such an overdetermined combination.

In a certain sense, one can even say that the Kantian system has already been shown to be contradictory. All the assumptions of Einstein's theory are so evident that they have as much right to be regarded as requirements of reason as do the principles of Euclidean geometry. The principle of the relativity of the co-ordinates is, as all Neo-Kantians agree, a requirement of the transcendental philosophy. But the principle of relativity leads to *non-Euclidean* geometry on the basis of certain empirical facts alone, in particular, of the equivalence of inertial and gravitational mass.[20] The system of self-evident principles of reason thus leads to contradictions, contradictions that cannot be recognized within the system itself but which become apparent when the system is applied to experience.

Reason has the task of adjusting the principles of the understanding to experience. Although some constitutive principles must always be presupposed for the establishment of empirical knowledge, such an adjustment can be made by means of a 'method of successive approximations', a method in which previously employed principles are assumed to hold approximately. Thus it is possible to *change the constitutive principles themselves*. In this way, the *a priori* loses its apodictic character, but it retains the more important property of being "constitutive of objects".[21]

In this connection, I wrote: "The role of reason [in knowledge] consists not in the fact that there are invariant elements in the system of co-ordination, but in the fact that arbitrary elements occur in the system" (47, p. 85 [1920f]). Although in this quotation I appear to be on the side of conventionalism, I should not like to choose this name for my view. In the first place, conventionalism does not recognize, as Kant did, that these 'conventions' determine the concept of object, that the particular thing or law is defined only by their help and not by reality alone. Secondly, the term 'convention' overemphasizes the arbitrary elements in the principles of knowledge; as we have shown, their *combination* is no longer arbitrary.[22] I agree with Schlick, who drew my attention to this matter, that Poincaré, the father of conventionalism, would acknowledge the restricted character of the combination of the principles and that his stress on the arbitrariness of the principles themselves was due to the historical context in which he wrote. For modern philosophy of science, however, it is important not only to detect the *arbitrary* principles of knowledge, but also to determine the totality of admissible combinations. That part of our scientific knowledge which stems from reason must be distinguished, by a sort of invariant theoretical method, from the objective content of science, a content which, in the present form of science, is no longer clearly visible.

In order to draw this distinction, I have attempted to formulate the presuppositions of the theory of relativity in axiomatic form. So far, there exists only a brief report (53 [1921d]) of these investigations. I shall therefore take this opportunity to mention some further results which will serve to clarify the role of fictions in the theory of relativity; I shall restrict myself to the special theory of relativity. Such an axiomatization must begin by distinguishing between *axioms* and *definitions*. In this context, in contrast to mathematics, an axiom is a factual statement. However, experimental evidence for this statement is not required; it may be provisionally asserted as a physical hypothesis. In any case, the axioms are not fictions. The axioms are supplemented by *definitions* which contain rules concerning the co-ordination of certain empirical phenomena to certain mathematical concepts. This distinction is, of course, to a certain extent arbitrary, but that is unimportant. What matters is the rigorous adherence to this distinction once the arbitrary stipulations have been given. Definitions are not fictions either. They do not claim to be either true or false, whereas fictions are *falsehoods* deliberately employed (cf. Vaihinger's remarks above, p. 5). A fiction is a statement which is capable of being true, but which, in fact, is not, whereas it would be nonsense to apply the predicate 'true' to a definition. A systematic construction of the theory of relativity can be achieved by means of the two auxiliary concepts of axiom and definition. Fictions appear only occasionally as pedagogical devices; they can always be eliminated.

Einstein's famous simultaneity equation is a *definition*. It specifies a rule for determining what time indices are to be attached to events happening at different space points. It would be a mistake to believe that the definition of simultaneity given in the special theory of relativity claims to be 'more correct' than any other definition of simultaneity. Clearly, Einstein does not make this claim since in the general theory of relativity he abandons the definition of simultaneity given by reference to light rays. Even in gravitation-free spaces to which the special definition of simultaneity could be applied, it is not necessary to do so; according to the general theory, any other definition of simultaneity will lead to an exhaustive description of nature and even to the same physical laws. It is only *advantageous* to use the special definition of simultaneity in this case. By 'advantageous' we do not mean 'economical'; we can state precisely what the advantage is, namely, that the defined synchronism becomes *transitive*.[23] One can also use a non-transitive synchronism. But the use of a transitive synchronism has a further advantage: a moving observer passing a system 'at rest' will never find that he is moving backward in time when he looks at clocks in the system 'at rest'. (This result is due to the fact that no physical system, and therefore no observer, can travel faster than light.) Thus, the law of causality will not be violated. It is possible to describe the situation in such a way that the law of causality is violated, and the resulting physical laws can even be expressed by the same mathematical equations that express the usual physical laws. But it is, of course, preferable not to violate the law of causality.

What does the theory of relativity assert about absolute time? Above all, that it, too, is merely introduced by *definition*, and therefore has the same epistemological status as relative time. Furthermore, the theory asserts that it is not possible to discover a natural simultaneity. Of course, one might say that the time of a given co-ordinate system is absolute, and demand that everything be measured with reference to this time, but such a time is an 'empty' absolute time.[24] The problem is to find *a rule whose uniform application leads, in every co-ordinate system, to one and the same simultaneity*. Such a simultaneity could be defined by transported clocks, if the time of the clocks were not dependent upon the velocity of the transport (cf. 54 [1922d]). The theory of relativity states that *there is no such rule*. This is a factual statement. (One could call it an axiom, but it turns out to be more expedient to divide it into several axioms.) This assertion may be either true or false; according to the present state of our knowledge, it seems to be true. If this assertion were false, absolute simultaneity would still rest upon a *definition*, albeit a definition with certain advantageous properties. Its advantage would consist in the fact that the same mechanism could be employed to define simultaneity in every co-ordinate system. But the possibility of defining absolute time would not exclude the use of relative time; the latter would be a time with *different* advantageous properties.

The theory of relativity makes two assertions about absolute time. First, that there is no absolute time; second, that if there were an absolute time, it would not be absolute.

In the course of the axiomatization, a strange result occurred. In order to define a physical metric, one must establish its relationship to certain physical things; the definition of congruence, for instance, relates this concept to rigid measuring rods. It is possible to define the entire space-time metric of the special theory of relativity in terms of light signals alone. Imagine a domain of space in which mass points travel freely. In this space, light signals can be used to define the conditions under which two mass points are at rest relative to each other; one can speak of a uniquely defined 'rigidity of light'. Uniform time need not be assumed; it, too, can be uniquely defined by reference to light signals. Furthermore, the equality of line segments can be established by light signals. A pure light-geometry can therefore be used as a foundation of the space-time metric. Einsteinian kinematics rests on the hypothesis that light geometry is identical with the geometry of rigid rods and natural clocks. This hypothesis is asserted by a special group of axioms.[25] It is the advantage of the light geometry that it avoids defining the metric by reference to such complicated things as rigid bodies and clocks; these things should be introduced at the *end* of a physical theory, not at its beginning, since a knowledge of their mechanisms presupposes a knowledge of all physical laws.

SUMMARY

Four prominent viewpoints can be distinguished in the literature on relativity.

1. The 'As If' philosophers regard the assertions of the theory of relativity as fictions. Their view is in opposition to the physical theory and to experience because they attack the factual assertions of the theory. They have no justification for their interpretation. Philosophical considerations cannot decide whether the factual statements of the theory are correct; only experience can do so.

2. The conceptions influenced by Mach contradict one another. Although Petzold is an adherent of the theory, his interpretation is influenced by his philosophical position and is therefore in part incompatible with the physical theory.

3. Among the Neo-Kantians, some representatives maintain that Kantianism is not affected by the theory of relativity. Cassirer, however, proposes a modification of Kant's doctrine of 'pure intuition'. Only in this way can

Kant's theory be reconciled with Einstein's, but then the transcendental method seems to lose its apodictic character.

4. For the relativists, the philosophical significance of the theory consists in the modification of certain fundamental epistemological concepts. The relativists deny the existence of pure intuition. Furthermore, they deny the apodictic character of the *a priori*, and they do not regard general principles of knowledge as necessary but as arbitrary. (This is the position of conventionalism.) The relativists leave the decision concerning the admissible combinations of these principles to experience, and they believe that a continuous change of these principles is possible.

NOTES

[1] This is the most detailed exposition; it contains, in addition to a statement of the theory of relativity, an elementary survey of the physical problems which gave rise to it. In the new edition, the section on the theory itself has been extended, and this section demonstrates a clear understanding of the theoretical problems involved.

[2] Spinning electrons are the best clocks and therefore all such experiments are concerned with them. The transversal Doppler effect would be a confirmation of the retardation of clocks, but has not yet been investigated because of experimental difficulties. The red shift of spectral lines on the sun is also an empirically testable statement about real clocks.

[3] Compare, for instance, the chemical equation:

2 litres of hydrogen (+) 1 litre of oxygen = 2 litres of water vapour

which is also a contradiction if the plus sign is construed algebraically.

[4] It is strange that Kraus continuously charges Einstein with logical confusion, but of course, he only does so when he misunderstands Einstein's theory. He asserts, for instance, that the general theory of relativity contradicts the special theory (27, p. 476). He justifies this assertion (which is refuted by the fact that the special theory holds in the infinitesimal domain) by saying that huge, even astronomical dimensions would have to be called infinitely small. He does not understand that when the mathematician says that certain conditions hold in the infinitesimal domain, he means that these conditions constitute a limit which is approached as the domain decreases. And why are some astronomical dimensions not small compared to other, larger astronomical dimensions? It would be interesting to find out at what order of magnitude the philosophy of 'As If' starts using the concept of the infinitesimal; perhaps at the order of dimensions of a pin head?

[5] At least this is true for the special theory of relativity of which Lipsius speaks. In the general theory, matters are more complicated.

[6] This, however, is Lange's only contribution to the question of relativity; he is not aware of dynamic relativity even though he was familar with Mach's statements, nor does he take into account the influence of the masses of the fixed stars in his definition of the inertial system.

[7] Schlick wrote this paper before Einstein's final publication of the general theory of relativity, and it contains a mistake that came from Einstein (pp. 170–1). At first, Einstein thought he could prove the impossibility of a completely general relativity; he later proved the contrary. Schlick's presentation can easily be corrected, and its philosophical content is not affected by this circumstance.

[8] V. Laue, on the other hand (32), does not wish to exclude causal forces; according to him, a causal force is merely a function of the co-ordinate system, and may therefore exist for one reference system and not for another. But the difference between v. Laue and Petzold is probably only a terminological one.

[9] He writes: "In the present book, I have been able to incorporate the theory of relativity into the widest framework . . . because here we are concerned with the ultimate question, namely, what is the nature of the universe" (41, Preface).

[10] Petzold makes a further mistake with respect to the so-called clock paradox; let me refer in passing to the correct and clear presentations by Thirring (66) and Bloch (2, pp. 71 and 102). Petzold makes a fundamental mistake in his analysis: a coincidence, i.e. a *point event*, does not lend itself to different interpretations, but is the same for all observers; if two clocks stand side by side, and one is slow when directly compared with the other, it is slow for every observer. The objectivity of coincidence is an assumption of the theory of relativity which Petzold himself acknowledges. Another confusion is involved in Petzold's criticism of the mechanical models of the theory of relativity. He rejects these models for epistemological reasons since they are incompatible with his view that it is impossible to go beyond the observations in one system. In these passages (46), he always speaks of the different spaces of different observers, whereas the issue concerns differences of simultaneity in the *same* space. The shortcomings of such models consist in the fact that the measured distances are so small that one can see them at a glance; in this way, the simultaneity in one system becomes identical with the experienced simultaneity, and the simultaneity in the other system appears to be 'false'. For large dimensions, there is no experienced simultaneity. If one disregards immediate perception, possible in small dimensions, one can easily visualize the conditions in large dimensions. Of course, in the model, clocks and measuring rods must be corrected artificially, whereas such a correction is not necessary when light is used as a signal since once the zero point of time is determined, the correction takes place automatically. This fact illustrates the unique character of the velocity of light.

[11] L. Ripke-Kühn does not understand very much of the theory of relativity, but she is at least honest enough to reject the theory outright (55).

[12] Schlick has likewise explained (58, p. 162) that Kant's absolute time is not a logical presupposition of Einstein's time. I should like to refer, therefore, to this little known publication by Schlick written as early as 1915.

[13] This idea is contained in the interesting publication by Bollert (3, pp. 62–5) which shows a clear understanding of physics. Bollert admits that his interpretation deviates from Kant's specific doctrine of space (cf. Preface); I must, however, object to his view that the differential uniqueness of Euclidean space constitutes an *eternal necessity*, i.e. a synthetic judgment *a priori*. It is conceivable that Euclidean geometry may lose its unique position some day; perhaps there are already signs of this development in quantum theory. I have previously rejected this apology for apriorism (47, pp. 30 and 76 [1920f]); I must also reject Bollert's defense of the concept of substance, a defense which I had predicted (47, p. 75 [1920f]). It is interesting to note Bollert's presentation of the steps of objectification.

[14] Since Cassirer expresses this conception so clearly in the quoted passages, it is hard to understand why he states elsewhere (5, p. 75) that the theory of relativity refers only to *empirical* intuition, and why he even agrees with Sellien. If, in view of the theory of relativity, pure intuition must be changed, it is not possible to restrict this theory to empirical intuition.

[15] One must not forget that, in another sense, the theory of relativity gives up the ideality of space; cf. below p. 35.

[16] Schlick has also made this objection against Cassirer (57) and for a detailed exposition I refer the reader to his work as well as to my book *Relativitätstheorie und Erkenntnis apriori* (47 [1920f]). My book was written without knowledge of Cassirer's work and therefore I raised this objection directly against Kant.

[17] The Michelson and the Fizeau experiments; cf. for instance the presentation by Bloch (2).

[18] I have expressed this idea earlier in a similar formulation (47, p. 86 [1920f]); but the earlier passage is not quite clear, because I forgot to add the statement about the definition of the metric by rigid bodies. The above analysis may therefore be considered a correction of my previous statement; I hope to have Schlick's consent to it.

[19] With respect to this statement, cf. my above remarks on Dingler (p. 21) and on Cassirer (p. 29).

[20] The presentation of this issue in my book (47, pp. 27–8 [1920f]) is not quite correct; I have given a clearer exposition in (52 [1921c]).

[21] Whether one still wishes to call such an extension of the doctrine of the *a priori* 'Kantian' is a terminological question. Schlick is against doing so (57, pp. 98 and 111) because he regards the combination of these two properties of the *a priori* as Kant's most important assertion. I feel, however, that the discovery of the constitutive component is such an eminent philosophical achievement that one should not divorce Kant's name from it. In this discovery lies Kant's superiority to Hume, for Hume did not know what to do with his discovery of non-empirical principles in knowledge, and could only characterize them as 'habit'.

[22] One must not underestimate the restrictiveness of this combination. As I shall show in the near future, this restrictiveness means that the absolutist *cannot* use Euclidean space; otherwise he would have to sacrifice essential requirements of absolutism at points of singularity in space.

[23] This means: if clock A is synchronized with clock B, and clock B is synchronized with clock C, then A is synchronized with C.

[24] The absolute time of the Lorentz theory is 'empty' in this sense since it is nothing but Einstein's 'absolute' time of a given co-ordinate system.

[25] For a more detailed analysis I refer to (53) and a forthcoming publication [*Axiomatik der relativistischen Raum-Zeit-Lehre*, 1924h.]

BIBLIOGRAPHY OF BOOKS AND ARTICLES QUOTED IN THE PRECEDING ESSAY

1. Adler, Friedrich, *Ortszeit, Systemzeit, Zonenzeit und das ausgezeichnete Bezugssystem der Elektrodynamik*, Vorwärtzverlag, Wien, 1920.
2. Bloch, Werner, *Einführung in die Relativitätstheorie*. (Aus Natur und Geisteswelt, vol. 618.) Teubner, Berlin and Leipzig, 2nd ed. 1920.

3. Bollert, Karl, *Einsteins Relativitätstheorie und ihre Stellung im System der Gesamterfahrung,* Steinkopf, Dresden, 1921.
4. Born, Max, *Die Relativitätstheorie Einstein's und ihre physikalischen Grundlagen,* Springer, Berlin, 2nd ed. 1920. [First English translation, *Einstein's Theory of Relativity* (Methuen, London, 1924); revised edition by the author in collaboration with Günther Leibfried and Walter Bien (Dover reprint, New York and Constable, London, 1962) – Ed.]
5. Cassirer, Ernst, *Zur Einsteinschen Relativitätstheorie,* Bruno Cassirer, Berlin, 1921. [English translation by W.C. Swabey and M.C. Swabey as *Einstein's Theory of Relativity*, printed as a supplement to the author's *Substance and Function* (Open Court, Chicago and London, 1923) – Ed.]
6. Dingler, Hugo, *Die Grundlagen der Physik,* Vereinig. wiss. Verleger, Berlin and Leipzig, 1919.
7. Dingler, Hugo, *Physik und Hypothese,* Vereinig. wiss. Verleger, Berlin and Leipzig, 1921.
8. Drexler, Joseph, 'Grundwissenschaftliches zur Einsteinschen Relativitätstheorie', *Grundwissenschaft. Philosophische Zeitschrift der Johannes Rehmke-Gesellschaft* **2,** (Leipzig, 1921).
9. Ehrenfest, Paul, *Zur Krise der Lichtätherhypothese,* Springer, Berlin, 1913.
10. Einstein, Albert, *Relativity: The Special and the General Theory*, transl. by Robert W. Lawson, Hartsdale House, New York, 1947.
11. Einstein, Albert, 'Ernst Mach', *Physik. Ztschr.* **17,** 101–4 (1916).
12. Einstein, Albert, 'Dialog über Einwände gegen die Relativitätstheorie', *Die Naturwissenschaften* **6,** 697–702 (1918).
13. Einstein, Albert, *Sidelights on Relativity: I. Ether and Relativity,* transl. by G. B. Jeffery and W. Perret, Methuen, London, 1922.
14. Einstein, Albert, *Geometrie und Erfahrung,* Springer, Berlin, 1921. The quotation in the paper refers to the English ed. *Sidelights on Relativity: II. Geometry and Experience,* transl. by G. B. Jeffery and W. Perret, Methuen, London, 1922.
15. Einstein, Albert, 'Über eine naheliegende Ergänzung des Fundamentes der allgemeinen Relativitätstheorie', *Preussische Akademie der Wissenschaften, Sitzungsberichte,* pt. I, 261–4, (1921).
16. Eddington, A. S., 'A Generalization of Weyl's Theory of the Electromagnetic and Gravitational Fields', *Proc. Royal Soc.* A., **99,** 104, London, 1921.
17. Freundlich, Erwin, *Die Grundlagen der Einsteinschen Gravitationstheorie,* Springer, Berlin, 1920.
18. Gehrke, E., 'Die Relativitätstheorie: eine wissenschaftliche Massensuggestion', *Schriften a.d. Verl. d. Arbeitsgemeinsch. deutscher Naturforscher zur Erhaltung reiner Wissenschaft* e.V., no. I, Berlin, 1920.
19. Geiger, Moritz, *Die philosophische Bedeutung der Relativitätstheorie,* Lecture, gehalten im I. Zyklus gemeinverständlicher Einzelvorträge, veranstaltet v.d. Univerität München, Niemeyer, Halle, 1921.
20. Geissler, Kurt, *Gemeinverständliche Widerlegung des formalen Relativismus.* Hillmann, Leipzig, 1921.
21. Hamel, G., 'Zur Einsteinschen Relativitätstheorie', *Sitzungsberichte der Berliner Mathematischen Gesellschaft* **19,** (Göttingen, 1921).

22. Höpfner, L., 'Versuch einer Analyse der mathematischen und physikalischen Fiktionen in der Einsteinschen Relativitätstheorie', *Annalen der Philosophie* **2,** no. 3, 466–74 (Leipzig, 1921).
23. Höpfner, L., 'Zur Analyse der philosophischen Ausdrucksform der Einsteinschen Relativitätslehre', *Annalen der Philosophie* **2,** no. 3, 481–7 (Leipzig, 1921).
24. Kopff, August, *Grundzüge der Einsteinschen Relativitätstheorie*, Hirzel (Leipzig, 1921).
25. Kopff, August, 'Das Rotationsproblem in der Relativitätstheorie', *Die Naturwissenschaften* **9,** no. 1, 9–15 (Berlin, 1921).
26. Kraus, Oskar, 'Fiktion und Hypothese in der Einsteinschen Relativitätstheorie', *Annalen der Philosophie* **2,** no. 3, 335–96 (Leipzig, 1921).
27. Kraus, Oskar, 'Die Verwechslungen von "Beschreibungsmittel" und "Beschreibungsobjekt" in der Einsteinschen speziellen und allgemeinen Relativitätstheorie', *Kantstudien* **26,** nos. 3–4, 454–86 (Berlin, 1921).
28. Kraus, Oskar, 'Die Unmöglichkeit der Einsteinschen Bewegungslehre'. *Die Umschau*, no. 46, 681–4, Frankfurt and Leipzig, 1921.
29. Kries, J. V., 'Ueber die zwingende und eindeutige Bestimmtheit des physikalischen Weltbildes', *Die Naturwissenschaften* **8,** 237–47 (Berlin, 1920).
30. Lange, L., *Die geschichtliche Entwicklung des Bewegungsbegriffes*, Leipzig, 1886.
31. V. Laue, Max, *Das Relativitätsprinzip*, vol. I and II, Vieweg, Braunschweig, 1913.
32. V. Laue, Max, 'Die Lorentz-Kontraktion', *Kantstudien* **26,** nos. 1–2, 91–5 (Berlin, 1921).
33. Lenard, P., *Ueber Relativitätsprinzip, Aether, Gravitation*, Hirzel, Leipzig, 1920.
34. Linke, Paul, 'Relativitätstheorie und Relativismus', *Annalen d. Philosophie* **2,** no. 3, 397–438 (Leipzig, 1921).
35. Lipsius, Friedrich, 'Die logischen Grundlagen der speziellen Relativitätstheorie', *Annalen der Philosophie* **2,** no. 3, 439–65 (Leipzig, 1921).
36. Mach, Ernst, *The Science of Mechanics*, transl. by Thomas J. McCormack, The Open Court Publishing Co. (La Salle and London, 1942).
37. Mach, Ernst, *The Principles of Physical Optics*, transl. by John S. Anderson and A. F. A. Young (Dover Publications, New York, 1953).
38. Pauli, Wolfgang, 'Relativitätstheorie', *Enzykl. d. math. Wiss.* (Leipzig and Berlin, 1921).
39. Petzold, Joseph, 'Mechanistische Naturauffassung und Relativitätstheorie', *Annalen der Philosophie* **2,** 447–63 (Leipzig, 1919).
40. Petzold, Joseph, 'Das Verhältnis der Machschen Gedankenwelt zur Relativitätstheorie'. Appendix to Mach, *Die Mechanik in ihrer Entwicklung*, 8th ed. Leipzig, 1921.
41. Petzold, Joseph, *Das Weltproblem*, 3rd ed., Wissenschaft und Hypothese, vol. XIV, Teubner (Leipzig and Berlin, 1921).
42. Petzold, Joseph, *Die Stellung der Relativitätstheorie in der geistigen Entwicklung der Menschheit*, Sibyllenverlag (Dresden, 1921).
43. Petzold, Joseph, 'Die Relativitätstheorie der Physik', *Zeitschrift für positivistische Philosophie* **2,** no. 1 (Berlin, 1919).
44. Petzold, Joseph, 'Kausalität und Relativitätstheorie', *Zschr. f. Phys.*, p. 467 (1920).
45. Petzold, Joseph, 'Die Relativitätstheorie im erkenntnistheoretischen Zusammenhange des relativistischen Positivismus', *Verh. d. dtsch. Phys. Ges.* **14,** 1055 (1912).

46. Petzold, Joseph, 'Die Unmöglichkeit mechanischer Modelle zur Veranschaulichung der Relativitätstheorie', *Verh. d. dtsch. Phys. Ges.*, p. 495 (1919).
47. Reichenbach, Hans, *Relativitätstheorie und Erkenntnis apriori.* [1920f].
48. Reichenbach, Hans, 'Die Einsteinsche Raumlehre. [1920a].
49. Reichenbach, Hans, 'Die Einsteinsche Bewegungslehre'. [1921f].
50. Reichenbach, Hans, 'Entgegnung auf O. Kraus'. [1921g].
51. Reichenbach, Hans, 'Erwiderung auf Herrn Th. Wulffs Einwände gegen die allgemeine Relativitätstheorie'. [1921e].
52. Reichenbach, Hans, 'Erwiderung auf H. Dinglers Kritik an der Relativitätstheorie'. [1921c].
53. Reichenbach, Hans, 'Bericht über eine Axiomatik der Einsteinschen Raum-Zeit-Lehre'. [1921d].
54. Reichenbach, Hans, 'Relativitätstheorie und absolute Transportzeit'. [1922d].
55. Ripke-Kühn, Lenore, 'Kant kontra Einstein', *Beiträge zur Philosophie des deutschen Idealismus,* Beiheft No. 7 (Erfurt, 1920).
56. Schlick, Moritz, *Raum und Zeit in der gegenwärtigen Physik,* 2nd ed., Springer (Berlin, 1919). [English translation by H. L. Brose, *Space and Time in Contemporary Physics* (Oxford University Press, Oxford and New York, 1920) – Ed.]
57. Schlick, Moritz, 'Kritizistische oder empiristische Deutung der neuen Physik?' *Kantstudien* **26,** nos. 1–2, 96–111 (Berlin, 1921). [English translation by P. Heath in Moritz Schlick, *Philosophical Papers, 1910–1936* (*Vienna Circle Collection,* Reidel, Dordrecht and Boston, forthcoming) – Ed.]
58. Schlick, Moritz, 'Die philosophische Bedeutung des Relativitätsprinzips'. *Ztschr. f. Philosophie u. philos. Kritik* **159,** no. 1, 129–75 (Leipzig, 1915). [English translation, *ibid.* – Ed.]
59. Schlick, Moritz, 'Anmerkungen' in Helmholtz, *Schriften zur Erkenntnistheorie*, Springer (Berlin, 1921). [English translation by M. Lowe in Hermann von Helmholtz, *Epistemological Writings*, ed. by R. S. Cohen and Y. Elkana (*Boston Studies in the Philosophy of Science*, Reidel, Dordrecht and Boston, 1977) – Ed.]
60. Schlick, Moritz, *Allgemeine Erkenntnislehre*. Springer (Berlin, 1918). [English translation of 2nd German edition by A. E. Blumberg, *General Theory of Knowledge* (Springer-Verlag, New York and Vienna, 1974) – Ed.]
61. Schneider, Ilse, (Rosenthal-) *Das Raum-Zeitproblem bei Kant und Einstein,* Springer (Berlin, 1921).
62. Schouten, J. A., Lecture given on the Mathematikertag in Jena, 1921. *Mathematische Zeitschr.*
63. Sellien, Ewald, 'Die erkenntnistheoretische Bedeutung der Relativitätstheorie'. *Kantstudien,* Ergänzungsheft, no. 48 (Berlin, 1919).
64. Sommerfeld, Arnold, 'Die Relativitätstheorie' *Südd. Monatshefte* **18,** no. 2, 8–15.
65. Thirring, Hans, *Die Idee der Relativitätstheorie,* Springer (Berlin, 1921).
66. Thirring, Hans, 'Ueber das Uhrenparadoxon in der Relativitätstheorie', *Naturwissenschaften* **9,** 209 (1921).
67. Thirring, Hans, 'Erwiderung auf E. Gehrke'. *Naturwissenschaften* **9,** 482 (1921).
68. Vaihinger, Hans, *Die Philosophie des Als Ob*, Meiner (Leipzig, 1922). The quotations in the paper refer to the English ed. *The Philosophy of 'As If',* Kegan Paul, London, 1924.
69. Weyl, Hermann, *Raum. Zeit, Materie,* Springer, Berlin, 1921. [English translation of 4th edition by H. L. Brose, *Space, Time, Matter* (Dover reprint, New York, 1922) – Ed.]

45. THE THEORY OF MOTION ACCORDING TO NEWTON, LEIBNIZ, AND HUYGHENS*

[1924d]

I

Newton's theory of motion has a considerably greater influence upon the historical development of the problem of motion than the doctrine of his opponents, Leibniz and Huyghens. Nevertheless, it is ironic that Newton, who enriched science so immensely by his physical discoveries, at the same time largely hindered the development of its conceptual foundation. However fertile his optical discoveries, his emission theory of light delayed the acceptance of the wave theory, formulated with great insight by his contemporary Huyghens, by about a century. However far reaching Newton's discovery of the law of gravitation, his theory of mechanics arrested the analysis of the problems of space and time for more than two centuries, despite the fact that Leibniz, who was his contemporary, had demonstrated a much deeper understanding of the nature of space and time. Only today, when physics has finally abandoned Newton's point of view in optics and mechanics, can we do justice to the two men whose unfortunate fate it was to have possessed insights that were too sophisticated for the intellectual climate of their times. A man's historical influence depends not only upon the profundity of his ideas, but also upon his having the good fortune to offer ideas which are in tune with the spirit of the times. It seems that the faculty of abstraction in Newton's time had not reached a sufficiently advanced stage to permit the ascendancy of the discoveries of a Leibniz or a Huyghens concerning the relativity of motion. Even Kant's theory of space and time, formulated almost a century later on the basis of Newton's mechanics,[1] constitutes a regression in comparison with the discernment and precision of Leibniz' formulations. Leibniz had in the main advanced beyond the employment of such vague terms as 'the ideality of time and space', although he used them occasionally. Even Kant's early essay, 'Neuer Lehrbegriff der Bewegung und Ruhe' (1758), which contains the most extensive elaboration of the concept of relative motion and which makes his critical doctrine of space appear like a return to Newton, falls short of the level achieved by Leibniz.

The injustice of history goes even further; for there is no direct historical connection between the relativistic conception of motion developed in our

*From *Modern Philosophy of Science: Selected Essays,* ed. and tr. by Maria Reichenbach, Routledge & Kegan Paul, London, 1959, pp. 46–66.

day and the ideas of the two earlier relativists. On the contrary, their theories seem to have been forgotten by the scientists and to have been preserved only by the historians of philosophy, and the latter have never influenced the development of philosophy in any remarkable way. When physicists finally solved the problem of motion, they turned back more exclusively to Newton, although this time in order to object to his theory. In their opposition to Newton, physicists of our day rediscovered the answers which Newton's two contemporaries had offered in vain. Mach, the first relativist of the new era, developed his conception of the problem of motion through a criticism of Newton's *Principles*.[2] He did not know anything of Leibniz' well-grounded objections to Newton, and made only a few naïve comments about him.[3] Even Einstein's solution, going far beyond Mach's, took Newton's 'classical mechanics' at its starting point without any reference to Leibniz and Huyghens.[4] It almost looks as if a curse has been cast upon a discovery that was not in step with its own time. The later acceptance of relativistic views stems not from the gradual penetration of the earlier discovery, but from new and independent creative work stimulated by a reconsideration of Newton's theories, which in this way have once more become immensely fruitful.

II

Newton certainly believed that his mechanics was a complete justification of the theory of absolute motion. Not only did he give (in the introductory section of the *Principia Mathematica Naturalis Philosophiae*) definitions of absolute space and absolute time which have become the classical formula of absolutism, definitions characterizing space and time as independently existing entities which supply the measure for all space-time events[5] but he also specified methods for establishing relations between empirical events and these absolute entities. It is well known that he assigned the decisive role in these methods to the centrifugal force. The core of Newton's doctrine consists in the fact that he considered the occurrence of a force as evidence for a state of motion. Of course, Newton knew that it makes no sense to speak of absolute motion if motion is regarded purely kinematically, i.e. as a change of distance between masses. Kinematically speaking, there is no difference between the views of Ptolemy and of Copernicus; both describe the same fact: the relative motion of the celestial bodies. Yet for one of the views, the Copernican, Newton found a dynamic explanation in his law of attraction, whereas for the other view no such explanation seemed possible. The confirmation of absolute motion was therefore left to dynamics. In the introduction to his *Principia*, Newton gave two illustrations of this idea: the famous pail experiment, and the two spheres connected by a string which rotate around their common center of gravity. The tension of the string permits us to discover the velocity of rotation and to determine it quantitatively even if no other possibilities for such a determination were given. Newton also proposed a most ingenious method for recognizing the direction of the rotation. At first this result seems to be paradoxical because the quantity of the centrifugal force does not depend upon the direction of the rotation. But if one changes the velocity of the rotation of the spheres by imposing additional forces in one direction or the other, one discovers, through the increase or decrease in the tension of the string, the direction of the original rotation. These considerations were of major interest to Newton. The declared purpose of his book was to provide methods which permit inferences concerning the true motions of bodies when their apparent motion, i.e. their relative or kinematic ones are given. The introduction ends with these words: "But how we are to obtain the true motions from their causes, effects and apparent differences, shall be explained more at large in the following treatise. For this end it was that I composed it."[6]

Had Newton's *Principia* no other use than that claimed by its author, the work would not possess the importance which it actually has. Perhaps no other passage so clearly demonstrates the tragic influence of Newton's philosophy upon his judgment, an influence which led the great physicist to regard his epoch-making achievement, the physical discovery of the fundamental laws of mechanics, as being of secondary importance. After two centuries of error, 'philosophia naturalis' has finally abandoned Newton's philosophical views, but this does not detract from Newton's contributions to physics. Fortunately, all empirically grounded knowledge is independent of the interpretation of its discoverers, but the price of this independence is that such knowledge holds only approximately.

III

The reason for Leibniz' opposition to Newton's theory of motion must be sought in the philosophical differences between the two thinkers. Newton was a consistent scientist of invincible persistence, who left the test of an hypothesis to experience; he withheld his theory from publication for twenty years because it did not agree with astronomical observations, though he presented it with renewed conviction when more precise observational material had been gathered. Leibniz, on the other hand, had a flexible mind; he arrived at his conclusions by abstract manipulation, for he considered the logical analysis of a problem to be a better guide to its solution than a reliance upon empirical data since such data are always open to various interpretations. For Leibniz, the solution to the problem of motion grew out of a conception of space derived from philosophical considerations. Although the means of modern physics were not available to him, he remained faithful to his philosophical position in the face of Newton's serious counterarguments. Although Leibniz' views finally triumphed, he himself was unable to completely refute Newton. The decisive answer to Newton's argument concerning centrifugal force was given by Mach. (As is well known, Mach's answer is based upon the fact that centrifugal force can be interpreted relativistically as a dynamic effect of gravitation produced by the rotation of the fixed stars; therefore, no absolute state of motion is dynamically discernible.) One is inclined to concede greater recognition to the adherent of a mistaken view which has the advantage of consistency; the reason that Leibniz' position failed to influence the development of science evidently lies in the fact that he had not yet found Mach's answer. On the other hand, we must not forget that it was a profound insight into the nature of space which forced Leibniz into opposition; he, too, can claim the merit of a consistency, which would not permit him to relinquish a cogent epistemological position.

The best known presentation of Leibniz' theory of space is contained in his correspondence with Clarke, which was widely read in the eighteenth century, when the reality of space was discussed with a vehemence similar to that with which it was discussed at the time when Einstein presented his theory.[7] In the correspondence, the theologian Clarke appears as the defender of Newton's view. To a certain extent, however, we may regard his defense as authoritative since he drafted his answers to Leibniz with Newton's help. The correspondence reads like a modern discussion of the theory of relativity. The relativist tries in vain to convince an opponent so enmeshed in his absolutistic ideas that he does not notice how often his arguments presuppose the doctrine which he wants to prove, nor to what extent his allegation that the relativistic position is contradictory rests upon his own tacit assumption of an absolutistic conception. Before dealing with this correspondence, let us examine the rigorous foundation of Leibniz' theory of space and time in his 'Initia rerum mathematicarum metaphysica',[8] which was written during the period of the correspondence.

The work begins with a very profound explanation of time and space. Only physical objects and their states are presupposed as data and it is on the basis of certain relations between them that we later construct the order of time and space. Causality is the physical relation which leads to the order of time. If two physical states stand in the relation of cause and effect, the cause is *defined* as the earlier state, the effect as the later one. The temporal sequence is therefore the order of causal processes; logically, causality is primary, time secondary. Causal connections are discovered, but time is defined. In addition to the relation of temporal sequence, there exists the relation of simultaneity. It holds for physical states that are not causally connected. Leibniz notes, however, that this condition is not sufficient. He adds the condition that a uniform definition of simultaneity for all events must not contradict the temporal sequence as determined by the order of causal processes.[9] Evidently, Leibniz meant to refer to this condition in the following quotation when he remarks about "the events of past years . . . not co-existing with those of this year" and further about "the connection of all things". I shall quote the entire passage: "Given the existence of a multiplicity of concrete circumstances which are not mutually exclusive, we designate them as *contemporaneous* or *co-existing* (dicentur existere simul). Hence, we regard the events of past years as not co-existing with those of this year, because they are qualified by incompatible circumstances.

"When one or two non-contemporaneous elements contains the ground for the other, the former is regarded as *antecedent* (prius), and the latter as the *consequent* (posterius). My earlier state of existence contains the ground for the existence of the later. And since, because of the connection of all things, the earlier state in me contains also the earlier state of the other thing, it also contains the ground of the later state of the other thing, and is thereby prior to it. All existing elements may be thus ordered either by the relation of *contemporaneity* (co-existence) or by that of being *before or after in time* (succession). (Et ideo quicquid existit alteri existenti aut simul est aut prius aut posterius.)"

"*Time is the order on non-contemporaneous things*. (Tempus est ordo existendi eorum quae non sunt simul.) It is thus the *universal* order of change in which we ignore the specific kind of changes that have occurred."[10]

This passage seems to me to represent a depth of insight into the nature of time that has not been equalled in the whole classical period from Descartes to Kant. Even Kant's theory of time (as formulated, for instance, in the 'Second Analogy of Experience') does not attain Leibniz' perspicacity. Leibniz' characterization of time as the general structure of causal sequences is superior to Kant's unfortunate characterization of time as the form of intuition presupposed by causality. This difference emerges clearly only from the vantage point of an axiomatically grounded theory of knowledge: I must therefore refer the reader to my book on *Axiomatik* [1924h].

Leibniz' explanation of space is closely connected with his explanation of time. Since there are many things which exist simultaneously, there must exist an order of co-existing things, space.

"*Space is the order of co-existing things*, of the order of existence for all things which are contemporaneous."[11]

Subsequently, Leibniz makes an important attempt to define the topological order of space and time. This problem concerns the relation *between*. If a point P is between P_1 and P_2, P is nearer to P_1 than P_2 is. Thus the metrical order of a continuum, the quantitative comparison of distances, is based on a topological order, and we can speak of a *topological distance* which no metrical determination may contradict. Therefore, when Leibniz speaks of 'propinquity or remoteness' in the following passage, this must be understood topologically. However, such a topological propinquity is bound to a linear continuum, i.e. P is nearer to P_1 than P_2 to P_1 only along a certain line. Therefore, Leibniz introduces the concept of a straight line, which is actually a metrical concept. He is faced with the problem of defining under what conditions P lies between P_1 and P_2; this definition must be constructed in qualitative terms without the use of geometrical concepts. Leibniz attempts this construction with the help of the concepts of simplicity and of 'utmost definiteness'. However imperfect Leibniz' result, the attempt to specify the topology of space and time by means of definitions is highly important. We quote the whole passage in translation: "In each of both orders – in that of time as that of space – we can speak of a *propinquity* or *remoteness* of the elements *according to whether fewer or more connecting* links are required to discern their mutual order. (Secundum utrumque ordinem (temporis vel spatii) propiora sibi aut remotiora censentur, prout ad ordinem inter ipsa intelligendi plura paucioraque correquiruntur.) Two points, then, are nearer to one another when the points between them and the structure arising out of them with the utmost definiteness present something relatively simpler. Such a structure (of utmost definiteness) which unites the points between the two points (interpositum maxime determinatum) is the simplest, i.e. the shortest and also the most uniform, *path* from one to the other; in this case, therefore, the straight line is the shortest one between two neighbouring points."[12]

This passage is somewhat difficult to understand. If, for example, I want to determine the time order of two events E_1 and E_2 which do not follow each other closely enough to permit a direct judgment, I insert a third event E between them. If E_1 is earlier than E and E is earlier than E_2, I can infer that E_1 is earlier than E_2. For the purpose of determining the order of E_1 and E_2, we require a connecting link. The time interval will be longer or shorter depending on whether more or less elements are needed between E_1 and E_2. In order to avoid appealing to the concept of a connecting link, Leibniz speaks of 'the simplest structure of utmost definiteness' determined by two points, and this structure in fact contains a connecting link. The concept of the simplest structure is to be regarded as defining the relation *between*. The shortcoming of this definition consists, of course, in the vagueness of the terms 'simple' and 'utmost definiteness'.

It is noteworthy, furthermore, that in the construction of this theory of space and time Leibniz starts with the order of time, and considers the order of space only after he has given a determination of simultaneity, a procedure which turned out to be fruitful for later axiomatization of the theory of space and time.[13]

Having constructed his own rigorous theory of space and time, Leibniz was confronted with the natural philosophy of Newton. Newton begins with very precisely formulated empirical statements, but adds a mystical philosophical superstructure. Leibniz had to oppose a theory which regards space and time as autonomous entities existing independently of things, nor could he help but consider his own theory superior, feeling as one who had emancipated himself from the primitive notions of everyday life and who was obliged to assume the role of expositor. In the detailed fifth letter to Clarke, Leibniz tries to convince his opponent by describing the origin of the idea of space. "I will here show, *how* Men come to form to themselves the Notion of *Space*. They consider that many things (choses) exist at once, and they observe in them a certain *Order* of Co-Existence, according to which the relation of one thing to another (rapport des uns et des autres) is more or less simple. This Order, is their *Situation* or Distance (situation ou distance). When it happens that one of those Co-Existent things changes its *Relation* to a Multitude of others, which do not change their Relation among themselves; and that another thing, newly come, acquires the same Relation to the others, as the former had; we then say, it is come into the *Place* (place) of the former; And this Change, we call a *Motion* in That Body, wherein is the immediate Cause of the Change."[14] Leibniz correctly remarks that he does not define 'place' but 'same place' (la même Place);[15] and he recognizes that this procedure is admissible. We may supplement the foregoing quotation by remarking that a change in the relation of one thing A to other things,

C, E, F, G, . . . is ascertainable by observation, for instance, by observing the increasing brightness of a thing *F* which had previously been in *A*'s shadow. If the former state recurs in such a way that another thing *B* now has that relation to the things, *C, E, F, G, . . .*, which *A* formerly had, so that *B* now casts a shadow on *F*, we say that *B* has taken the place of *A*. From small differences in the position of *B*, we can discover that *B* is now the cause of the shadow, and not *A*. *Place of A* is a concept which characterizes a certain state of things *A, C, E, F, G, . . .*

Leibniz continues: "And That which comprehends *all those Places*, is called *Space* (ce qui comprend toutes ces places, est appellé Espace). Which shows, that in order to have an Idea of *Place*, and consequently of *Space*, it is sufficient to consider these *Relations*, and the Rules of their Changes, without needing to fancy any absolute Reality *out of* the Things (aucune réalité absolue hors des choses) whose Situation we consider."[16] If we hypostatize space, we go beyond what is meant by the concept of place; place, too is nothing that exists as such, but is rather a state of things. Strangely enough, we are inclined to hypostatize only a spatial place, although there exist other instances of the same logical relation where no one ventures a hypostatization. Leibniz cites genealogy as an example of the latter.[17] In a genealogical order, every individual has a 'place', and exactly as in a spatial order the place of an individual indicates nothing but certain relations he bears to other individuals. In this context, nobody thinks of 'place' as something absolute, as having any meaning apart from the relations between individuals. Although here, too, we can conceive of 'that which comprehends all those places', nobody interprets it as an independently existing entity. The genealogical order schematizes the structure of ancestral relations between individuals, and is not something else existing in addition to this structure.[18] This example is very striking, indeed, and provides an excellent elucidation of the problem of space. The parallelism extends even further than Leibniz may have realized. According to the investigations by K. Lewin,[19] we may regard the genealogical order as the space-time order of biological evolution, in exactly the same sense as we have to regard the causal order as the space-time order of physics. Here, as in many other areas, – I will mention only the *analysis situs* and the geometric characteristic – Leibniz' ideas have undergone an unforeseen extension and elaboration in our days, for only now is it possible to understand them fully.

I should like to make just a brief remark concerning the superiority of Leibniz' theory of space and time over Kant's.[20] Leibniz speaks of the 'ideality of space and time' only in connection with more rigorous formulations.

And he makes it clear enough that space and time constitute a framework in which things are ordered, but that they themselves are not things. Kant held essentially the same view but his terminology lends itself to dangerous misinterpretation. In spite of their ideality, space and time are no less anchored in the nature of things than any other conceptual schema of science. Although genealogy constitutes a conceptual order, it nevertheless describes an actual state of individuals. One must, however, avoid the familiar epistemological mistake of taking 'description' to mean 'copy of reality'. Space and time have a similar function; although they are conceptual schemata, they describe actually existing states of things. Recent axiomatizations have shown that the empirical content of the description can be formulated much more rigorously. Leibniz, of course, did not yet see the possibility of such an extended construction in which space and time represent nothing but the empirically given order of causal sequences in the world. Yet his space-time theory leads directly to this result, whereas Kant's theory does not. Due to his doctrine of the dual nature of space and time, their transcendental ideality and empirical reality, Kant dealt quite inadequately with the objective character of space and time.[21]

The answer which Clarke gives to Leibniz' presentation of the problem of space manifests such narrow-mindedness that it cannot contribute significantly to a clarification of the question. Convinced of his superiority, Clarke declares, with the complacency of a person not inhibited by any capacity for further enlightenment, that the relevant passages in Leibniz' letter "do not contain serious *Arguments*".[22] Furthermore, he attempts to point out contradictions in Leibniz' views; he thinks it is a contradiction to say that an order of situations is a quantity.[23] Leibniz had previously explained that the serial order, the existence of preceding and succeeding elements, provides the conditions of measurability.[24] Clarke evidently holds the view that quantity is an absolute concept. How little Clarke understands Leibniz' train of thought is borne out in an earlier letter. Clarke argues that were the earth and sun and moon placed where the remotest fixed stars now are, while preserving their relative positions, then, according to Leibniz' view, not only would they retain the same effects upon each other, but they would also occupy their present places since Leibniz considers place to be a mutual relation, which would be a manifest contradiction.[25] Clarke does not see that a contradiction arises only for the absolutist, not for the relativist. If there existed no other bodies in the world than the earth, the sun, and the moon, these bodies would indeed retain their places because then the shift to another position would be completely fictitious. But if, in addition, the fixed stars existed, then, on Leibniz' view, the earth, the sun, and the moon would change their places since their positions relative to the fixed stars would have changed, and their places are determined not only by their mutual relations, but also by their relations to all other bodies.[26] Leibniz still defends kinematic relativity brilliantly. Clarke had objected that the fact of motion cannot be dependent upon observation; a ship is moving even though this motion is not observable to a person inside the cabins.[27] Leibniz gives the conclusive answer: "Motion does not indeed depend upon being *Observed*; but it does depend upon being *possible to be Observed*. There is no *Motion* when there is no *Change that can be Observed*. And when there is no *Change that can be Observed*, there is *no Change at all*."[28]

The principle of the identity of indiscernibles, as much as we are inclined to grant it, raises great difficulties for Leibniz' relativistic viewpoint as soon as motion is no longer considered kinematically, but dynamically. Although Clarke's philosophic interpretation of space and time is quite inferior,[29] he avails himself of one argument concerning the problem of motion which Leibniz cannot refute decisively. Newton had suggested that centrifugal force could be used as a criterion for distinguishing absolute motion from relative motion. When Clarke presents Newton's argument,[30] Leibniz must admit that the characterization of a motion as absolute is at least possible dynamically, the principle of the identity of indiscernibles to the contrary notwithstanding. "However, I grant there is a *difference* between *an absolute true motion of a Body,* and a *mere relative Change in its Situation with respect to another Body*. For when the immediate Cause of the Change is in the Body, That Body is truly in Motion; and then the Situation of other Bodies, with respect to it, will be changed consequently, though the Cause of that Change be not in Them".[31] Leibniz might argue that the centrifugal force of the earth's rotation, noticeable, for instance, in the flattening at the poles, shows that the driving forces of the rotation inhere in the earth and not in the fixed stars; therefore, the rotation of the earth must, in this case, be called "an absolute true motion", whereas the rotation of the fixed stars is "a mere relative change of situation". Such an admission constitutes a serious break in Leibniz' theory of motion because the distinction between absolute and relative motion is equivalent to an acceptance of Newton's view. Leibniz lacks Mach's argument which alone can defend the relativity of dynamic motion. From here on, Leibniz' argumentation becomes inadequate. Clarke had declared:[32] "The *reality of Space* is not a *Supposition*, but is *proved* by the fore-going Arguments, to which no Answer has been given." Of course, Clarke does not realize which of his arguments is decisive because none of the arguments to which he refers mention centrifugal force; so far his arguments have consisted merely of inadmissible absolutistic contentions. It would seem that at this point Clarke's invisible mentor, Newton himself, has intervened, for only now does Clarke mention the problem concerning rotation. From here on, Leibniz' reasoning exposes itself to attack. He does not think that his admission of a distinction between true and relative motion adversely affects his conception of space. This view is probably to be understood in the following way: For Leibniz, space is not the cause of centrifugal forces; rather, these forces are located in the bodies themselves. If one speaks of true motion, one no longer regards motion as a process in space because such a process is always relative, but as a dynamic process which is not spatial.

Leibniz therefore needs two different concepts of motion, one applying to a process in space, the other to a process in the interior of things. This conception is hardly defensible because it implies that motion as an internal process enables us to distinguish kinematically equivalent motions. Unfortunately, Leibniz passes over these difficulties very rapidly. Immediately after the quoted passage about true motion he writes: "Thus I have left nothing unanswered, of what has been alledged for the absolute reality of Space".[33] One must concede that Clarke is right in not being content with this explanation; in his answering letter, he reverts to the subject of centrifugal force.[34] He concludes that, according to Leibniz' point of view, the rotating sun would lose its centrifugal force if all matter around it were destroyed. This conclusion is actually correct, and modern physics admits it. Unfortunately, we do not know what Leibniz would have answered because at this point the correspondence was interrupted by his death. If we were to carry through Leibniz' view consistently, we would have to say that it would be possible for centrifugal forces to occur on a sun not surrounded by matter and therefore on a sun not surrounded by fixed stars; but in such a case, we should not regard centrifugal force as a criterion of rotation because the motion of an isolated body is a meaningless fiction. For an isolated body, the concept of motion as a process in space loses its meaning, but we can draw no conclusions concerning its dynamic state.[35] It is, of course, futile to pursue Leibniz' viewpoint further than he himself did. We must regret that the problem of rotation is not treated more extensively in this correspondence because Leibniz might have been forced to give a more rigorous analysis which might have saved him from the inconsistency of accepting dynamic absolute motion. Under the circumstances, we must be content with Leibniz' consistent defense of the relativistic interpretation of space as a schema of order, even though he did not develop a relativistic theory of motion.[36]

IV

The correspondence with Clarke could not lead to a solution of the problem of motion because the intellectual level of the disputing opponents was too uneven. Leibniz had to answer the theologian Clarke with theological arguments that were supposed to settle the question about the nature of space on the basis of the dignity and perfection of God. On the other hand, Leibniz, too, was so accustomed to theological arguments, it was so natural for him to frame epistemological considerations in theological terms, that a great deal of the discussion becomes unfruitful for us. We may expect a greater reward, therefore, when Leibniz faces an opponent who is his peer and to whom he can communicate his ideas in a more sophisticated fashion. For this reason, his correspondence with Huyghens is one of the most important documents for the evaluation of his theory of motion.

Even though the theory of motion which Leibniz formulates in his correspondence with Huyghens is more profound than that which he developed in his letters to Clarke, one must not regard the more profound theory as the later one. Chronologically, this correspondence preceded that with Clarke by about twenty years, and Leibniz based his answers to Clarke upon the previously developed theory of motion. Apparently, the belief that he could expect a complete understanding of his ideas from his former mathematics teacher encouraged Leibniz to be more open in his communications with Huyghens, whereas Clarke's hostile attitude compelled him to pass over unsolved problems, and, instead, to present the epistemological foundations of his conception in a more detailed manner. The above considerations explain the method used in our presentation of Leibniz' theory of motion, which begins with the chronologically later source and then turns to the chronologically earlier one.

Discussions among like-minded scholars who are working in the same field are usually briefer in their expositions than are presentations intended for a wider audience. What has to be explained to the student or opponent by means of prolix and detailed illustrations and psychological devices can be conveyed to a co-researcher in a single remark. Concentration on the same subject-matter creates a common atmosphere in which only the barest means of expression are needed for the sharing of knowledge. This circumstance may make it difficult for later readers to understand a correspondence between experts; however, advances in science may enable those who come later to see a problem in the same light as, or perhaps even in a brighter light than, its discoverers did. Thus, we who have been granted the privilege of an acquaintance with Einstein's solution of the problem of motion can appreciate the great significance of the correspondence between Leibniz and Huyghens. Incidentally, the discussion of the problem of motion constitutes the smallest part of this correspondence. The interesting passages on the problem of motion, in which the two great mathematicians inform each other briefly about their respective views, are to be found only in the last year of the correspondence,[37] and these passages are scattered among long expositions of the integral calculus (in which Leibniz tried to explain the advantages of his calculus to his teacher who at first hesitated to accept it), discussions of Newton's and Huyghen's optics, of atoms, and of many other matters.

The lucid style of this correspondence reveals at once a most fortunate meeting of minds. The two correspondents are not engaged in a controversy over world views; on the contrary, they agree in their fundamental approach to the problem. Both Leibniz and Huyghens are opponents of Newton's philosophy. They are repelled by the supernaturalistic tendencies of this physicist who leans upon Henry More and who always turns into the mystic and dogmatist as soon as he leaves the boundaries of his special field. Both know that the crude conception of a substantial space which Newton shares with the epistemologically untutored layman cannot be correct. However much they respect Newton's physical discoveries, they reject the philosophical consequences he draws from these discoveries. Huyghens even conjectured that Newton would revise his interpretation of motion in the next edition of the *Principia*, so secure did he feel in his rejection of Newton's absolutism.

The discussion of motion begins when Huyghens asks Leibniz about his concept of real motion. In his answer, Leibniz does not enter into an analysis of the concept of space, as he does in the letters to Clarke, but forthwith explains his conception of motion. This conception exhibits the same duality that we noted above. Motion as a process in space is relative, but there must exist a subject of motion in which the impelling force occurs, and this subject is in true motion. Leibniz thinks that if there were no such subject, the impelling force would not be real. Nevertheless, he does not regard this conception as a violation of relativity because the geometrically observed processes do not furnish any means of determining the subject of the motion. "If *a* and *b* approach each other, I admit that all phenomena will be the same, no matter to which of them one ascribes motion or rest; and even if there were 1000 bodies, I agree that the phenomena will not give us (nor even the angels) an unfailing criterion for determining the subject or the degree of the motion (pour déterminer le sujet du mouvement ou de son degré), and that each of them could just as well be conceived as being at rest."[38]

Leibniz gets into difficulties with this conception. He must assert "that actually each body has a certain degree of motion, or, if you wish, of force, notwithstanding the equivalence of the assumption".[39] Although he admits that the true force cannot be ascertained, it is supposed to exist. Since this contradicts his principle of the identity of indiscernibles, he must look for a method of determining the true force. He believes that he can escape his difficulties by finding a non-geometrical method of ascertaining the true force. He is faced with a problem in dynamics, and his concept of true motion compels him to assert that dynamics contains a non-geometrical metaphysical principle, and that 'motion' has a metaphysical, as well as a spatial significance. Because the subject of the motion can be distinguished, Leibniz concludes "that there exists in nature something that geometry cannot determine. And this is not the least of the reasons of which I avail myself in order to prove that besides extension and its changes (variations) which are something purely geometrical, one has to recognize something more important, namely, force".[40] Here we have in much purer form than in the answer to Clarke a formulation of the metaphysical concept of motion. Motion in the metaphysical sense is not 'change of place', but occurrence of forces.

In this passage, modern and obsolete ideas are strangely mixed. Leibniz realizes that it is impossible to recognize true motion by direct observation. Nor does dynamics supply direct evidence for the occurrence of a force since one can invent a dynamic hypothesis for each of the equivalent interpretations of motion. A given interpretation will be preferred only because of its simplicity ('la simplicité de l'hypothèse'). Thus, the Copernican system is preferred to the Ptolemaic system because of its simplicity; it is merely the most adequate system for explaining the phenomena (ad explicanda phaenomena aptissima).[41] Yet Leibniz believes that the simplest explanation is to be regarded as the true one[42] (qu'on peut tenir la plus simple, tout considéré, pour la véritable), and thus his investigation of the logical equivalence of the dynamic explanations leads him to reaffirm the metaphysical distinctness of one of the motions. It is his metaphysical system which influences him at this point; his *Monadology* is closely connected with his absolutistic concept of force, and in this way his total system becomes the obstacle which prevents him from making a consistent analysis of this particular problem.

Leibniz presents, for the first time, one argument which certainly does constitute a serious objection against Newton's theory of motion, and which has exerted great historical influence. This argument demonstrates the inconsistency inherent in Newton's special treatment of uniform rectilinear motion. If space is conceived as something real, and if it is possible through dynamic criteria to ascertain a motion with respect to space, then it must also be possible to characterize one of the uniform motions as a state of rest. The dynamic relativity of uniform translation is just as inconsistent within the framework of Newton's absolutistic theory of motion as is the postulation of a metaphysical state of motion within Leibniz' relativistic theory. It would seem that Newton himself was not aware of this inconsistency; at least, I know of no passage which would lead one to believe he was.

Leibniz, who was much more perceptive about epistemological issues, must have felt uneasy about the discrepancy in his theory of motion. By admitting 'true forces', Leibniz had singled out one of the kinematically equivalent motions, and he was therefore induced to attempt, once again, to extend kinematic relativity to dynamic motion by means of a number of assumptions about the subject of motion. Newton had claimed that circular motion was a counter-instance to dynamic relativity. Leibniz, however, declared that "in this respect circular motion has no prerogative" and that "all assumptions are equivalent".[43] One would expect Leibniz at this point to present a solution along the lines of Mach's argument. Mach contended that the origin of the centrifugal force need not be restricted to the earth as the subject of motion, but may be ascribed, just as well, to the fixed stars; thus the distribution of the forces upon the masses may be adjusted to the selected kinematic description of the motion. Unfortunately, Leibniz merely hinted at another explanation based on certain peculiar views about the nature of a body; we shall probably never know what he meant by this remark.

V

In comparison with Leibniz, Christian Huyghens – or Hugens de Zulichem, as he signed his name in the French version under these letters – reveals himself as the more consistent relativist. Although he admits that he had previously accepted the special role of circular motion in Newton's sense, he says that he has abandoned this view in the meantime. He is firmly convinced that there is no real motion, but only relative motion; nor does he let himself be "diverted by Newton's reasoning and experiments in his *Principles of Philosophy* which I know are erroneous" (sans m'arrester au raisonnement et expériences de Mr. Newton dans ses Principes de Philosophie, que je scay estre dans l'erreur).[44] He also rejects Leibniz' concept of true motion; "I cannot agree with you that a number of bodies which are in motion relative to each other have each a certain degree of true motion or force" (de mouvement ou de force véritable).[45] Evidently he perceived the weakness of Leibniz' conception, and from this hint alone, which focuses on the major error in Leibniz' theory, we may conclude that Huyghens had a more discriminating conception of dynamic motion. Yet these letters do not convey anything about the content of Huyghens' theory; he merely says that he has found "two or three years ago the more correct view[46] concerning circular motion. The last letter Huyghens wrote to Leibniz, four months later, does not refer to the problem of motion; then his death interrupted the correspondence forever.

Thus, this correspondence concludes with the assurance by both mathematicians that they possess a solution to the problem of circular motion which does justice to the relativistic point of view even in its dynamic aspects. Although in Leibniz' case we must be content merely with his assurance without having any grounds for believing that he had in fact found a solution, we are in a more fortunate position as far as Huyghens is concerned. Among his posthumous writings, which were carefully preserved in Leiden, several manuscripts have recently been discovered by J. Korteweg, and these have now been published by J. A. Schouten.[47] After more than two centuries, we have finally received the answer to the most important question left open in the correspondence with Leibniz. These manuscripts reveal that Huyghens was the first relativist to attempt to solve the problem of the dynamics of circular motion, though his solution is not correct.

There are four separate sheets on which Huyghens made notes concerning the problem of circular motion. The first and earliest shows Huyghens still holding Newton's view that the velocity of rotation must be calculated by means of centrifugal force. The second sheet, which Korteweg dates 1692 on the basis of a remark in one of Huyghens' letters to Leibniz, attests to the decisive change in Huyghens' interpretation of circular motion, and, together with the third sheet dating from the same year, indicates Huyghens' solution of the problem. The fourth sheet, which must be dated later, contains a summary of this solution.

Huyghens' solution is based on a discovery in kinematics: change in mutual distance is not the only criterion of relative motion. If we look at a rotating disk, two diametrically opposed points always move in opposite directions, i.e. they move relative to each other without any change of distance between them.[48] In this case, a change of distance is prevented by the rigid connection between the points. Therefore, Huyghens concludes that one cannot infer from the constancy of the distance between two bodies that they are at rest relative to each other; such an inference is justifiable only if, in addition, there is no physical connection between the bodies. It is true that free bodies, but only such bodies, will change their distances when they are in relative motion. If two spheres connected by a string are pushed in opposite directions, perpendicular to the connecting line, they will start moving relative to each other; because of the connecting string, this relative motion will not be accompanied by a change in their mutual distance but will be confined to rotation. The customary definition of 'rest' must therefore be changed: "one must know that bodies are at rest relative to each other only if they retain their position relative to each other while being free to move separately, and being neither tied nor held together (il faut donc scavoir que l'on connoit que des corps sont en repos entre eux, lorsqu'estant libres a se mouvoir séparément et point liez ni detenus ensemble, ils gardent leur position entre eux)".[49] The condition that the bodies be disconnected is the novel feature in Huyghens' definition.

On the basis of this definition, Huyghens arrives at an interpretation of circular motion as relative motion, as the motion of the parts of a circle relative to each other, not as an absolute motion with respect to space. It is only because this relative motion is not accompanied by a change of distance that its relative character has been overlooked. Hence dynamic relativity has been established: From the occurrence of a centrifugal force, the occurrence of a rotation can be inferred, but this dynamic effect again confirms only a relative motion, that of the individual particles to each other. "For a long time I thought that one had a criterion for true motion in centrifugal force. Actually, with respect to all other phenomena, it makes no difference whether a disk or a wheel rotates where I am standing, or whether I circle the stationary disk. But if a stone is put on the perimeter, it will be thrown outward when the disk is rotating, and therefore I used to think that the circle does not rotate relative to any other body (nulla ad aliud relatione). Nevertheless, this phenomenon only shows that the parts of the wheel are driven in opposite directions relative to each other (motu relativo ad se) through the pressure exerted upon the perimeter. Circular motion is thus a relative motion of the parts which are driven in opposite directions, but held together by a string or connection (ut motus circularis sit relativus partium in partes contrarias concitatarum sed cohibitus propter vinculum aut connexum)".[50]

This solution of the problem of rotation is as new as it is interesting. It represents a consistent interpretation of dynamic relativity, of the idea that even from the occurrence of forces, only a relative motion of bodies can be inferred. This solution abandons Leibniz' concept of true motion, and Huyghens carefully emphasizes that circular motion does not help us to ascertain which of the rotating parts have received the first impact, i.e. which of them are the subjects and are in true motion in Leibniz' sense. One might perhaps consider this solution the first theory of relativity – even though it cannot be maintained in view of the modern theory.

Today we would say that the relative motion of the parts of the circle can be 'transformed away', that there is a stationary system of co-ordinates in which it vanishes, namely, the system which rotates along with the parts. In this respect, rotational relative motion differs from other motions; translational relative motions, for example, are accompanied in every stationary co-ordinate system by differences in the velocities of the individual bodies. With regard to dynamic relativity, the following requirement must be made: if a dynamic effect due to the motion of a body is observable from a co-ordinate system, the body must also be in motion relative to this co-ordinate system. This condition is realized by motions which 'cannot be transformed away', where the force occurs in every co-ordinate system (i.e. the force is a tensor); in the instance of circular motion, however, we should expect the centrifugal force to vanish in the accompanying rotating system. Yet actually this is not the case; a body moving with a rotating disk is subject to the pull of the disk, and the centrifugal force cannot be interpreted as an effect of the relative motion of the parts of the disk since this effect also occurs in that system in which the relative motion is equal to zero.

Although we must reject Huyghens' solution as incorrect, its historical merit remains undiminished. It manifests consistency in interpreting the problem of motion never achieved before Huyghens and only attained by Mach and Einstein after him. This solution is the answer of a persistently searching mind who was guided by his insight into the nature of motion and who, thus motivated, took up the fight against the impressive system of Newton. Both Leibniz and Huyghens are sustained in their approach to the problem of motion by a conviction which constantly draws their theories in a single direction, just as a magnetic field draws the needle of a compass. If, in this pursuit, Leibniz was more rigorously trained in epistemology and therefore better able to give a logical analysis of the problem, Huyghens was better grounded in applied mathematics and, being less hindered by an epistemology of his own, was more consistent in his solution of the problem of rotation. Huyghens must have had a very vivid feeling of superiority towards Newton's theory, and the final words from Newton's introduction to the *Principia*, which we quoted in the beginning of this paper, probably appeared tragic even to the contemporary of the great dogmatist of absolutism. On one of the separate sheets where true motion is called a 'false idea' and a 'chimera', we find the following note in the margin: "Mr. Newton dit qu'il a escrit tout son traité pour connoitre le vray mouvement".[51] Although Huyghens knew that the goal which Newton had set for his work was unattainable, he could not prevent the recognition accorded to the *Principia*. It is incumbent upon *us*, however, who have witnessed the final abandonment of Newton's theory of motion, to bestow our belated recognition on the two unfortunate relativists of the classical period of absolutism.

NOTES

[1] See Ernst Cassirer, *Substance and Function and Einstein's Theory of Relativity*, (Chicago–London, 1923), p. 410f.

[2] In this context, compare my presentation in [1922f], p. 328f. I gave a detailed analysis of the Mach-Newton controversy in [1922c]. See also Schlick, 'Die philosophische Bedeutung des Relativitätsprinzips', *Ztschr. f. Philos. Kritik* **159**, 129 (1915).

[3] He writes (*Die Mechanik in ihrer Entwicklung*, Brockhaus, 8th ed., Leipzig, 1921, p. 431): "There is no need to spend any time on Leibniz, the inventor of the best of all possible worlds and of the pre-established harmony; these inventions have found the treatment they deserve in Voltaire's apparently funny, actually very serious novel *Candide*. As is well known, he was just as much a theologian as he was a philosopher and scientist". [My translation. M. R.]

[4] See for instance Einstein, 'Grundlage der allgemeinen Relativitätstheorie', Section 2, *Annal. d. Physik* **49** (1916) [English translation by W. Perrett and G. B. Jeffrey as Chap. 7 in *The Principle of Relativity* by Einstein, Lorentz, Minkowski and Weyl (Methuen, London, 1923; Dover reprint, New York n.d.) – Ed.]

[5] These definitions are quoted and translated [into German] in H. Scholz, 'Zur Analysis des Relativitätsbegriffes', *Kantstudien*, p. 388 (1922). The same work provides a fairly detailed exposition of Newton's conception of relative motion, correctly identifying it with changes in mutual distances.

[6] *Sir Isaac Newton's Mathematical Principles*, translated into English by Andrew Motte in 1729, ed. by Florian Cajori, (Berkeley, 1947), p. 12.

[7] See Cassirer, *Erkenntnisproblem in der Philosophie und Wissenschaft der neueren Zeit* vol. II, p. 471. (Berlin, 1911–20).

[8] [The following quotations are taken from *Leibniz, Selections*, ed. by P. P. Wiener, New York, 1951, rather than from sources used in the original German paper – M. R.]

[9] Leibniz does not notice, of course that even this condition does not necessarily lead to a unique definition of simultaneity; such considerations could only originate in the context of Einstein's theory of time. A detailed investigation of this problem is presented in my book [1924h]. The theory of time contained in this book may be regarded as a continuation of Leibniz' ideas, although at the time of writing I was not aware of this connection since I did not then know the relevant passages by Leibniz.

[10] Wiener, *op. cit.*, pp. 201–2.
[11] Wiener, *op. cit.*, p. 202.
[12] *Ibid.*, p. 202. [I inserted 'of utmost definiteness' because Wiener does not translate 'maxime determinatum'. – M. R.]
[13] In my book [1924h] the topology of space is constructed in a similar way; in this axiomatization, the concept *spatially nearer than* is reduced to the concept *temporally earlier than* by means of a special kind of causal sequence: first signals.
[14] Samuel Clarke, *A Collection of Papers, Which passed between the late Learned Mr. Leibnitz and Dr. Clarke, in the Years 1715 and 1716*, (London, 1717), pp. 195–7. [In this edition, the original French correspondence is printed on the left-hand pages, Clarke's own English translation on the right-hand pages of the book. I have substituted this edition for the Cassirer edition used in the German essay. – M. R.]
[15] *Ibid.*, p. 202.
[16] *Ibid.*, p. 197–9.
[17] *Ibid.*, p. 201.
[18] Leibniz carries the parallelism between spatial order and genealogical order a step further so that he can also apply the concept of motion within the genealogical order. "And if to this one should add the Fiction of a *Metempsychosis*, and bring in the *same* Human Souls again; the Persons in those Lines might change Place" (Clarke, *op. cit.*, p. 201). Metempsychosis functions as an analogy to motion in space and time; one individual replaces another one, assuming that relation to the whole which the first individual had before him. Yet strictly speaking, this extension destroys the parallelism. The genealogical net has a time dimension, and is to be compared not solely to space, but to a space-time manifold of two dimensions, such as that used in diagrams of the Lorentz transformation. In this case there is no motion; there are only *world lines*. Every time line of the net which proceeds from grandfather to father to son, etc., represents a motion, namely, the transition from one generation to the next. Each generation corresponds to *space at a certain time*.
[19] Kurt Lewin, *Der Begriff der Genese in Physik, Biologie und Entwicklungsgeschichte*, (Berlin, 1922).
[20] A detailed comparison between Leibniz' theory and Kant's is to be found in Cassirer, *Leibniz' System*, Marburg, 1902, p. 271f. Cassirer makes use of the evaluation of Leibniz given by Kant in 'Metaphysische Anfangsgründe der Naturwissenschaft'. Cassirer is intent on pointing out the close relationship between the two doctrines, whereas it seems to me that in the light of the relativistic solution of the problem, Leibniz' theory must be ranked above that of Kant's.
[21] For a characterization of reality by means of conceptual systems, see my presentation in [1920f] p. 85f. However, the characterization of the invariant content of the metric given in this book needs correction, and is superseded by my [1924h].
[22] Clarke, *op. cit.*, p. 299.
[23] *Ibid.*, pp, 309, 347.
[24] *Ibid.*, p. 215.
[25] *Ibid.*, p. 75.
[26] Cassirer has already pointed out this *petitio principii* of Clarke's on p. 141 of his edition of *Leibniz' Werke*.
[27] Clarke, *op. cit.*, p. 133.
[28] *Ibid.*, pp. 211–13.

[29] The opinion that Clarke was not too successful in defending his position is widely accepted, even by those philosophers who have moved away from Kantianism. See, for example, Hans Vaihinger, *The Philosophy of 'As If'*, London, 1924, p. 229f. However, Vaihinger's attempt to reconcile the contrasting viewpoints of the opponents by using his notion of fictions (p. 230) is not tenable in the light of modern physics.
[30] Clarke, *op. cit.*, p. 133.
[31] *Ibid.*, p. 213.
[32] *Ibid.*, p. 135.
[33] Clarke, *op. cit.*, p. 213.
[34] *Ibid.*, p. 295.
[35] Cassirer presents a different interpretation in his edition of *Leibniz' Werke*; see p. 219, footnote 158.
[36] It has been pointed out a number of times, in discussions of the theory of relativity, that the distinction between *true* and *spatial* motion is also made by Kant in 'Metaphysische Anfangsgründe der Naturwissenschaft'. Kant calls circular motion 'true motion' and distinguishes it from 'apparent motion', but he does not characterize it as absolute motion in contrast to relative motion. This conception of Kant's was anticipated by Leibniz. One advantage of Kant's interpretation – which, incidentally, like our own elaboration of Leibniz' ideas treats the problem of an isolated body by reference to a rotation recognizable through Coriolis forces – seems to me to be that he does not regard the distinction between true and apparent motion as an objection to Newton. The vehemence with which Leibniz opposed Newton is due, apart from the historical situation which made him Newton's rival in the controversy over the invention of the integral calculus, rather to their different conceptions of space than to their disagreement about motion. Leibniz' theory of space is superior to Newton's, and, as we have tried to show, also to Kant's, but he emphasized the differences between his theory of motion and Newton's more strongly than is warranted in view of his inability to demonstrate dynamic relativity.
[37] Recognizing the importance of this discussion E. Cassirer has included these passages of the correspondence in his Leibniz edition.
[38] [I am translating the original text from *Leibnizens mathematische Schriften*, ed. by C. I. Gerhardt, Berlin, 1850, vol. II, p. 184. – M. R.]
[39] Gerhardt, *op. cit.*, p. 184.
[40] *Ibid.*, p. 184.
[41] From the 'Promemoria' which Leibniz presented to the Vatican in Rome. See Gerhardt, *op. cit.*, vol. VI, p. 146 n.
[42] One is inclined to agree with Leibniz' assertion since today, more than at any other time, simplicity is accepted as a criterion of truth. But just in this case the argument is not correct. I must refer the reader to my book [1924h] (§ 2).
[43] Gerhardt, *op. cit.*, vol. II, p. 199.
[44] *Ibid.*, p. 177.
[45] *Ibid.*, p. 192.
[46] *Ibid.*
[47] *Jahresbericht d. deutschen Mathematiker-Vereinigung* **XXIX**, 136–44 (Leipzig, 1920).
[48] In mathematical language: the difference of their velocity vectors is not equal to zero.
[49] *Op. cit.*, p. 139.
[50] *Ibid.*, p. 142.
[51] 'Mr. Newton says that he has written his whole treatise only in order to find the true motion.' *Ibid.*, p. 141.

46. THE RELATIVISTIC THEORY OF TIME

[1924f]

While, from the very first, the theory of relativity emerged in quite a complete form as regards its physical aspects, its philosophical foundations were uncovered only gradually. To be sure, Einstein's work characteristically rests upon a totally lucid insight into basic philosophical concepts, but the creator of this great theory possessed this insight more as intuitive knowledge than in the form of a conceptually complete epistemology. Above all, his attention is directed so intently upon the physical development of the theory that dwelling at length upon the philosophical foundations would only obstruct his progress. His all-too-brief philosophical comments have caused some opponents to accuse him of a lack of philosophical understanding and to attempt to restrict his achievements to a merely mathematical significance – a misunderstanding that can only appear laughable to anyone well acquainted with Einstein's physical theory, particularly with his mode of thinking. Is it possible to regard as a philosophical layman a physicist who so successfully returns to epistemological problems in resolving physical difficulties? Surely it would be more fitting to endeavour to establish through modest philosophical work what Einstein has actually achieved by way of philosophical knowledge, while leaving to him the right to continue expanding his new physics untroubled by its philosophical consequences. It has frequently occurred during the course of the history of philosophy that the natural sciences have produced many more philosophical ideas than the philosophy of the scholars. We thank the exact sciences for the conquest of scholasticism, – such men as Copernicus, Kepler, and Galileo, without whom others, such as Descartes and Leibniz – themselves half-mathematicians – would not have been able to lay the foundations of their epistemology. Will we perhaps witness in our own day another turning point, whereby mathematics and physics will again recognize the way of knowledge better than the judges appointed for that purpose?

Thus we should not try to *judge* – it is first of all necessary to *comprehend.* What is the philosophical significance of Einstein's discoveries? We will be able to answer this question only if we allow the same exactitude that is the hallmark of the mathematician to predominate in our philosophical methodology. Better to limit ourselves to one special problem than to discourse about the whole in propositions of empty generality: let us, for the time being, bend our attention only to the meaning of Einstein's first innovation, the *relativization of time.*

I

We will straight away correct an error of which the great majority of popular presentations of the theory of relativity must be accused. We commonly read such statements as, "What appears to one observer as simultaneous appears to another observer who is moving in relation to him as not simultaneous." Indeed, the writers of these articles work diligently to present the relativity of *time* as a relativity of the *observer,* in the apparent belief that the differences in question are comparable to the variation in the spatial perspectives of two remotely situated observers and originate in the subjectivity of sense perception. But this shows a misunderstanding of the logical character of the Einsteinian theory of time, which pertains only to the *logical* conditions of knowledge, not to the *psychological.* It does not query the perceptions of the observer, but rather the knower's schema for ordering his knowledge. It discloses presuppositions of knowledge, not of knowing as a psychological act.

For if we just take a good look at the sense perceptions of an observer, we see that they are *unable* to give any report regarding simultaneity at different locations, as it is treated in Einstein's theory. The observer is bound to his location in space; he merely receives reports, signals from distant points, and all that he can *perceive immediately* is the simultaneity of their arrival at his position. This position is, to be sure, not a precise mathematical point, but we may regard it as practically infinitesimal in comparison to the distances that light traverses in a very few seconds, and which are the subject of relativity theory. The arrival of a signal at the observer is to be regarded as a coincidence, or point event; we shall take over the concept of *simultaneity at the same position* unaltered from classical physics. The *logical* problem that arises over and above sense perception is this: how is the observer able to come to a *temporal ordering of spatially distant events?*

"By means of physical measurements", is the first answer to come back. The observer measures the spatial distance and divides it by the velocity of the signal, thus getting the time taken to traverse the path. If, for instance, a light ray from Sirius arrives at the earth simultaneously with a light ray from the sun, we can calculate from the distance of these stars and the speed of light the time at which each of the light rays departed; and then we have the comparative times on the sun and on Sirius.

This is certainly correct. But we need to know the speed of light. How are we to measure it?

There is basically only *one* method for measuring the velocity of a signal, which can be roughly schematized in the following way. Imagine that clocks are set up at two distant points. A signal leaves the first point at, say, 12:00 o'clock. It arrives at the second point at 12:05; thus the signal took five minutes to traverse the path, which we shall measure, and its velocity is established by means of division. This is the only possible method for measuring the velocity.

But is this really true? Was the velocity of light not measured in a completely different fashion by Fizeau? He sent a light ray to a distant point, from which it was reflected and returned. Thus he needed to measure only the time of departure, without regard to the time of its arrival at the mirror. But he got only the sum of the times for the journey there and the journey back. He was not able to measure precisely what interests us: *the velocity in a single direction.* Our previous assertion, then, is true.

We must note that our measurement of velocity leads to a difficulty. In order to measure the velocity, we require *two* clocks, at different locations. In order for the difference in time that we read off from them to be meaningful, the clocks obviously must be *compared;* i.e., we must establish whether the clocks *simultaneously* point to the same numbers. Yet our whole measurement of velocity has been instituted solely for the purpose of achieving a means of establishing simultaneity at distant positions. We find ourselves caught in a circle: in order to establish simultaneity of distant events we must know a velocity, and in order to measure this velocity we must be able to judge the simultaneity of distant events.

Einstein has shown the way out of this *logical circle:* we cannot *know* the simultaneity of distant events at all, but can only *define* it. Simultaneity is *arbitrary;* we can lay down whatever definitions we wish concerning it, without giving rise to an error. For if we subsequently make measurements, we will invariably reach the result of the same simultaneity that we inserted by definition in the first place; this process can never lead to a contradiction.

The question will be raised whether this is really Einstein's theory. Does the special theory of relativity not contain a stipulation according to which clocks are to be set, arranging them in such a way that the velocity of light is the same for the journey in each direction? Certainly – but it would be a mistake to suppose that this stipulation claims to be *truer* than any other.

It merely leads to simple numerical relations, but these, too, are valid only if certain objective conditions are realized, namely the absence of a gravitational field. And in such a case the simplicity of this particular definition is lost. Furthermore, even if this definition is simpler, a more complicated definition of time is also admissible in gravitation-free space, for it too would lead to an unambiguous quantitative description of all natural events, and this achievement is sufficient to satisfy the demand for *truth*. Only in the general theory of relativity did Einstein succeed in implementing complete arbitrariness in the choice of time [definitions]. According to this theory, all ways of defining time are equally legitimate. Thus Einstein's general theory offers the first real solution to the epistemological problem of simultaneity[1].

All this demonstrates that Einstein was, in fact, treating the *logical* problem of time order, not the *psychological* problem. The question of defining time exists for every observer in the same way, and the individual observer actually has all the possibilities for temporal definition at his disposal. We simply get a better illustration if two separate definitions of time are assigned to two different observers. Yet the observer 'at rest' can define time just as well as the observer 'in motion', although, to be sure, the velocity of light will not be constant for his system, but will be different for the two directions along a single path. For we have here not a difference in standpoints but a difference in the logical presuppositions of measurement, which must be arbitrarily established before any measurements can be made at all.

II

Only when we possess this conception are we able to tackle the epistemolocical questions entailed by the relativistic theory of time. The problem must first be presented with logical clarity before we can answer the further question: is Einstein's solution *admissible*?

It is occasionally admitted that it is impossible to measure simultaneity; but in reply the objection is raised that no conclusions concerning real relations are to be drawn from this fact. There are many cases in which physicists come up against the bounds of measurement, but are they therefor to conclude that the quantities to be measured do not exist at all? It is utterly impossible, for instance, to determine the precise number of molecules in a cubic centimeter of air, and we may state with the greatest assurance that we will never succeed in actually counting every single molecule. But may we conclude that this number does not exist? Quite the contrary; we must maintain that there always is a whole number that precisely characterizes this quantity. Einstein's error – for his opponent – consists in confusing the *impossibility of measuring* with *objective indeterminacy.*

Those who raise this objection fail to note a vital distinction. Sometimes there is an impossibility of measurement because of the limitations of our technical instruments; in these instances, I shall speak of *technical impossibility.* But in addition, there is *impossibility-in-principle* on *logical grounds.* Even if we possessed a perfect and complete experimental technology we would not be able to circumvent this impossibility-in-principle. It is, for instance, impossible in principle ever to discover whether the meter housed in Paris is really a meter. Not even the very greatest refinement of our geodetic instruments can shed any light on this matter – simply because it is impossible to establish absolutely what a meter is. That is why we call the measuring rod in Paris the *definition* of the meter; we arbitrarily declare it to be a unit, and the question of whether it really is this unit loses all meaning. The situation is just the same with simultaneity, which involves, not technical limitations, but a logical impossibility. Setting out to measure simultaneity without first laying down instructions for determining it is just as senseless as trying to measure a distance in meters without first stipulating that a certain rod shall be the unit. The comparison with the measurement of the number of molecules is false, for this number is defined precisely, and it is merely a technical impossibility that we encounter. The number that cannot be measured for technical reasons can at least be *approximated* with increasing precision – but this process fails in the case of logical impossibility. We cannot even *approximate* the simultaneity of distant events in the absence of any definition of simultaneity. Furthermore, the technical impossibility remains even after the stipulation is made. Even Einsteinian simultaneity can only approximately be determined in practice.

The objection, then, is untenable. The relativity of time asserts the existence of an arbitrary element in the logical foundations of our process of measurement and has nothing to do with technical limitations. It asserts that the nature of simultaneity is not given in objective events, but instead is introduced into descriptions of nature through the form of our thinking. It asserts that we must first lay down a definition before we can apply the arrangement of natural events into the four coordinates of space and time. There is no argument against *this* part of Einstein's theory.

III

But another objection arises at this point. Granted, the simultaneity of distant events is in principle not measureable, granted, even, that it must be handled by means of a definition: is this definition therefore *arbitrary?* Is it not, rather, tied to certain restrictions that originate in the nature of our *thinking?* That there are mental presuppositions in natural science is, indeed, a recognized fact. But are these preconditions not in turn subject to restrictions of a particular kind that stem from the fact that we, as essentially rational, must grasp intuitively the object of our knowledge[2]?

This objection is invariably raised by those who intellectually acknowledge the Einsteinian theory of time but are unable to comprehend it intuitively. "I can't imagine all that" is a retort often heard, precisely on the part of thorough thinkers. There appears to be a mental compulsion to retain the absolute meaning of simultaneity, which we shall now examine.

Let us take two events: I knock on the wall, and you knock on the wall over there. Did these two events occur simultaneously? No, I heard clearly that you knocked later than I did. Einstein asserts that we could call these two events simultaneous? Impossible! Here intuition fails; here my thinking is subject to other restrictions.

But what, then, am I asserting when I say that two events are not simultaneous? I probably wish only to say that my *perceptions* of the events contain a quality that I call 'not simultaneous'. But we proceed from the perception to the events only by means of a *chain of inference.* If I observe the psychological compulsion more precisely, I find that it exists only with respect to the *perception;* in speaking of an intuitive non-simultaneity of events, I am quite illegitimately transferring this compulsion to *things.*

Let us now implement the arbitrary definition of time, calling the two events simultaneous. What peculiar consequences this has! At the moment at which I knocked – say, at 1:25 – a sound wave also departed which reached the place where you knocked earlier than your knocking began. Earlier? Yes, earlier, for a clock set up at your position would have registered my sound wave sooner than your knocking. Nonetheless, the moment at which you knocked registers on your clock at 1:25, for this is our arbitrary stipulation. Thus the sound wave left my position at 1:25 and arrived at yours at 1:23! But then it must have run backwards through time – a paradox indeed!

But what does that matter? The sound wave reached you just as you were lighting your cigarette. That *also* occurred at 1:23, so the two events still belong together. The the sound wave returned to me, having been reflected from your wall. It reached me just as my hand was about to leave the wall: at 25 minutes and 1/10 second after one o'clock. The moment of its arrival, then, also registers as 25 minutes and 1/10 second after one o'clock. The return journey of the wave thus took two minutes and 1/10 second – a very slow sound! Measured at my position, the time of the journey there and back taken together equals 1/10 second, yet the time for the return journey alone is two minutes and 1/10 second. Nonetheless, the calculation is correct, for the time for the journey there was negative, minus two minutes, and the resulting sum is 1/10 second.

Thus the measured velocities are very peculiar. But no matter how disturbing we feel this to be, we note that it does not in any way alter our *perception.* You perceived that the sound wave arrived at the very moment when you were lighting your cigarette – and this fact is preserved in this peculiar time order. What *I* perceived – the return of the sound wave to my position as my hand was still touching the wall – is also preserved. The psychological compulsion that led me to recognize the coincidence is not disturbed in the least. And note: our definition of time pertains only to the *interpretation* of our perceptions, not to the perceptions themselves. What is immediately given is not itself called into doubt; modification is made only in the *ordering* of what is given in relation to the physical world.

For this very reason, an *a priori* ability cannot teach us anything concerning the justification of our definition of time. What we actually experience is not disturbed by our altered ordering of phenomena. The arbitrary element in the choice of simultaneity is just great enough so that it *never* leads to a false assertion respecting what is actually observed. Therefore our definition of time is also never false, and never inconsistent with the psychological force of perception.

IV

Nevertheless, is it not unbelievable that the velocity of sound becomes negative in one direction? For is that not impossible? Can a signal run backwards through time? It is very tempting to rule out this possibility *a priori.*

Rule it out? – But it happens! For we have here the very sound wave that runs backwards through time! What is real is also possible – that we may surely assert.

But – comes the reply – that this sound wave runs backwards through time stems only from the awkward definition of simultaneity. Had we defined it in the ordinary way the negative velocity would never have occurred. It only *appears* to be a backwards running wave; in *reality* it runs forward in time.

Appearance and reality? – Let us set these concepts aside for the time being. Still, it must be admitted that negative velocity stems solely from the awkward definition of time. In *this* instance it is avoidable, and it is quite proper to give preference to the more natural definition of time. But is negative time *always* avoidable?

That is a different question altogether. Is it *always possible* to give a *univocal* definition of simultaneity such that any signal or moving body will register a positive time difference on the clocks that it passes? – This is not a question to be answered hastily. For if we answer it in the affirmative, we are asserting something essential about nature: an assertion about the *order of causal sequences* in the universe.

This question has been investigated by the present author in a more comprehensive recent work [1924h]. It can only be resolved through an *axiomatic system of time* which discloses the presuppositions upon which an affirmative answer to it is based. Whether or not these presuppositions are fulfilled in nature is a separate question. We might note that they are fulfilled in the special theory of relativity, and the Einsteinian definition of time, which establishes the speed of light as constant, is at the same time to be determined in such a way that it satisfies the demand of causality for the positive temporal duration of all causal sequences. These presuppositions are usually *also* fulfilled in the general theory of relativity, but it remains an open mathematical question whether they are fulfilled for all theoretically admissable gravitational fields.

But one point is in any case certain: whether or not the question is to be answered affirmatively does not depend upon our mental capacity. And no matter how staunchly our intuitive faculties resist the notion of the possibility of a situation in which the temporal order does not uniformly conform to the requirements of causality, the answer to this question depends upon the *facts,* not upon *ourselves.* Axiomatic formulation has made a clear discernment of these facts possible. They pertain to the type of order of causal sequences in the world and are to be assessed independently of any definition of simultaneity.

Thus analysis of the problem of simultaneity leads to a surprising separation between *arbitrariness* and *necessity.* The definition of simultaneity, held by the absolute theory of time to be necessary, turns out to be arbitrary. On the other hand, carrying through a temporal order corresponding to causality does not lie at our discretion; whether or not it is possible depends upon empirical facts. Here again the absolute theory of time holds the opposite belief, i.e., that it is always possible to fulfill the requirement of causality respecting the temporal order of cause and effect. But this theory fails to take account of the axiomatic presuppositions contained in this claim; only experience can tell whether these occur.

V

One last question remains to be answered. Let it be granted that we are not logically compelled to impose causality upon the temporal order, that it cannot be imposed. But *if* it can be, does it not entail a certain definition of simultaneity? Can we perhaps find here a means of avoiding the relativity of simultaneity? We stipulate – in cases where it is possible – that simultaneity must be so defined that all signals possess a positive temporal duration. This is a justifiable demand, implementation of which gives the concept of time a very special meaning. Is simultaneity not thereby univocally laid down, in a manner free from any arbitrary elements?

This question, too, is to be answered only with the help of an axiomatic system, and here the response is that it could be the case that the relativity of simultaneity would thereby disappear. But that again depends upon the validity of certain axioms which indicate the characteristics of a causal sequence, and, as before, it is experience that must judge these axioms. Let us formulate the axiom in question. I send a signal from A to B and, directly afterwards, a signal from B to A; there will elapse, at A, some time between the departure of the first signal and the arrival of the responding signal. Is it possible to reduce this time indefinitely through a proper choice of signals? Note that the answer to this question is in no way dependent upon any definition of simultaneity at distant points or any metric of time, but involves instead a *topological axiom of time.* The univocality of simultaneity is determined by whether the answer is affirmative or negative.

But experience has already rendered a decision concerning this question. It teaches us that the time at A for a given B cannot be made indefinitely small. The fastest signals are light signals, and even these require a certain amount of time for a round trip. And there are no faster signals, for experiments with electrons have demonstrated that the kinetic energy of moving substances increases with respect to velocity at a rate exceeding the fourth power and would become infinitely great for the speed of light. This speed sets an objective limit to all actual transfer. This is, of course, not apparent *a priori,* but the reverse also cannot be asserted *a priori.* Experience decides in favor of the *former* possibility. But with this assertion the possibility of a univocal determination of simultaneity vanishes. It can, instead, be defined within a certain interval, while the causal requirement that all signals have a positive temporal duration remains nonetheless fulfilled. Then, to be sure, the sort of definitions of time (which we exemplified in Section III) will be out of the question, and for an area as small as a room the arbitrary interval will be reduced to an indiscernibly short time. That is why, for small areas, there is virtually no difference between absolute and relative time. This difference becomes noticeable only when we are dealing with astronomical dimensions or with very rapidly moving systems, such as electrons.

VI

We have shown that the relativity of simultaneity is epistemologically necessary, that it is not ruled out by any sort of psychological constraint, and that it is also not nullified by the demand for causality. Let us now investigate one last question: what is the relation between Einsteinian time and the time of daily life? Can the ordering of phenomena set up by the theory of relativity be called time at all?

The concepts of daily life are not formulated so precisely that they may be compared directly to scientific concepts. A variety of meanings are intertwined in the concept of 'time'. Sometimes we are thinking of the psychological experience. We feel time going by, sometimes slower, sometimes faster; we sense it in our consciousness as a linear sequence. But beyond this we also invariably order the events in the external world in accordance with such a sequence, and the very existence of the clock signifies the replacement of the subjective temporal sequence by an objective mechanism the course of which serves as the standard of the order and measure of time. And in setting up clocks, we invariably follow the fundamental principle of constructing the definition of time in such a way that the cause never occurs later in time than the effect. This leads for practical purposes to a quite definite determination of simultaneity, for (as we discussed in Section V) the interval remaining open is far too small for us to actually notice. Thus the objective time order of daily life came into being, and it is quite sufficiently unambiguous.

But at the same time the idea arose that this order of time in the only one possible. All primitive thinking claims to be the only admissible way of thinking – a stage that is overcome only by scientific education. Thus it is only now that we notice which peculiarity of the causal sequence made possible the implementation of such a simple temporal order. It is the fact that the velocity of light is so high; this causes the time of the light signal *ABA*, measured at *A*, to be extremely short and the admissible interval of simultaneity to be correspondingly small. If the limiting velocity were substantially less, being, say, of the order of magnitude of the velocity of sound, the arbitrary element in the choice of a definition of simultaneity permitted by the demand for causality would long since have been noticed. For even in daily life, the setting of clocks plays a major role; think, for instance, of the way in which the clocks in every railway station are set every morning in accordance with signals. Were the limiting velocity smaller, a railway engineer would long ago have discovered the theory of relativity without having to wait for Einstein, and by now people would long since have ceased to assert that the relativity of time is inconceivable, for they would frequently have experienced it in practice.

And for this reason we shall also set aside the distinction between appearance and reality. It has no place here; it is as meaningless to assert that one definition of time is really the right one, while the other is merely apparent, as it is to claim that the real linear measurement is established by the metric system, while other units lead only to apparent linear measurements. The schema into which we fit phenomena in order to describe them quantitatively is neither real nor apparent. It is a conceptual system, and only in the determination of whether it is applicable and whether it is unambiguous does any characteristic of reality find expression. The discovery of the presuppositions underlying the applicability is the task of philosophy – but the significance of this task does not lie in a glib defense of the notions of daily life. For these notions will in the end have to arrange themselves in accordance with what is first recognized by the intellect with broader horizons; in the future, the relativity of simultaneity will no more be inconceivable than the spherical shape of the world, which we have never actually seen, is to us today.

NOTES

[1] The general theory of relativity makes physical assertions going beyond the epistemological relativity of time. Cf. my comments in 'La signification philosophique

de la théorie de la relativité', [1922c], where I also discuss a definition of simultaneity not carried out through signals. A very thorough investigation of this question is to be found in my [1924h], Sections 2–3 and 21–8.

[2] This indeed is the conception of the Kantian philosophy, a thorough critique of which is to be found in my [1920f]. I have also given a presentation of other philosophical conceptions of the theory of relativity in [1922f].

From: *Sitzungsberichte der mathematisch-naturwissenschaftliche Abteilung der Bayerischen Akademie der Wissenschaften zu München* (Nov. 1925), pp. 133–175.

47. THE CAUSAL STRUCTURE OF THE WORLD AND THE DIFFERENCE BETWEEN PAST AND FUTURE

[1925d]

[Introduced by C. Carathéodory to the session of 7 November 1925 of the Bavarian Academy of Sciences]

1. DETERMINISM AND THE PROBLEM OF THE 'NOW'

It has become the custom to regard the hypothesis of causality in physics as so self-evident a necessity that no one even thinks of subjecting it to critical scrutiny. The extent to which this hypothesis represents extrapolation beyond the factual situation known by experience is seldom noticed; the usual defense of this standpoint is exhausted by the assumption that no exact natural sciences would be possible without it. We propose to demonstrate in the following essay that a quantitative description of natural phenomena is possible without the hypothesis of strict causality: a description that accomplishes everything that is achievable by physics and that furthermore possesses the capacity to solve the problem of the difference between past and future, a problem to which the strict causal hypothesis has no solution.

Before proceeding with this investigation, we must lay down a distinction that in itself reveals the problematic nature of the causal hypothesis. The first form of the causal hypothesis is present in physical *laws*, i.e., assertions of the form, 'If *A*, then *B*.' Let us call this the *implicative form* of the hypothesis. The second form, on the other hand, goes further and makes an assertion about the course of the universe as a whole; it asserts, that is, that this course is unalterably determined, that the past and the future are completely determined by any single cross-section of the four-dimensional universe. Let us call this assertion, also known as determinism, the *deterministic form* of the causal hypothesis. Obviously, the second assertion goes much further than the first, and for natural science to have made the step from the implicative form, which is at least plausible, to the claim that the course of the entire universe is controlled seems extraordinarily bold. It is generally justified by drawing a connection between the two forms, but no notice is taken of the fact that a second assumption is added to the implicative form of the causal hypothesis in the process, an assumption that, moreover, makes a highly dubious assertion in the light of the contents of experience. The auxiliary hypothesis will be revealed upon closer examination of the transition from the implicative to the deterministic form.

While the implicative form asserts that the cause A always has the effect B, it nevertheless makes this assertion only with regard to the case in which the cause A is really strictly present. But this very condition is recognized as never fulfilled, so that for every application of the implicative form a second hypothesis is required, relating to the remainder of the factors that are present in addition to A. This auxiliary hypothesis is usually formulated as the assumption that the remaining factors have only a quantitatively small influence. But this is not quite accurate. The assumption really runs as follows: the *remaining factors exert their influence in accordance with the laws of the probability calculus*. Relatively large deviations may occur from time to time, but over a number of recurring cases the deviations correspond to a statistical law. This is the assumption that the deviations conform to a continuous probability function, as I have demonstrated elsewhere.[1] This principle of probability comes into play whenever the causal hypothesis in its implicative form is applied to reality. It is not in any way deducible from the implicative form, constituting instead an independent assumption without which the implicative form would have no value; without it, the implicative form could in no case be applied to reality. Physical knowledge, then, rests upon *two* principles: the *principle of causal connection* and the *principle of probable distribution*.

How do we get from these considerations to the deterministic form? In order to establish the connection, let us formulate the deterministic form of the causal hypothesis for a universe filled with a continuous field of matter. In such a universe we would have no need to speak of individual phenomena, for we could give a complete description of the world by indicating the field distribution.

The deterministic hypothesis would then run as follows. If, for a cross-section of the four-dimensional universe (t = constant), the field distribution and also the first and second derivatives of the field quantities with respect to time are given, then the past and the future are fully determined. We can envisage the field distribution in such a way that, for instance, the Einsteinian tensor T_{ik} is strictly given as a function of selected spatial coordinates.

Comparison of this assertion with the contents of experience reveals the very difference that we already indicated. On the one hand, the state of the field is never given with total rigor; on the other hand, the earlier and later states calculated from it are invariably established only with probability. On the basis of experience, then, the deterministic hypothesis can be derived only by means of a transition to a limit that transforms the approximate field distribution into a strict distribution and probability into certainty. Uncritical advocates of determinism generally overlook the difficulties connected with this transition to a limit.

Let us suppose, for instance, that the distribution of matter within the terrestial sphere is given by its density σ as a function of the coordinates. In astronomy the stipulation $\sigma =$ constant is adequate. In geology, σ will vary in accordance with the strata. Physics goes much further, wanting to establish the position of every single molecule, i.e., to establish a density function to show much more minute spatial variations than geological density. Each of these levels of precision gives a prediction of events, i.e., of future conditions of density; the certainty of the result calculated increases with increasing precision. The deterministic hypothesis assumes that there exists a function (perhaps through dividing the scalar σ in a tensor T_{ik}) that determines the result with certainty.

Let it be granted that the degree of probability can approach 1 without limit – the deterministic hypothesis nonetheless contains the assumption that the series of field functions, ordered with increasing certainty, has a limit. Given the contents of experience, we can only make the assertion that for any field function there exists a more precise one possessing a higher degree of probability. To maintain the existence of a final function in this series possessing the probability 1 goes far beyond this assertion. This is the extrapolation that is contained in the deterministic hypothesis.

Of course, the extrapolation cannot simply be declared false without further investigation, but it can be asserted that everything that is explicable *with* it can also be explained *without* it. For it is only the fact of a precision capable of approaching 1, and never the existence of the limiting function itself, that is used for all verifiable physical propositions. Therefore the deterministic hypothesis is completely empty for physics; and while it cannot be directly refuted, there is also nothing to be said in its favor. For this reason we will omit this hypothesis in what follows and show that the causal structure of the universe can be comprehended with the help of the concept of *probable determination* alone.

The deterministic hypothesis has been regarded as having the merit of eliminating the concept of probability from the explanation of natural phenomena. From this point of view, the principle of probability is nothing but a helpful expedient that we use so long as we are ignorant of the exact determinants of an event; once we have a complete and precise knowledge of all the factors involved, this expedient becomes superfluous. But this argument for the deterministic hypothesis overlooks the fact that the concept of probability is eliminated only at the limit; it is still necessary for every proposition that, from a practical standpoint, can actually be put forward in the natural sciences. For even if the limiting function exists, it is never known in the strictest sense. Yet the fact remains that, for imprecise descriptions of natural events, the probability laws, at least, are valid, and this verifiable fact can only be explained if the probability laws do not represent the expedient of an imperfect state of knowledge but, rather, a characteristic of natural phenomena. If, then, the concept of probability were eliminated, the probability laws would have to be demonstrated to be a consequence of causal laws; but surely no such proof is possible.

Thus the deterministic hypothesis does not in the least succeed in making the concept of probability superfluous. Hence there can be no objection to pursuing the opposite course, abandoning determinism and setting up the concept of probability as the basic concept of knowledge. For, after all, the hypothesis of strict causality is only intended to express the idea that every deviation from strict lawful regularity must in turn be subject to a causal explanation, and this very idea can be retained even without the hypothesis of a limit. We accept, then, the validity of the implicative hypothesis, and not only in the form, 'If *A*, then *B* follows,' but also in the reverse form, 'If *B*, then *A* has occurred previously.' But we add to this assumption another assumption about probability relating to the factors not accounted for in *A* and *B*, the import of which is that these factors find expression according to the rules of the probability calculus. Both assumptions should apply at every level of precision, and we reject the claim that the second assumption ultimately becomes superfluous. In place of the unified hypothesis of determinism, then, we rest content with two assumptions that stand on a par: the assumption of a causal connection between the determining factors in natural events and the assumption of a probability distribution to account for the influence of the remaining factors. Surely we are conforming to the demand for a description of nature of the greatest possible accuracy in preferring this duality to a unified assumption so little capable of justification as is determinism.

And yet it is not even necessary to place the two assumptions separately side by side. The division of phenomena into a causal part and a probability part is of purely formal significance and can be replaced by the single assumption that a connection of a probabilistic nature exists between cause and effect. Whether *A* *would* with certainty bring about *B* *if* no further factors were present is a matter of indifference; as this case never occurs, we will rest content with the single assumption, 'If *A* exists, it determines a *B* in accordance with the laws of probability.' The *degree* of probability can be increased arbitrarily close to 1 through the most precise possible determinations of the participating factors[2] – here the idea is expressed that for every deviation in the effect another cause can be found – yet the connection between cause and effect retains, for every attainable level, the character of a probability. Thus we envisage a universe in which all dependencies are of the same nature as the relation between the appearance of one side of a die and the throw of the die; every step in the course of events is a throw of the die, and it is only the high probability of certain sequences that has misled us into reading absolute certainty into them. With this conception, then, we have also reached a unified assumption concerning the character of natural events, but we have omitted the causal assumption rather than the probability assumption. A universe of this nature possesses, in every one of its elements, only a probability relation.

It is the demand for a minimum number of presuppositions that forces us to reject strict causality. Yet we will discover, in the course of developing the theory of probability relation, that at the same time we gain one definite advantage over determinism which constitutes a strong justification: a clarification of the concepts of past and future.

That time order can be established on the basis of certain features of a causal structure has recently been made clear through the investigations of K. Lewin,[3] R. Carnap,[4] and the present author.[5] The meaning of 'earlier' and 'later' can be defined through causal series, for it is only because phenomena can be connected to one another by means of causal series that they possess a temporal relation. The characteristics of causal series necessary for this order can be formulated as axioms, among which the axiom excluding any closed causal series going in a single direction plays an important role. Thus it is possible to establish a topology of time in which the basic concepts 'earlier', 'later', and 'simultaneous' are defined. But what this topology has *not* been able to resolve up to the present, is the problem of the 'now'.

What does 'now' mean? Plato lived earlier than I, Napoleon VII will live later. But which of these three lives *now*? I undoubtedly have a distinct feeling that *I* am living now. But does this assertion have an objective meaning over and beyond my subjective experience? Its meaning might be given exhaustively by a description of my psychological condition. But is it not possible, after all, to give it an objective meaning?

To begin with, we will attempt to find this objective meaning in an assertion concerning relations of simultaneity. According to this reading, the proposition, 'I am living now', is identical to propositions of the form, 'I am living simultaneously with Mr. A' or 'I am living simultaneously with such and such a phenomenon.' If this is correct, there is no special 'now'; rather, the meaning of this word can be reduced to the concepts 'earlier', 'later', and 'simultaneous'. But is the meaning of 'now' thereby exhausted?

If the now can be reduced to simultaneity, the meaning of the question, 'What is happening now?' would consist in the following. This question itself presents an event F having a position in the course of the universe; what is being asked is, what is simultaneous with F? Yet while it is correct, this answer is not exhaustive. For it does not lie within my power to select the position of this F, which automatically takes its place at the time point 'now'. If the objection is made that I do indeed have the localization of F in my control inasmuch as I can wait before asking the question, the following reply must be given: certainly, I cannot choose just any point in time, but only future ones. But the point in time at which these eligible points are divided from the ineligible ones is the 'now'. For we simply cannot escape by means of such attempts the compulsion whereby the 'now'-point is absolutely distinguished for us as the experience of the boundary between the past and the future.

The problem, then, may also be formulated as the question of the difference between past and future. For determinism, no such distinction exists. If the future is already completely determined for any given temporal cross-section, it makes no difference whether it has already taken place or has yet to take place. The actual occurrence brings with it nothing new: that which will happen one hundred years from now is given in just the same sense as the events of the past war, and I could make observations in basically the same way about the wars of Napoleon VII as about the Battle of Verdun. With respect to the 'now,' there is no difference between Plato and me; I might just as well say that Plato is living now and I belong in the future. I could, to be sure, then assert that Plato lived *earlier* than I did, for determinism, too, recognizes an 'earlier' and a 'later'. But there is no 'now', no distinctive point in time, and the feeling that my existence is real, whereas Plato's life merely casts its shadow into reality, must be a mistake. But these considerations contradict our entire attitude toward our existence; we regard the future in a completely different manner from the past, and unless we are prepared to conceive of every single one of our actions, every thought attendant upon the shaping of our daily lives, as one immense error, determinism must be mistaken.

This is not to say that determinism *is* false, but we must be quite clear as to the contrast at hand. If determinism is correct, then we cannot in any way justify undertaking an action for tomorrow but not for yesterday. No doubt it is true that it is not even possible for us to give up our *intention* to act tomorrow and our belief in freedom – we surely cannot. The point is that, given determinism, our behavior would be senseless, for then tomorrow would be already past in the same sense that yesterday is. It is only determinism that compels us to draw this conclusion; if we reject it, we can avoid this claim which is so contradictory to our basic vital instincts. To be sure, we cannot permit a feeling of this kind to be decisive if our understanding speaks convincingly to the contrary. Rather, we must first analyze our understanding to establish whether its judgment is necessary – and, as it turns out, it is not.

For once we espouse the theory of probability relations, we are able to draw precisely that distinction between the past and the future that corresponds to our instinctive feeling. If events are not totally determined, it cannot be claimed that the future is already established. It is always possible that the opposite of what has been calculated will occur. The past, on the other hand, is definite, and the present is that threshold over which the universe steps in going from an indefinite state to a definite one. In the state of the universe, then, there is one distinctive cross-section that we designate as the present; the 'now' has an objective meaning. There would be a 'now' even if there were no more living persons; 'the present state of the planetary systems' is as definite a designation as 'the state of the planetary system at the time of the birth of Christ'.

A distinctive cross-section of this kind does not exist in the four-dimensional model of the world, as used, for instance, in the theory of relativity. But that is only because this model omits an essential feature. Any assertions about the universe, about past or future events, must be connected by means of chains of inference to certain perceived events, and the connecting points must all lie in a *single* cross-section, namely, the cross-section of the present. To take an actual case: if I wish to know in what year Charlemagne was born, I must open a history book; the perception of the number is the present perception that constitutes the beginning of a chain of inference to the proposition that it signifies the year of the birth of Charlemagne. (This inferential chain includes, for instance, the assumption that the book is a sufficiently reliable work of history.) If I wish to calculate the time of an eclipse of the sun, my starting points must be, as before, printed numbers in a book that I am reading 'now', or else present observations of the sun and the moon. Numbers that I do not look up, but remember instead, must be known 'now'. Here the experience of remembering is comparable to perception and again leads only by means of a chain of inference (e.g., by checks on the certainty of what is remembered) to the fact asserted. For any given state of the universe,

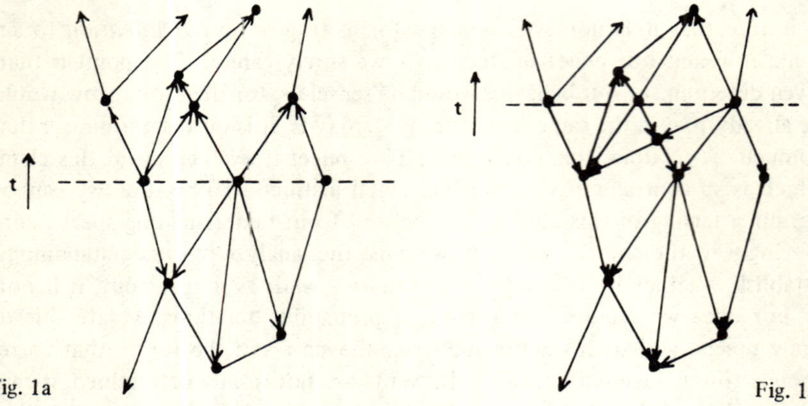

Fig. 1a Fig. 1b

Fig. 1. The course of the universe represented as a sequence of structures, the cross-section of the present being indicated.

then, there is a cross-section such that all propositions, both about the past and about the future, must be connected to it.

We have characterized this cross-section as being such that all assertions about the universe must be connected to it, but we have not thereby defined it subjectively. For we are compelled to select it not by our own constitution but precisely by the state of the universe. For every given state of the universe there is a unique and distinctive cross-section. Figure 1a is an accurate representation of the world-picture of the theory of relativity, and the course of the world consists in its transition from the state depicted in Figure 1a to that in Figure 1b, and so on.[6] The entire course cannot be shown in a single picture, but only in a sequence of pictures such as those in Figure 1. We are merely simplifying – which for many purposes is, of course, admissible – when we leave out the cross-section and all the arrowpoints and replace the sequence by a single picture.

Although we speak of a distinctive cross-section, we are not thereby asserting the existence of any absolute simultaneity. Rather, we must correct our assertion so that it conforms to relativity theory: the direction of the distinctive cross-section is arbitrary within the space of a certain interval. This description remains admissible if we define the present by means of the subjective experience 'now'. The experiences of a single person appear in the causal schema as a cross-section of such limited breadth that it may be regarded as practically a point. Every section of simultaneous events that is admissible in relativity theory then becomes, through this point-event, an admissible now-section. Thus the now-section can be defined by means of a point, which also accords with the definition of the now-section by the reversal of the direction of the arrows, shown in Figure 1. The causal cone $ds^2 = 0$ extending forwards and backwards from a point-event P itself divides the universe in such a way that all causal lines which coincide in P can be fitted out with arrow points, as in Figure 1. Only the points in the intermediate range (in Minkowski's sense) are not ordered in this process, and this range is filled by the admissible now-section through P. When, in what follows, we refer to a distinctive cross-section, we mean, more precisely, any given one of the distinctive cross-sections, and Figure 1 is to be interpreted accordingly.

Past and future are distinguished from one another by the manner in which they are determined by the distinctive cross-section. We will now demonstrate this by employing the *theory of probability relations*, clarifying in the process the sense in which the past may be called 'objectively determined' and the future 'objectively undetermined'. However, we abandon the assumption that the universe is filled by a continuous field and think instead of individual events (the connecting points in Figure 1) which are joined to one another by inference chains. This conception enables us to base the system of relations in the universe upon the topological characteristics of a net structure. The extension of the theory to cover continuous fields is bound up with difficulties that, for the time being, cannot be overcome.

II. TOPOLOGY OF PROBABILITY IMPLICATION

The relation that replaces a strict causal connection between events is called *probability implication*. We observe that when an event A occurs, the event B also occurs with a certain regularity. B need not *always* occur, but the instances of the occurrence and non-occurrence of B are regulated in accordance with the laws set forth in the probability calculus. The laws in question include not only the regularity of the *frequency* relation of occurrence and non-occurrence, but also the regularity of the *deviations* from this relation, i.e., the laws of distribution. We say

$$A \twoheadrightarrow B$$

which is put into words, as "A implies B with probability" or "A determines B." This says nothing as to the *degree* of probability, which can lie anywhere between 0 and 1 (inclusive). Thus the relation $A \twoheadrightarrow B$ applies not only in those instances in which we would say, in common parlance, that B is "made probable" by A, but also in those which B is "made improbable" by A. The symbol '$\multimap$' for probability implication is derived from the symbol $\supset$, for strict [*streng*] (logical) implication, by the addition of the cross stroke. Strict implication then emerges from probability implication as a limiting case, occurring when the probability = 1.

Two objections will be raised against the introduction of probability implication. First, how is it possible to assert the regularity of the frequency between A and B for all cases, given that it has only been observed for a finite number of cases? We answer that we will not here investigate this question, which presents the problem of induction, but will presuppose it to be both possible and meaningful to draw a probability inference about all observations upon the ground of a finite number of observations. This presupposition is made not only by our probability theory but also by every scientific investigation of nature, and we will therefore simply assume that it is justified. Second, the question will be raised what meaning it can have to ascribe a probability to the occurrence of the individual event B if this number does not signify anything for the individual case, but only for arbitrarily long series of repetitions. Again we reply that we will simply assume this assertion to be meaningful – and this assumption, likewise, applies not only to our theory but is also invariably made in science and in daily life. Criticism of this assumption – which must take due note of the fact that the problem is basically no different for the individual case than for any finite number of cases to be predicted – is a very important part of epistemology, but we will not concern ourselves with it here.

For present purposes, then, we regard probability implication as a basic concept, just as implication may be introduced in logic as a basic concept that cannot be derived from any other concept. $A \multimap B$ means, "If A occurs, then, with probability, B occurs." Or: "If A occurs, with probability, then B occurs, with probability". But this means that we take "B occurs, with probability" as a basic concept that cannot be further analyzed. The relation of probability implication cannot be placed between just any events, but only between certain events; experience teaches us exactly which ones these are. $A \multimap B$, then, is a factual proposition.

The most striking characteristic of probability implication, as opposed to implication, consists in the fact that $A \multimap \bar{B}$ is always given with $A \multimap B$, where $\bar{B}$ (in words: not-B) is the absence of the event B. From the standpoint of probability calculus this is self-evident: if p is the measure of probability governing B, then $1-p$ is the corresponding value for $\bar{B}$. Along with the regularity of the *occurrence* of B, a corresponding regularity is given for the *absence* of B, i.e., for the occurrence of $\bar{B}$. This basic characteristic can be symbolized as follows:

$$(A \mathrel{-\!\!\!\ni} B) \supset (A \mathrel{-\!\!\!\ni} \bar{B}), \tag{1}$$

where $\supset$ signifies strict implication.

It is important to distinguish the assertion,

$$\overline{A \mathrel{-\!\!\!\ni} B}.$$

from the assertion $A \mathrel{-\!\!\!\ni} \bar{B}$. The former says the assertion $A \mathrel{-\!\!\!\ni} B$ is false, which means that no regular relation exists between the occurrences of A and B of the kind represented by the laws of probability. From statement (1) we immediately derive

$$\overline{(A \mathrel{-\!\!\!\ni} B)} \supset \overline{(A \mathrel{-\!\!\!\ni} \bar{B})}.$$

There are certain cases in which $B \mathrel{-\!\!\!\ni} A$ is valid, in addition to $A \mathrel{-\!\!\!\ni} B$. In these cases the probability implication is convertible. Only experience can teach us whether or not convertibility is possible: again, it is a factual assertion. The measure of probability is generally different for the two directions.

To take some examples: The rising of the barometer implies with probability that the weather will be good. Conversely: If the weather is good, it follows with probability that the barometer has risen. On the other hand: If I meet Mr. X on Y street, it follows with probability that Mr. X is going to Z. But the converse is not valid: If Mr. X is going to Z, it does not follow, even with probability, that I will meet him on Y street.

Following is a compendium of laws of probability implication which claims neither to be exhaustive nor to represent a table of independent axioms. We may, however, consider that it encompasses the most important laws. We will avail ourselves of Russell's notation for mathematical logic, except for replacing Russell's negation sign with the more visible over-stroke.

The notation is to be interpreted as follows.

$a \supset b$	*a* implies *b*
$a \mathrel{-\!\!\!\ni} b$	*a* implies *b* with probability
$a \cdot b$	*a* and *b*
$a \vee b$	*a* or *b* or both (inclusive 'or')
$\bar{a}$	not-*a* (negation)

Laws of Probability Implication

1*.	$(a \Rrightarrow b) \supset (a \Rrightarrow \bar{b})$	Equivocality
2*.	$(a \Rrightarrow b \,.\, c) \supset (a \Rrightarrow b) \,.\, (a \Rrightarrow c)$	Dissolution of final 'and'
3*.	$(a \vee b \Rrightarrow c) \supset (a \Rrightarrow c) \,.\, (b \Rrightarrow c) \,.\, (a \,.\, b \Rrightarrow c)$	Dissolution of the initial 'or'
4*.	$(a \Rrightarrow b \vee c) \supset (a \Rrightarrow b) \vee (a \Rrightarrow c)$	Dissolution of the final 'or'
5*.	$(a \Rrightarrow b) \supset (a \,.\, c \Rrightarrow b)$	Initial factor
6*.	$(a \Rrightarrow b) \,.\, (a \Rrightarrow c) \supset (a \Rrightarrow b \,.\, c)$	Final multiplication
7*.	$(a \Rrightarrow c) \,.\, (b \Rrightarrow c) \supset (a \vee b \Rrightarrow c)$	Initial addition
8*.	$(a \Rrightarrow b) \vee (a \Rrightarrow c) \supset (a \Rrightarrow b \vee c)$	Final addition
9*.	$(a \Rrightarrow b) \,.\, (b \Rrightarrow c) \supset (a \Rrightarrow c)$	Transitivity
10*.	$(a \Rrightarrow b) \,.\, (b \supset c) \supset (a \Rrightarrow c)$	Transitivity for the partial limiting case

The laws of probability implication are exactly analogous to those of implication. But it is important to note that implication occurs alongside probability implication in these laws; we cannot, for instance, take a proposition that is correct for implication and replace the symbol '$\supset$' by '$\Rrightarrow$' throughout, but must instead do this only in particular places. Probability implication, then, is based upon implication, and strict implication cannot be rendered superfluous by probability implication. Conversely, even though implication is a limiting case of probability implication, we cannot demand that all laws remain correct if '$\Rrightarrow$' is replaced by '$\supset$' throughout. This substitution, too, can be made only in certain places. If, for instance in 1*, the initial '$\Rrightarrow$' is replaced by '$\supset$', the second '$\Rrightarrow$' may not be so replaced. For the measure of the second probability implication becomes equal to 0 when that of the first becomes equal to 1. (The meaning of certain of these laws will first become clear through their application below.)

Using probability implication, we will now endeavour to formulate assertions about the causal structure of the universe. These will be *topological* propositions, as probability implication and the laws for it which have been state so far have not yet made any use of the *measure* of probability. We imagine certain events as given in the cross-section of the present and infer from them the existence of other events with the help of natural laws. All the natural laws are of the form $a \ni b$. How we happen to have knowledge of these individual laws and how, in particular, it is possible to establish such laws between *temporally successive* events when only *simultaneous* events are given – all this we shall leave aside for the present. We will, then, simply take the laws as given, and attempt to show that the method of inference differs topologically according to whether we are inferring the existence of past or future events.

We shall use the following procedure in order to disclose the structural difference of the direction of time. We make the initial assumption that we know from some other source whether the inference in question concerns past or future events and are thereby able to characterize the mode of inference. Conversely the distinctive feature of the mode of inference may be used to define the direction of time.

The simplest ordering of events is the *undivided chain*.

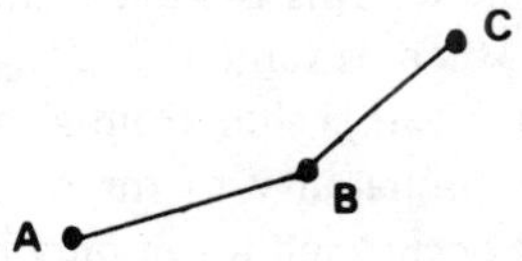

Fig. 2. The undivided chain.

Here the following inferences are valid.

$$A \ni B \quad B \ni C \quad A \ni C \qquad (2)$$

$$C \ni B \quad B \ni A \quad C \ni A$$

No direction is distinguished here. The undivided chain, then, offers no indication of the direction of time, which can first be gained with the emergence of *connecting points*. In this way we are led to base the temporal order upon the characteristics of a *net structure*.

The simplest basic form of a net structure as given in Figure 1 is a fork. We will first consider a fork with the point directed toward the future. We shall call it a *pointed fork* or, alternatively, a *conjunctive pointed fork*, in order to distinguish it from the disjunctive pointed fork that is to be discussed later on.

The following inferences are valid for the pointed fork.

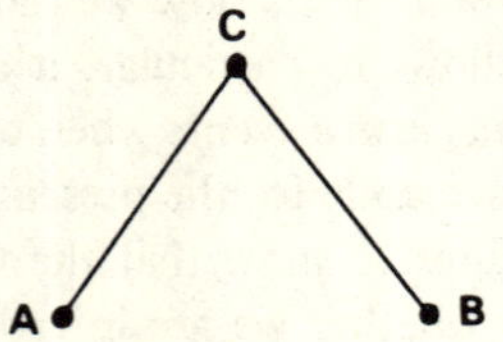

Fig. 3. The pointed fork.

$$\begin{array}{lll} A\,.\,B \ni C & C \ni A\,.\,B & \\ \overline{A \ni C} & C \ni A & \overline{A \ni B} \\ \overline{B \ni C} & C \ni B & \overline{B \ni A} \end{array} \qquad (3)$$

The characteristic feature here is the initial 'and' in the first proposition on the left. As in strict implication, it cannot be dissolved; i.e., only A and B taken together can determine C. This is what characterizes an inference concerning the future. $\overline{A \ni C}$, then, is valid.

For example: One billiard ball is shot from A and one from B, and C represents their collision. The probability of the occurrence of C is given only when *both* events A and B occur, and it can only be calculated from the *two* separate probabilities for the arrival of the balls at place C. If nothing is known about the start of the billiard ball at B, i.e. there is no probability for the event of this ball occurring at the place C, then there is no probability relation between the departure of the billiard ball at A and the event C.

The proposition $A\,.\,B \ni C$ is convertible: from it we may infer $C \ni A\,.\,B$. The last 'and' can be cancelled out, in accordance with 2*, giving us $C \ni A$ and $C \ni B$. With these we are able to construct the rule for the direction of time:

The law of direction: If probability implication is valid in only one direction, then the antecedent is the temporally later event.

In symbols this runs:

$$(C \ni A)\,.\,\overline{(A \ni C)} \supset (A < C) \qquad (4)$$

where $A < C$ means "A is earlier than C."

Because the probability implication between C and A is valid in only *one* direction, no implication can be established between A and B in *any* direction; for $A \ni B$ would, by 9*, imply $A \ni C$ and $C \ni B$, and, likewise, $B \ni A$

would presuppose $B \ni C$ and $C \ni A$. The pointed fork, then, is *intransitive*.

The pointed fork is completely determined by (3). If any three events are given, and if the relations (3) are known to hold between them, then these three events must constitute a pointed fork. Which of them represents the point directed toward the future can be seen from the non-symmetrical occurrence of these events shown in the relations (3). If, in (3), A is exchanged with B, the identical system of propositions results. But if C is exchanged with A or B, different propositions result. Thus the intransitive fork has a *topologically distinguished angle*.

We shall give the name *saddle fork* to the fork having its point directed toward the past, and here, too, we shall eventually distinguish a *disjunctive* saddle fork from the *conjunctive saddle fork* now under discussion, for which the following relations are valid:

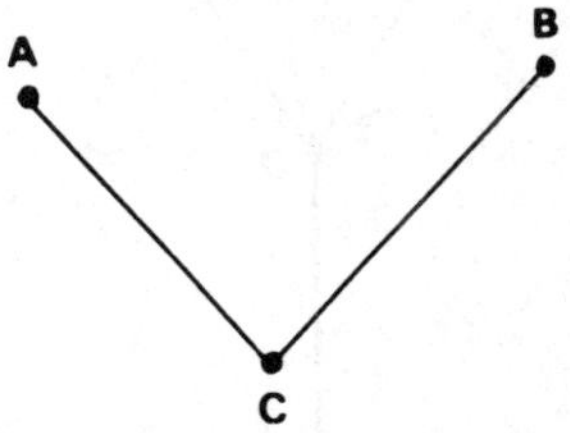

Fig. 4. The saddle fork.

$$\begin{array}{lll} A \vee B \ni C & C \ni A \,.\, B & \\ A \ni C & C \ni A & A \ni B \\ B \ni C & C \ni B & B \ni A \end{array} \quad (5)$$

What is characteristic here is the initial 'or' in the first proposition on the left. By 3*, it can be cancelled out and thus leads, in contrast to the pointed fork, to $A \ni C$ and $B \ni C$. By 9*, then, $A \ni B$ and also $B \ni A$; thus the saddle fork is *transitive*.

For example: Let A and B once again stand for shooting one billiard ball each, but now C will stand for a cause common to them, perhaps the signal at which both balls are released. If I only observe A, I may directly conclude with probability that the signal has been given. Even if B fails to take place at all, we may draw a probability inference from A to C; the discharge mechanism might have failed to operate for B. The addition of the observation of B to the observation of A simply has the effect of *increasing* the probability of

C. And if I observe *A* while knowing nothing as to the occurrence of *B*, I may infer *B* from *A* via the common cause *C*.

We recognize here the decisive difference between past and future. The pointed fork and the saddle fork are symmetrical with respect to the cross-section of the present; if the mode of inference into the past were the same as the mode of inference into the future, then relations (3) and (5) would have to be identical. But, in fact, they differ in a very essential point: inference into the future requires an antecedent 'and', while inference into the past requires only an antecedent 'or'. Nothing short of the *totality* of all causes is required for inferences into the future, but inference about the past can be made on the basis of a *partial action* [*Teilwirkung*].

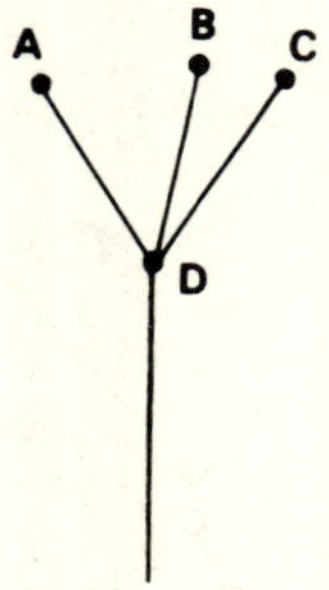

Fig. 5. A three-pronged fork.

It is, then, precisely through the intransitive fork that the direction of time is distinguished, just as we formulated it in the rule of direction; it can *not* be distinguished by means of the transitive fork. For because of its very transitivity, this fork has no topologically distinguished angle. If, in (5), we exchange *A* with *B*, or *B* with *C*, or *C* with *A*, we obtain either the propositions (5) or else such propositions as can be inferred from them according to the stated laws of probability implication. It is therefore impossible to deduce from (5) which angle is temporally earlier, and it is impossible to infer at all that any one angle must lie in the past. If we imagine three causal chains starting from an event *D* and extending towards the future (Figure 5), eventually leading to the events, *A*, *B*, and *C*, then it is precisely the relations (5) that will pertain between these as well. The undivided chain (Figure 2) also leads to these same relations, for the relations (2) are identical with (5). And for this reason we cannot infer from the existence of the relations in (5) that a saddle fork is present. The temporal direction of such events, which are related through (5), can only be determined by their relation to pointed forks in the network of the structure.

The significance of the saddle fork, on the other hand, consists just precisely in its transitivity, which makes possible the establishment of a probability implication between events which are not bound together by means of a continuously ascending or continuously descending causal chain. Probability implication between events belonging to the same cross-section of the present can therefore only be established via *past* events, not *future* events, for only the fork directed toward the past is transitive. It is only the common *cause*, not the common *effect*, that produces a probability relation between simultaneous events.

The practical significance of the saddle fork for experimental physics is extraordinarily great. The vast majority of all inferences, even those concerning future events, are drawn by way of the saddle fork. We have an instance of a saddle fork whenever, for example, the temperature in an electric oven is controlled by the strength of the heater current. We observe a deflection of the indicator, from which we infer the strength of the electric current as its cause, and, in turn, the second effect of the current, the production of heat. It is this principle upon which all instruments of measurement rest. The observed partial effect A stemming from the cause C is selected in such a way that the relation $A \ni C$ possesses a high degree of probability, C being thereby securely established. If, in turn, B is observed, the relation $C \ni B$ may be established experimentally. In most cases B is not observed directly, but only a partial effect D, stemming from B, which gives a high probability to the inference $D \ni B$. The pattern of inference corresponds to Figure 6, in which the chains possessing a high probability are indicated by the heavy lines. If, for instance, we wish to calibrate the temperature of an electric oven, we would assign the following meaning to the symbols in Figure 6.

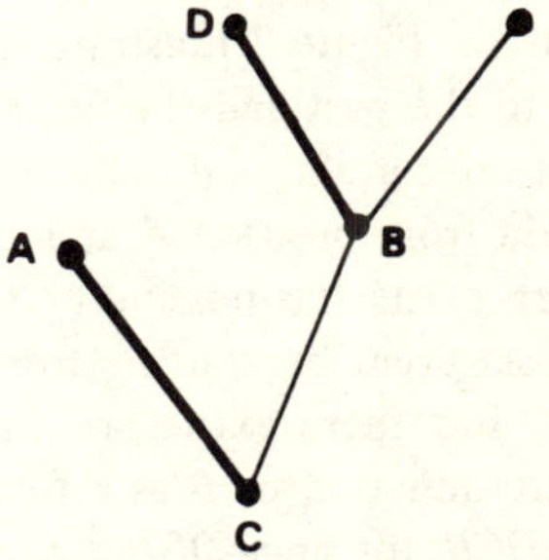

Fig. 6. An inference to events in another chain using the saddle fork.

A: deflection of the indicator on the ammeter
C: current strength
B: temperature
D: numbers given on a thermometer

An analogous inferential procedure in a completely different sphere is seen in the use of circumstantial evidence in jurisprudence. In a case where complicity in a murder is proven by means of circumstantial evidence, the symbols in Figure 6 might stand for:

A: *X*'s fingerprints, discovered at the scene of the crime
C: presence of *X* at the scene of the crime
B: murder
D: signs of the murder

In this inference, $C \mathrel{-\!\!\ni} B$ is assigned a high probability, while $B \mathrel{-\!\!\ni} C$ is assigned a low one. That is, the participation of *X* cannot be inferred simply from the signs of the murder, but may indeed be inferred once the fingerprints are also taken into account.

The transitivity of the saddle fork also supplies the answer to a question that we touched upon earlier. The cross-section of the present is all that is ever given, and hence all that we ever observe are probability inferences between simultaneous events. How is it possible to establish assertions of the form $C \mathrel{-\!\!\ni} A$ if *C* is the cause of *A*? Here, again, the saddle fork helps us out; saddle forks giving a high degree of probability to one branch are those most commonly used. Yet it is, in fact, important to be clear as to the fact that the totality of the past is a network construction joined together entirely by probability implications between simultaneous events.

The property of transitivity belonging to the saddle fork frequently makes it possible to reach a conclusion about the past with far greater certainty than an inference about the future. Figure 7 illustrates a linking together which is symmetrical with respect to the past and the future in which the chain *AB* is supposed to be particularly uncertain, and in both directions. As a result, the prediction of *B* is uncertain from position *A* and likewise the retrodiction of *A* from *B*. However, there exists the possibility of making a more certain determination of *A* as a past event by starting from another event *C* and following the route *CDEFA*. But there exists no corresponding possibility of making a more certain determination of *B* as a future event by starting from *F* and taking the route *FEDCB*, for since *DCB* is a pointed fork, no inferences can be drawn via *C*. Hence the *prior* determination of events is very much

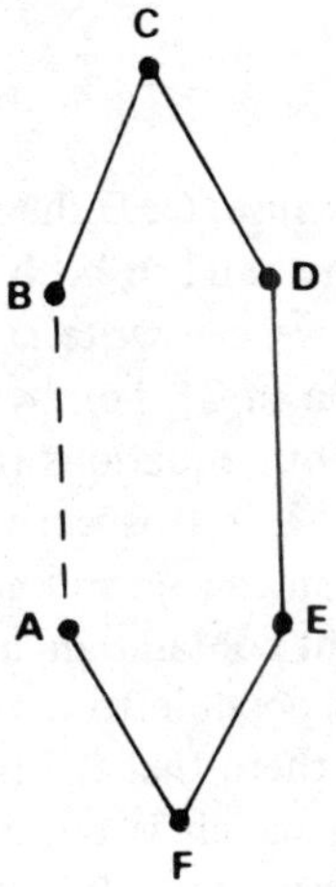

Fig. 7. Sequence of a particularly uncertain chain.

affected by *one* uncertain chain, while the retrospective determination in the corresponding case may be quite certain.

Yet it is not the higher *degree* of probability that is the distinctive feature of inference into the past. Indeed, there are instances in which such inferences become uncertain. It is, rather, the *mode of inference* that distinguishes retrospective inference from predictive inference. Let us look at one more structure in order to make this quite clear. The double fork in Figure 8 is symmetrical with respect to its connections for past and future, but its probability relations prove to be asymmetrical.

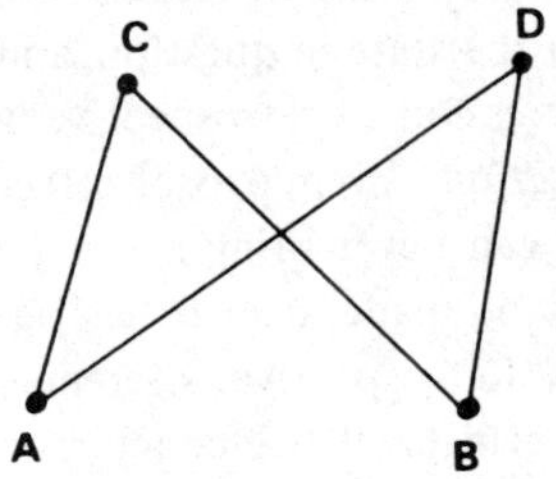

Fig. 8. A double fork

They are as follows:

$$A \,.\, B \ni C \,.\, D$$

$$C \vee D \ni A \,.\, B \tag{6}$$

$A \,.\, B$ is the totality of the cause, $C \,.\, D$ the totality of the effect. Inferences from the whole to the part are valid in both directions as a result of a basic feature of implication which we can write as $a \,.\, b \supset a$. It is expressed in the laws of probability implication in 2*, i.e., by means of the dissolution of the final 'or', which is valid for both equations (6). On the other hand, we may infer the whole from the part in retrospective inference – this is the significance of the antecedent 'or' in the second equation – but not in predictive inference, where the antecedent contains an 'and'.

Given this knowledge, it is possible to determine a concept which we discussed at the outset. We said there that the future was objectively indeterminate, in contrast to the past, which is objectively determinate. Now, what does objective determinateness mean? It is tempting to give the following definition: a state is objectively determinate if the probability with which it can be subjectively determined can approach 1 without limit. This definition has the immediate disadvantage of using the degree of determinability in order to define objectivity, but it becomes completely untenable once we abandon the assumption that there exists a limiting function describing a universal cross-section with absolute certainty. For the probability 1 then corresponds to no defined state of the universe at all; the limiting case has degenerated and cannot be used to define 'objective'.

However, there is another way in which we can determine the concept 'objectively determinate'. We will call the past objectively determinate because it may be inferred simply from a partial effect. For an inference from the part to the whole presupposes that the whole has already been independently established. In the concept 'objectively determinate', we are pursuing the idea that we can no longer alter the state in question, and this is the very characteristic that finds special expression in retrospective inference, which represents *evidence* rather than *causation*. Even partial effects may be adduced as *evidence*, but a partial cause can never bring about an event. Thus an assertion about the future can only be made after it has been established that all the partial causes are present. Retrospective inference, on the other hand, does not require all the partial effects. It is characteristic of past events that they can be *recorded*. What was the temperature in this room the day before yesterday? We would confront great difficulties were we to attempt to infer this today from the remaining effects. But if there is a recording thermometer in the room, it is easy to find the answer; the causal chain connected with this instrument can be used in a retrospective inference possessing a high probablity for which the other effects are not required. No analogous arrangement is possible for the future, however. We cannot record the future; i.e., a single partial chain is not sufficient for its determination.[7]

The future, then, must be regarded as 'objectively indeterminate'. For if the limiting function does not exist, the totality of all partial causes is not a defined quantity. It cannot be claimed that it is simply a lack of technical means that presses the definiteness of the future state of the universe beyond the bounds of certainty. Indeterminacy is, rather, an objective characteristic of the causal structure.

Thus the difference between 'objectively determinate' and 'objectively indeterminate' may be reduced to a topological difference in probability implication. The difference between the two terms 'or' and 'and' is decisive not only for the difference between the past and the future, but also for the characterization of the objectively determinate, as opposed to the indeterminate. It may at first seem strange to determine a concept in this way, but once we have freed ourselves of the notion of characterizing what is objectively determined by the degree of determinacy, this new method appears far more auspicious. For it employs a qualitative rather than a quantitative difference as the mark of objectivity. After all, the determination must give expression to the fact that the past is beyond all causal influence, and this is a qualitative difference that cannot be embodied in a degree of probability.

We could also attempt to formulate the idea that causal influences can only be propagated in a temporally forward direction in the following way. If A is the cause of C and a slight *alteration* [*Änderung*] is induced in A, a slight

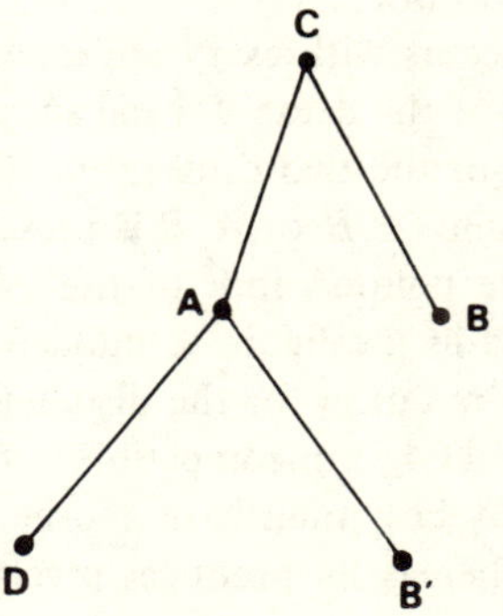

Fig. 9. Pointed forks displaying the unique direction of causal influence.

alteration will also appear in C. But if C is altered slightly, *no* change will arise in A. However, the formulation employing arbitrarily introduced alterations is open to objection, as it is completely impossible to introduce arbitrary changes from the standpoint of determinism. This defect can be avoided by using probability implication. A change in C means simply that a second causal chain, not stemming from A, arrives at this point, so that a pointed fork (Figure 9) is formed; that this auxiliary chain emanating from B has no influence upon A is expressed in the absence of a probability implication between B and A. If, on the other hand, the auxiliary chain coming from B' connects with A (Figure 9), then $C \twoheadrightarrow A$ and $A \twoheadrightarrow B'$ will entail $C \twoheadrightarrow B'$, according to 9*; i.e., the effect of B' is to be observed in C. The concept of probability implication thus permits an unexceptional formulation of the fact that causality is propagated only in a temporally forward direction, and never backwards.

III. THE RELATION BETWEEN PREDICTIVE PROBABILITY AND RETROSPECTIVE PROBABILITY

In the above, we have assumed that a probability exists for the direction 'temporally forwards' as well as for the direction 'temporally backwards'. We wish to reveal the presupposition contained in these assumptions and to demonstrate how retrospective probability can be calculated from predictive probability. For this purpose we must introduce a new kind of relation into the structural schema, a connection with *possible* causal chains.

1. *The Disjunctive Pointed Fork*

Let B be an effect that occurs with every appearance of the cause A_1 as well as with every appearance of the cause A_2; and suppose that B does not result through the *combination* of the two causes, but occurs only when one, and only one, of the two chains A_1B or A_2B is present. This distinguishes this case from the conjunctive pointed fork of the preceding section. We shall draw in an arc of an angle in the figure to indicate the disjunctive property.

We introduce a special notation for the disjunctive property. We may not say that the combination A_1A_2 is incompatible with the occurrence of B, for A_2 (or A_1, as the case may be) could have another effect Q_1 (as it implies B only with probability), whereas A_1 produces precisely B. It is only when no further effect Q_i from A_1 or A_2 is present that B is incompatible with the combination A_1A_2, for since we assume that a causal chain never comes to an

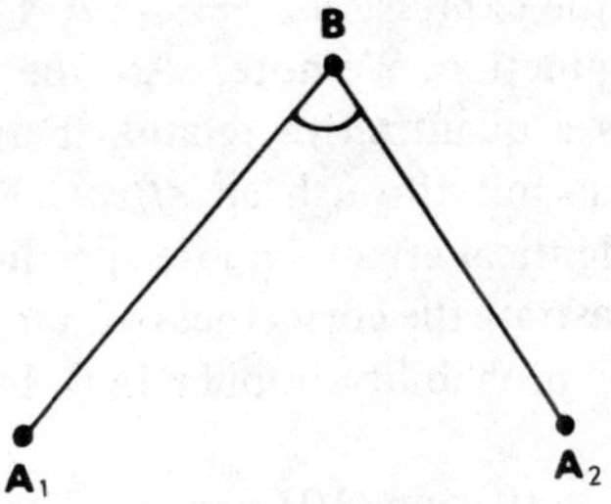

Fig. 10. The disjunctive pointed fork.

end, both chain A_1B and chain A_2B must be present, which by hypothesis is incompatible with B. Therefore we write:

$$B \twoheadrightarrow A_1 \wedge A_2 \;=\; \left\{ \begin{array}{ll} B \,.\, Pl(\overline{Q_i}) \twoheadrightarrow A_1 \,.\, A_2 & [0] \\ B \,.\, Pl(\overline{Q_i}) \twoheadrightarrow A_1 & [p] \\ B \,.\, Pl(\overline{Q_i}) \twoheadrightarrow A_2 & [q] \\ \quad Q_i \twoheadrightarrow A_1 \vee A_2 & \end{array} \right. \quad Df. \qquad (7)$$

Here $Pl(\overline{Q_i})$ signifies the logical product of all possible $\overline{Q_i}$, which is to say:

$$Pl(\overline{Q_i}) \;=\; \overline{Q_1} \,.\, \overline{Q_2} \ldots \overline{Q_n} Df. \qquad (8)$$

The letter or number in the square brackets '[ ]' signifies the *measure* of probability for the relevant implication. It is essential for the definition that this amount is equal to 0 in the first line. The expression standing to the left in (7), which is defined by the right-hand side, is to be read, "*B* alone determines A_1 or A_2." The symbol '∧' signifies the 'exclusive or', but it is important to note that we have not defined this symbol, but only the expression '$B \twoheadrightarrow A_1 \wedge A_2$', and that this expression still contains the meaning '*B* alone'.

We can now set down the relations for the disjunctive pointed fork:

$$A_1 \vee A_2 \twoheadrightarrow B \qquad B \twoheadrightarrow A_1 \,.\, A_2 \qquad B \twoheadrightarrow A_1 \wedge A_2 \qquad (9)$$

From (7), 2*, 3*, 5*, and 9*, it follows that:

$$\begin{array}{cccc} A_1 \twoheadrightarrow B & A_2 \twoheadrightarrow B & B \twoheadrightarrow A_1 & B \twoheadrightarrow A_2 \\ & A_1 \twoheadrightarrow A_2 & A_2 \twoheadrightarrow A_1 & \end{array} \qquad (10)$$

The 'or' in the antecedent of the first expression in (9) is immediately conspicuous, for this expression involves an inference into the future. This 'or' can occur only because the expression $B \twoheadrightarrow A_1 \wedge A_2$ is valid, i.e., because we have an instance of disjunction. We note, also, the sequence $A_1 \twoheadrightarrow A_2$ and $A_2 \twoheadrightarrow A_1$, which states a quantitative relation between events related not through a common cause but through an *effect*. This is, to be sure, not a *common* effect but an identical effect – more specifically, a *possible identical effect*. In order to demonstrate the correctness of our conclusions (and thereby also that of our laws of probability implication), let us analyze this case in greater detail.

Are all the assertions in (9) and (10) correct? The first assertion in (9) and the last two assertions in (10), in particular, appear dubious. Yet some probability implications must hold here. Those below follow from the meaning of the problem:

$$A_1 . \overline{A_2} \twoheadrightarrow B \quad [u] \tag{11}$$

$$\overline{A_1} . A_2 \twoheadrightarrow B \quad [v] \tag{12}$$

$$B \twoheadrightarrow A_1 . \overline{A_2} \quad [u'] \tag{13}$$

$$B \twoheadrightarrow \overline{A_1} . A_2 \quad [v'] \tag{14}$$

The first two equations represent the definition of the problem, the last two the assumption of the two corresponding retrospective probabilities. But one more equation must hold:

$$A_1 . A_2 \twoheadrightarrow B \quad [s] \tag{15}$$

For should A_1 and A_2 be present, we can interpret B as the point of a *conjunctive* pointed fork, the chains of which possess the probability u and $1 - v$ (or $1 - u$ and v), which gives us:

$$s = u(1 - v) + v(1 - u) = u + v - 2uv \tag{16}$$

We wish to demonstrate that the relations (9) and (10) follow in their entirety from the five equations (11)–(15); in doing so, we will not appeal to our laws of probability implication – by means of which this could be immediately proven – but will instead employ calculation. We will follow through all the possible combinations of the three events A_1, A_2, and B with respect to their occurrence and non-occurrence. For the sake of simplicity we shall assume that besides A_1 and A_2 no further causes are possible for B; this assumption places no restrictions on our assertions, but merely decreases the number of unknowns and equations. The following table lists the possible combinations with their frequencies, which we must imagine to have been observed by means of experiments:

$$\begin{array}{ll} A_1 . \overline{A_2} . B & n_1 \\ A_1 . \overline{A_2} . \overline{B} & n_2 \\ \overline{A_1} . A_2 . B & n_3 \\ \overline{A_1} . A_2 . \overline{B} & n_4 \\ A_1 . A_2 . B & n_5 \\ A_1 . A_2 . \overline{B} & n_6 \end{array} \tag{17}$$

Because of the aforementioned simplifying assumption, the combination $\overline{A_1} . \overline{A_2} . B$ does not exist; the combination $\overline{A_1} . \overline{A_2} . \overline{B}$ may be ignored, as it is not included in any of the relations (9) and (10). Now the following question arises: are the observed frequencies $n_1 \ldots n_6$ unchanged in their relation if the enumeration is extended over a larger number of events?

The following five equations are given through the five relations (11)–(15).

$$\frac{n_1}{n_1 + n_2} = u \tag{18}$$

$$\frac{n_3}{n_3 + n_4} = v \tag{19}$$

$$\frac{n_1}{n_1 + n_3 + n_5} = u' \tag{20}$$

$$\frac{n_3}{n_1 + n_3 + n_5} = v' \tag{21}$$

$$\frac{n_5}{n_5 + n_6} = s \tag{22}$$

These five equations are independent of one another. Equations (18), (19), and (22) simply represent the connection between n_1 and n_2, n_3 and n_4, and n_5 and n_6, and are independent of one another as well as of (20) and (21). These latter two are also independent of each other.

Equations (18)–(22), then, establish the relations between the six unknowns $n_1 \ldots n_6$; these relations must therefore remain constant if the relations (11)–(15) are valid. But in that case every other probability relation between the magnitudes [degrees of probability] of A_1, A_2, and B must be valid, for each of these relations is established through the enumeration of the frequencies $n_1 \ldots n_6$. For instance, the probability of $A_1 \twoheadrightarrow A_2$ becomes

$$t = \frac{n_5 + n_6}{n_1 + n_2 + n_5 + n_6} \tag{23}$$

which must remain constant if the relations between $n_1 \ldots n_6$ remain constant. We have hereby proved that the relations (9) and (10) follow from (11)–(15) without using in the proof the laws of probability implication.[8]

Let us now discuss the case in which A_1 as well as A_2 can have no effect other than B: viz., $A_1 \twoheadrightarrow B[1]$, $A_2 \twoheadrightarrow B[1]$. In this case, $A_1 . A_2 . B$ is impossible and $n_5 = 0$. However, (20) and (21) become mutually dependent, as $u' + v' = 1$. There are, then, four equations for five unknowns, and, once again, only the relations between n_1, n_2, n_3, n_4, and n_6 are determined.

We can now trace the consequences of the relations (9) and (10). $A_1 \twoheadrightarrow A_2$ and $A_2 \twoheadrightarrow A_1$ signify that the two events A_1 and A_2 occur with a regular frequency. Were we to survey all the events in the universe, we would discover that the frequency of A_1 and A_2 displays a constant relation. *Not only a common cause, but also a possible identical effect establishes a probability and frequency relation between events.*

This result follows from the existence of predictive *and* retrospective probability. If only retrospective probability existed, regular frequency would be found *only* between BA_1 and BA_2; that is, we could only enumerate the cases in which A_1 or A_2 are accompanied by B.

2. *The Disjunctive Saddle Fork*

Here the same relations hold as with the disjunctive pointed fork; thus the disjunctive fork gives no indication of direction.

$$B_1 \vee B_2 \twoheadrightarrow A \quad A \twoheadrightarrow B_1 . B_2 \quad A \twoheadrightarrow B_1 \wedge B_2 \tag{24}$$

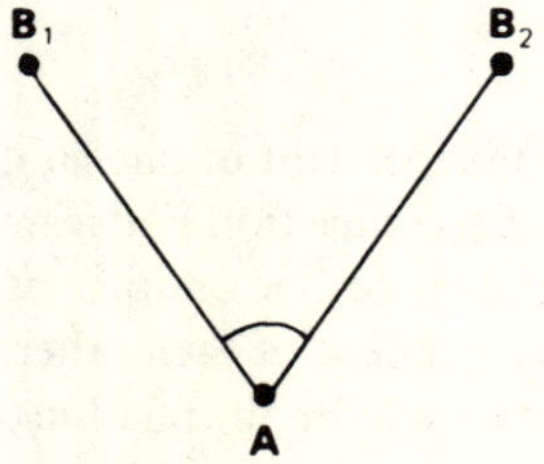

Fig. 11. The disjunctive saddle fork.

The expression $A \twoheadrightarrow B_1 \wedge B_2$ is defined here just as it is in (7), except that in this case we are to understand by Q_i the further possible *causes* of B_1

or B_2. The relations are the same as in the conjunctive saddle fork; thus the addition of the disjunctive relation does not signify any change.

The conclusion regarding the frequency of events, corresponding to the previous one, is: *A common cause establishes a frequency relation between events even when only one of the events can at any one time be the effect of the said cause.* Taken together, these propositions may be formulated as follows.

Distribution Law for Events in the Universe: *Events that are traceable to an identical or to a common cause or that are capable of producing an identical effect display a regular and reciprocal frequency in their occurrence in the universe.*

3. *The Conjunctive Pointed Fork*

We shall now turn, for a few cases, to a quantitative calculation of the retrospective probability on the basis of the predictive probability. Let us begin with the conjunctive pointed fork. In Figure 12, the predictive probabilities are shown as p_i and q_i; the corresponding retrospective probabilities will be shown as p_i' and q_i'. For the measure of probability in $A \ni B$, which was previously shown after the proposition in square brackets, we shall introduce the notation $W(A \ni B)$.*

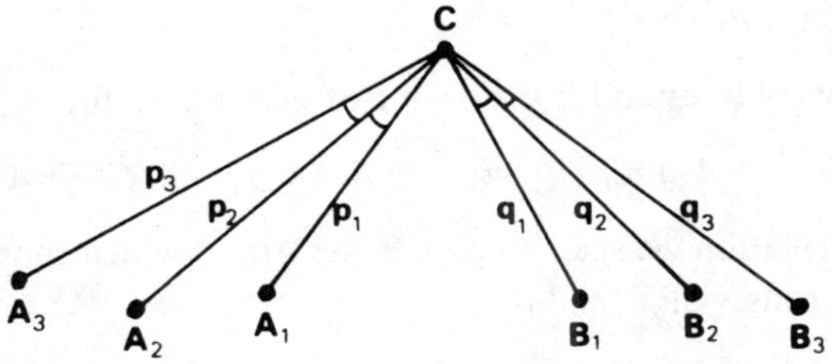

Fig. 12. A pointed fork with other possible causes indicated.

[* *W* is used here and throughout because it is the first letter of the German word 'Wahrscheinlichkeit' (probability). From the standpoint of perspicuity, *P* should be used in its place in English, but as *P* already has a different use in various formulas, this could only create ambiguity. Therefore the notation is left just as Reichenbach set it down. The same comment applies below to the use of *V* for '(Kausal) Verkettung' rather than *C* for 'causal chain'. – EHS]

Then, given that A_1 and B_1 are independent events, we have:

$$W(A_1 . B_1 \ni C) = r_1 = p_1 . q_1 \quad (25)$$

The quantities p_i and q_i do not indicate here, for instance, the probabilities $W(A_i \ni C)$ and $W(B_i \ni C)$, for, according to (3), these implications do not exist. They simply signify, instead, the probability of the occurrence in C of the *partial* effect stemming from A_i (or B_i, as the case may be). Thus we obtain the probability r_i only from its product.

In order to draw inferences about the retrospective probability from the predictive probability, we must make use of A_i and B_i $(i > 1)$, the other possible causes of C. Specifically, we must make some assumption about the totality of possible causes of C, for instance, one of the following two assumptions:

ASSUMPTION H. Any two events $A_i B_k$ can together produce C.

In this case,

$$W(A_i . B_k \ni C) = p_i \cdot q_k \quad (26)$$

ASSUMPTION J. For every A_i, a certain B_i must be added in order for C to be produced.

In this case,

$$\begin{aligned} W(A_i . B_i \ni C) &= p_i \cdot q_i \\ W(A_i . B_k \ni C) &= 0 \end{aligned} \qquad i \neq k \quad (27)$$

We will now try to establish the degree of probability for:

$$W(C \ni A_1) = p'_1 \quad W(C \ni B_1) = q'_1 \quad W(C \ni A_1 . B_1) = r'_1$$

For this calculation we shall employ Bayes' rule,[9] which runs: If $X_1 \ldots X_n$ are all possible causes of Y, and if

$$W(X_i \ni Y) = z_i \quad W(X_i) = \alpha_i$$

then

$$W(Y \ni X_i) = z'_i = \frac{\alpha_i z_i}{\sum_1^n \alpha_k z_k} \quad (28)$$

$W(X_i)$ is the so-called '*a priori* probability' for X_i, i.e., the relative probability of the X_i in relation to one another with respect to their occurrence in the universe. These degrees of probability α_i may all be identical, in which case:[16]

$$z'_i = \frac{z_i}{\sum_1^n z_k} \quad (29)$$

However, without the α_i, the problem is not fully defined and p_i' not calculable. The α_i are clearly the relative probabilities of our *distribution law*, and we shall therefore call them *distribution probabilities*, as the use of the term '*a priori*' in this context can be misleading. It is solely because of the existence of these probabilities that retrospective probability can be calculated on the basis of predictive probability. In school-book examples of Bayes' rule, the α_i are generally established by tracing the X_i back to a common cause X which produces the X_i disjunctively. However, this demonstration is unnecessary. If a retrospective probability exists, the α_i must exist, and this proof is sufficient for our purposes.

With Bayes' formula, equation (28) brings us back to our problem. Let $W(A_i) = \alpha_i$ and $W(B_i) = \beta_i$. Let the number of possible causes A_i be m and those of B_i be n. We may conceive of every combination $A_i B_k$ as a single event possessing the *a priori* probability $\alpha_i \beta_k$ and producting the effect C with the probability $p_i q_k$. Then, by (28), if we make assumption H:

$$
\begin{aligned}
p_1' &= \frac{\alpha_1 p_1 \cdot \sum_1^n \beta_i q_i}{\sum_1^m \sum_1^n \alpha_i p_i \beta_k q_k} = \frac{\alpha_1 p_1}{\sum_1^m \alpha_i p_i} \\
q_1' &= \frac{\beta_1 q_1 \sum_1^m \alpha_1 p_i}{\sum_1^m \sum_1^n \alpha_i p_i \beta_k q_k} = \frac{\beta_1 q_1}{\sum_1^n \beta_i q_i} \qquad (30) \\
r_1' &= \frac{\alpha_1 p_1 \beta_1 q_1}{\sum_1^m \sum_1^n \alpha_i p_i \beta_k q_k} = p_1' \cdot q_1'
\end{aligned}
$$

On the other hand, assumption J gives us (assuming $m = n$):

$$
p_1' = q_1' = r_1' = \frac{\alpha_1 p_1 \beta_1 q_1}{\sum_1^m \alpha_i p_i \beta_i q_i} \qquad (31)
$$

Thus, given assumption H, the retrospective probability r_1' is calculated from the individual retrospective probabilities p_1' and q_1' just as the corresponding predictive probability r_1 is calculated from the individual predictive probabilities p_1 and q_1. Given assumption J, on the other hand, the calculation changes for the retrospective probability; specifically, all three retrospective probabilities become the same. Whether assumption H or assumption J corresponds to reality depends, of course, upon the particular conditions of the problem; most cases will probably turn out to exemplify a combination of both. In such cases, r_1' will turn out to have a corresponding intermediate value.

4. *The Conjunctive Saddle Fork*

Here, too, we must employ another assumption if we are to calculate the retrospective probability from the predictive probability, as follows.

ASSUMPTION K. B_1 and B_2 necessarily have *a common cause*.

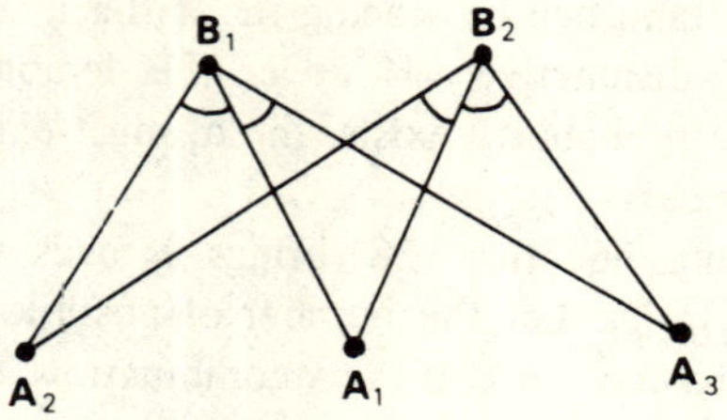

Fig. 13. A saddle fork with other possible causes indicated in accordance with assumption K.

Let $A_1 \ldots A_m$ in Figure 13 be the possible common causes, and let their distribution probability be $W(A_i) = \alpha_i$. We will then introduce the following additional stipulations.

$$\begin{aligned} W(A_i \ni B_1) &= p_i & W(B_1 \ni A_i) &= p_i' \\ W(A_i \ni B_2) &= q_i & W(B_2 \ni A_i) &= q_i' \\ W(A_i \ni B_1 B_2) &= r_i & W(B_1 B_2 \ni A_i) &= r_i' \end{aligned}$$

This gives us:

$$\begin{aligned} p_i' &= \frac{\alpha_i p_i}{\sum_1^m \alpha_k p_k} \qquad q_i' = \frac{\alpha_i q_i}{\sum_1^m \alpha_k q_k} \\ r_i' &= \frac{\alpha_i p_i q_i}{\sum_1^m \alpha_k p_k q_k} = \frac{p_i' \cdot q_i'}{\alpha_i \sum_1^m \frac{1}{\alpha_k} p_k' q_k'} \end{aligned} \tag{32}$$

In discussing these formulas, let us take the simple case in which the α_i are all of identical magnitude. Then the formulas in (32) can be reduced to:

$$p_i' = \frac{p_i}{\sum_1^m p_k} \qquad q_i' = \frac{q_i}{\sum_1^m q_k} \qquad r_i' = \frac{p_i' q_i'}{\sum_1^m p_k' q_k'} \tag{33}$$

The quantities p'_i, q'_i, and r'_i are, as probabilities, all < 1, as is easily seen. But, in addition:

$$r'_i > p'_i \cdot q'_i \tag{34}$$

because

$$\sum_1^m p'_k q'_k < \sum_1^m \sum_1^m p'_k q'_i = \sum_1^m p'_k \cdot \sum_1^m q'_i = 1$$

If the saddle fork were turned upwards, so as to become a pointed fork, we would obtain $r'_i = p'_i \,.\, q'_i$, given that p'_i, q'_i, and r'_i now stand for the corresponding predictive probabilities.[11] We now have reached the following proposition: *Retrospective inference from two events to one produces a higher probability than the corresponding predictive inference with the same probabilities*. The distinctive characteristic of retrospective inference is also expressed metrically as a proposition about *evidence*, as opposed to *effect*; each individual effect permits the drawing of a retrospective inference to the cause with a certain probability, and any additional effect serves simply as a *confirmation*, but not as a *necessary condition*, of this inference.

To be sure, this confirmation can also be of a negative nature, if the second effect happens to 'make' the supposed cause 'improbable'. In order to examine this case, let us compare r'_i with the individual probability p'_i. Then:

$$r'_i = p'_i \,.\, f \quad f = \frac{q'_i}{\sum_1^m p'_k q'_k} = \frac{q_i \sum_1^m p_k}{\sum_1^m p_k q_k} \tag{35}$$

According to whether $f \lesseqgtr 1$, $r'_i \lesseqgtr p'_i$. Thus:

$$\begin{aligned} r'_i &\lesseqgtr p'_i \quad \text{if} \quad q_i \sum_1^m p_k \lesseqgtr \sum_1^m p_k q_k \\ r'_i &\lesseqgtr q'_i \quad \text{if} \quad p_i \sum_1^m q_k \lesseqgtr \sum_1^m p_k q_k \end{aligned} \tag{36}$$

If, for instance, all q_k are of equal value, then $r'_i = p'_i$, i.e., addition of the event B_2 neither increases nor decreases the probability of A_i as calculated solely on the basis of B_1. In this case, all causes A_i are equiprobable causes of B_2. If q_i is the highest of all q_k, then $r'_i > p'_i$; if it is the lowest, then $r'_i < p'_i$. The probability increases or decreases, then, according to whether B_2 'makes' the cause A_i probable or improbable. Condition (36) states that this 'making probable' occurs when q_i exceeds a certain intermediate value derived from q_k with the assistance of p_k.

In the formulas (32) and (33), we should further note that it is possible to express the compound retrospective probability r_i', consisting of the retrospective probabilities p_i' and q_i' of the individual events and the distributive probabilities a_i, without including the predictive probabilities p_i and q_i. This is a peculiarity that does not hold for every case; it depends here upon assumption K. Let us now take a look at another case which also possesses this feature and yet is based upon a different assumption.

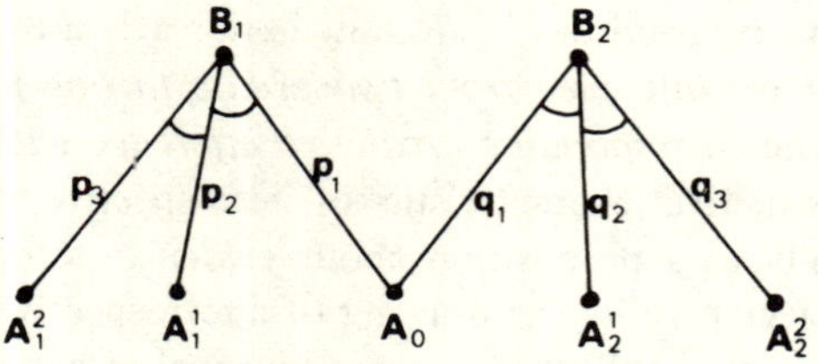

Fig. 14. A saddle fork with other possible separate causes indicated.

Let us assume that events B_1 and B_2 need no longer be explained by a single cause A_0 and admit instead for each the separate causes A_1^i and A_2^k (Figure 14). But what we are attempting to establish is precisely the probability of the existence of the common cause A_0. In order to simplify the problem, let us add yet another assumption, as follows:

ASSUMPTION L. If A_0 is present, then all $A_1^1 \ldots A_1^m$ and $A_2^1 \ldots A_2^n$ are excluded. Here

$$A_0 \supset \bar{A}_1^i \,.\, \bar{A}_2^k.$$

In reality, an assumption of this nature, like all the previous ones, will not hold with absolute certainty; i.e., the implication appearing in assumption L is only a probability implication of a very high degree. But in actual cases we will often be permitted to treat this high degree of probability as certainty. For instance, assumption L may apply to the many instances of proof by mean of circumstantial evidence in which it is a matter of tracing the various clues, each of which, taken alone, admits of independent causes, to a common cause. For purposes of calculation we will make the additional assumption that the α_i are all equal; otherwise the result will not be perspicuous. Using the same notation as above, we get:

$$p'_0 = \frac{p_0}{\sum_0^m p_k} \qquad q'_0 = \frac{q_0}{\sum_0^n q_k}$$

$$r'_0 = \frac{p_0 q_0}{p_0 q_0 + \sum_1^m \sum_1^n p_i q_k} = \frac{p_0 q_0}{\sum_0^m \sum_0^n p_i q_k - p_0 \sum_0^n q_k - q_0 \sum_0^m p_k + 2p_0 q_0}$$

$$= \frac{p'_0 q'_0}{1 - p'_0 - q'_0 + 2p'_0 q'_0} = \frac{p'_0 q'_0}{(1-p'_0)(1-q'_0) + p'_0 q'_0} \tag{37}$$

Here too, then, r'_0 can only be expressed through retrospective probability, and, at that, only the retrospective probabilities p'_0 and q'_0 are included, while the other retrospective probabilities cancel out. Once again, of course, $r'_0 < 1$, but also $r'_0 > p'_0 q'_0$, for the denominator in (37) is < 1. This result appears when we observe that $0 < p'_0 < 1$ and $0 < q'_0 < 1$. If we stipulate that $p'_0 = 1 - \delta$, and $q'_0 = 1 - \eta$, then the denominator will equal $\delta\eta + p'_0 q'_0 < (p'_0 + \delta) \times (q'_0 + \eta) = 1$. Again, calculation shows the retrospective probability to be higher than the corresponding predictive probability.

Let us inquire more closely into the case in which $r'_0 > p'_0$. We must also stipulate that:

$$\frac{q'_0}{1 - p'_0 - q'_0 + 2p'_0 q'_0} > 1$$

From which it follows that:

$$q'_0 > \tfrac{1}{2} \tag{38}$$

Accordingly, if $p'_0 > \frac{1}{2}$, then $r'_0 > q'_0$. We establish, then, that the second event B_2 increases the probability of A_0, calculated from B_1, if it implies only A_0 with a probability higher than $\frac{1}{2}$.[12]

5. *The Conjunctive Double Fork*

Finally, let us examine the case in which there is a double connection. Two independent causes A_1 and B_1 have two common effects C and D. We shall attempt to reduce this case to those we have already discussed.

The proof can be carried out in two different ways, which we shall schematize as follows:

a)		b)	
$C \ni A_1 B_1$	$[r'_1]$	$CD \ni A_1$	$[s'_1]$
$D \ni A_1 B_1$	$[R'_1]$	$CD \ni B_1$	$[S'_1]$
$CD \ni A_1 B_1$	$[w'_1]$	$CD \ni A_1 B_1$	$[w'_1]$

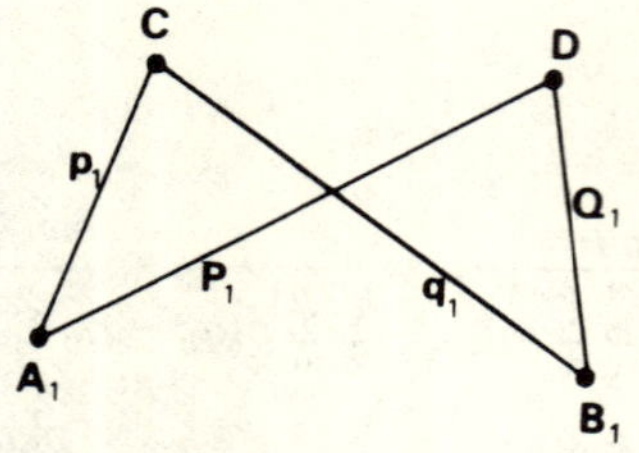

Fig. 15. A double fork with predictive probabilities indicated.

In method a), the first two lines stand for a retrospective inference in the conjunctive pointed fork; $A_1 B_1$ is then conceived of as a *single* event, and in the third line a retrospective inference is carried out in the saddle fork. In method b), the first two lines stand for a retrospective inference in the saddle fork; then *CD* is conceived of as a *single* event, and in the third line a retrospective inference is carried out for the pointed fork. Both methods of proof must lead to the same goal. But we must make assumptions of the sort made for the relevant single forks if the conclusion is to be defined. We shall make assumption K for the saddle fork and assumptions J and H, in succession, for the pointed fork. In order to keep the calculation relatively perspicuous, we will once again stipulate that the distributive probabilities a_i are of identical value.

Let us choose method a). With assumption J we get, according to (31):

$$w(C \twoheadrightarrow A_i B_i) = r_i' = \frac{p_i q_i}{\sum_1^m p_i q_i} \quad w(D \twoheadrightarrow A_i B_i) = R_i' = \frac{P_i Q_i}{\sum_1^m P_i Q_i}$$

Then, with assumption K, according to (33):

$$w_1' = \frac{r_1' R_1'}{\sum_1^m r_i' R_i'} = \frac{p_1 q_1 P_1 Q_1}{\sum_1^m p_i q_i P_i Q_i} \tag{39}$$

On the other hand, the predictive probability w_1 for $A_1 B_1 \twoheadrightarrow CD$ becomes:

$$w_1 = p_1 q_1 P_1 Q_1 \tag{40}$$

Let us define the values p_i', q_i', P_i', Q_i' as:

$$p_i' = \frac{p_i}{\sum_1^m p_k} \quad q_i' = \frac{q_i}{\sum_1^n q_k} \quad P_i' = \frac{P_i}{\sum_1^m P_k} \quad Q_i' = \frac{Q_i}{\sum_1^n Q_k} \tag{41}$$

which we may not, however, interpret in this case as retrospective probabilities corresponding to the unprimed quantities; as assumption J has, in accordance with (31), the effect of making these retrospective probabilities equal to r_i' or R_i', as the case may be. We may then rewrite (39) as:

$$w_1' = \frac{p_1'q_1'P_1'Q_1'}{\sum_1^m p_i'q_i'P_i'Q_i'} \tag{42}$$

Now let us again use method a), but with assumption H rather than assumption J. Then, according to (30), we get:

$$w(C \twoheadrightarrow A_iB_k) = r_{ik}' = \frac{p_iq_k}{\sum_1^m \sum_1^n p_iq_k} = p_i'q_k'$$

$$w(D \twoheadrightarrow A_iB_k) = R_{ik}' = \frac{P_iQ_k}{\sum_1^m \sum_1^n P_iQ_k} = P_i'Q_k' \tag{43}$$

in which we may now interpret p_i', q_i', P_i' and Q_i' as retrospective probabilities corresponding to the unprimed quantities, for this is admissible according to assumption H. With assumption K and (33), we obtain the further result:

$$w_1' = \frac{p_1'q_1'P_1'Q_1'}{\sum_1^m \sum_1^n p_i'q_k'P_i'Q_k'} = \frac{p_1q_1P_1Q_1}{\sum_1^m \sum_1^n p_iq_kP_iQ_k} \tag{44}$$

Thus we get a result that differs from (42) (or (39), as the case may be) in that the double sum appears in the denominator.

When method b) is used, assumption J cannot be carried out, for in the very first step different expressions appear for s' and S' – which, according to assumption J, must be the same. If, on the other hand, we use assumption H, we once again end up with (44).

We are now in a position to study the result. With respect to its chains, the double fork is symmetrical for past and future, but there is nonetheless a characteristic difference in the type of probability pertaining to each. The predictive probability comes out, according to (40), as $w_1 = p_1q_1P_1Q_1$; for the retrospective possibility, however, this expression must also be divided by a summation expression. This applies equally to assumption J according to (39) and assumption H according to (44). This lack of symmetry becomes particularly clear with assumption H, according to which it is admissible to interpret p_i', q_i', P_i', and Q_i' as individual retrospective probabilities. For if we conceived of the double fork as heading in the temporally opposite direction

– that is to say, of C and D as earlier and A_1 and B_1 as later events – and of the primed individual probabilities as predictive and the unprimed individual probabilities as retrospective probabilities – we would obtain the following result.[13]

$$W(CD \twoheadrightarrow A_1B_1) = w'_1 = p'_1 q'_1 P'_1 Q'_1 \tag{45}$$

$$W(A_1B_1 \twoheadrightarrow CD) = w_1 = \frac{p'_1 q'_1 P'_1 Q'_1}{\sum_1^m \sum_1^n p'_i q'_k P_i Q'_k} \tag{46}$$

The measure of the probability implication between the events A_1B_1 on the one hand and the events CD on the other would, that is, be different, although the probabilities between the individual events maintain their numerical value. If we imagine that it is not known in which temporal direction the double fork is to be conceived as pointing, but that at the same time the individual probabilities in both directions are known – that is, $p_1 q_1 P_1 Q_1$ and p'_1, q'_1, P'_1 and Q'_1 – and also the collective probabilities w_1 and w'_1, then the temporal direction of the fork can be determined by examining whether these values fulfil the relations (40) and (44) or the relations (45) and (46). A complete examination also requires knowledge of the probabilities with an index other than 1, which can, indeed, be determined in just the same way as the others, but even without these values it is possible to decide whether w_1 is correctly represented by (40) or w'_1 by (45). Thus the temporal direction can be determined by measuring probability implications, i.e., basically by the enumeration of statistical regularities.

On the other hand, no assertion is made as to whether the retrospective inference produces the greater or the smaller probability. According to (44) – that is, to assumption H –

$$w'_1 \gtreqless w_1, \quad \text{if} \quad \sum_1^m \sum_1^n p_i q_k P_i Q_k \lesseqgtr 1 \tag{47}$$

With assumption J, (39) gives us:

$$w'_1 \gtreqless w_1, \quad \text{if} \quad \sum_1^m p_i q_i P_i Q_i \lesseqgtr 1 \tag{48}$$

There is no general comment to be made concerning these two relations, as p_i, q_i, P_i, and Q_i represent unbound probabilities. We see here once again that it is not the *degree* of determinability, but the *manner* of determinability that distinguishes past and future. Under some conditions inference into the future may prove more certain than inference into the past, compared with respect to an identical causal chain.

On the other hand, (44) gives us:

$$w'_1 > p'_1 q'_1 P'_1 Q'_1, \tag{49}$$

for, according to (41):

$$\sum_1^{} \sum_1^{n} p'_i q'_k P'_i Q_k < \sum_1^{m} \sum_1^{n} \sum_1^{m} \sum_1^{n} p'_i q'_k P'_r Q'_s$$

$$= \sum_1^{m} p'_i \sum_1^{n} q'_k \sum_1^{m} P_r \sum_1^{n} Q'_s = 1$$

Therefore w'_1 as determined by (44) is higher than w'_1 as determined by (45), and we can say: If we compare a causal chain V_1 (as in Figure 15) with another chain V_2 which is a symmetrical mirror image of the first, mirrored at the cross-section through the present, we find that a higher degree of probability attaches to the retrospective inference in V_1 than to the predictive inference in V_2. It is only in this respect that the difference in the manner of inference also signifies a difference in the degree of determinability.

Thus the same result is confirmed for the double fork that was already shown to hold for the saddle fork, as compared to the pointed fork. In comparing these forks, we must take account of one other difference. The retrospective inference of the saddle fork is an inference from two events to one; it must be compared reflexively with the predictive inference (25) of the pointed fork, which gives in (33) and (37) the result that we formulated in (34). Here again, then, greater certainty pertains to the retrospective inference. In contrast, the retrospective inference of the pointed fork is an inference from one event to two events; it must be compared reflexively with the predictive inference of the saddle fork. In the latter, $r_1 = p_1 . q_1$; if we compare this with (31) and (30) by contrasting, in the latter formulas, the formations containing the primed values with the formation $p_1 . q_1$, we find a greater certainty for the retrospective inference attaching only to assumption J, while for assumption H the probabilities prove equal. This case, then, constitutes an exception to our rule.

The results established in Section III can be summarized in the following propositions:

1. The existence of predictive and retrospective probability presupposes a law of distribution for events in the universe.

2. The measure of retrospective probability cannot be calculated from that for predictive probability without auxiliary assumptions, which may vary for the different kinds of cases, and the presence of which must be established empirically.

3. It cannot be asserted that in a given structure of the universe the past can always be determined with greater certainty than the future. But in the cases discussed (with one exception), the distinctive topological feature of retrospective inference enables us to calculate the past from given individual retrospective probabilities with greater certainty than would be possible if these same individual probabilities were predictive probabilities and could be combined so as to form a reflexively symmetrical inference into the future.

NOTES

[1] [1915b] and [1916a]; also [1920c] and [1920d]. The continuous function is not always the Gaussian function, which applies only to certain cases.

[2] Whether the probability can in every instance actually approach 1 without limit or whether, instead, limits appear in certain places before 1 is reached is a matter open to dispute. These limits might also turn out to be, in fact, unattainable, so that the proposition that for every level of precision achieved a higher level exists would remain true. Justifiable as such an assumption – which would be confirmed if quantum theory abandoned all attempts at causal explanation and contented itself with the probability jumps of electrons – may appear, we shall not discuss it here; and everything that follows is also compatible with a probability capable of approaching 1 without limit.

[3] *Zschr. f. Phys.* **13**, 62 (1923).

[4] *Kantstudien* **30**, 331 (1925).

[5] [1924h], [1921d]. The 'Axiomatik der speziellen Relativitätstheorie [Axiomatic System of the Special Theory of Relativity] by C. Carathéodory (*Berl. Akad. Ber.*, 1924, p. 12) is closely related to my proposed axiomatic system.

[6] Note that, in Figure 1b, that portion corresponding to the future in Figure 1a is drawn somewhat differently, for the purpose of indicating that the future has turned out somewhat differently from what was calculated in 1a.

[7] That is why there is a science of history only for the past. The chronicle, i.e. the registration of events, is the characteristic mark of history.

[8] This proof can also be carried out if (13) and (14) are replaced by the relations $B \rightarrow\!\!\!\ni A_1$ and $B \rightarrow\!\!\!\ni A_2$ or also if (11) and (12) are replaced by the relations $A_1 \rightarrow\!\!\!\ni B$ and $A_2 \rightarrow\!\!\!\ni B$.

[9] Cf. any textbook on probability calculus, e.g., *Lehrbuch der Wahrscheinlichkeitrechnung* by Czuber, 1908, p. 175.

[10] $\Sigma_1^n z_i' = 1$, as can be seen from (28) and (29); therefore the z_i' are called 'bound probabilities'. The z_i, on the other hand, are 'unbound probabilities'; that is $\Sigma_1^n z_i \gtreqless 1$.

[11] If the α_i are not of identical magnitude then (34) need not be fulfilled. But in such a

case, $p_i'q_i'$ would also cease to signify predictive probability in the pointed fork, as the *a priori* probability of A_i enters in as well, and p_i' and q_i' would then not be independent probabilities. – Note, too, that we are not here comparing the retrospective probability with the predictive probability of the same case – not, that is, r_i' with r_i – but with the predictive probability of a corresponding case in which p_i' and q_i' signify predictive probabilities (namely, probabilities of partial effects in the sense indicated in the comments following (25)). That is, we are comparing r_i' with $p_i'q_i'$.

[12] If we had not stipulated that the α_i are equal, this slot would not be filled specifically by 1/2, but rather by some construct consisting of the α_i.

[13] Here the probabilities with an index other than 1 refer to implications between A_1 and B_1 on the one hand and events C_i and D_i on the other, which we have not previously considered. These probabilities, then, are not comparable to those in the preceding formulas. In contrast, all probabilities with the index 1 apply to the same implications as before, and in the same direction between the events; the only difference is that we now take the events to be reversed with respect to temporal direction, so that the primed values now pertain to the direction 'temporally forwards'.

48. THE AIMS AND METHODS OF PHYSICAL KNOWLEDGE

[1929g]

TABLE OF CONTENTS

Part (a) The General Theory of Physical Knowledge

1. THE VALUE OF PHYSICAL KNOWLEDGE

If we ask a physicist why he does physics, he will be hard put to it to give a really pertinent answer. He may attempt to justify his work by pointing out its value for science in general or its value for mankind, or perhaps he will appeal to the fact that history has shown us an ever-progressing

Translated from 'Ziele und Wege der Physikalischen Erkenntnis', *Handbuch der Physik, vol. 4: Allgemeine Grundlagen der Physik.*

development of the human mind, to which each of us is obligated to contribute his own little bit of brain power. But if he takes an honest look at himself, he will find that all these pretty-sounding goals are nothing but a well constructed mask, a sort of Sunday frock that he dons whenever he must appear in public. In fact, the process that turns an individual into a physicist is both more complicated and more crude. Talent; environment; often, too, material pressures; stumbling by chance into a stimulating circle of fellow physicists; a lucky creative inspiration that brings him recognition in the field, enabling him to establish himself as a physicist – in short, people are drawn into physics, as into other professions, by their experiences, without much relation to an idealistic concern for the development of mankind. And precisely those physicists who have had to struggle with material want, making their way laboriously in science despite the attractions of a more practical occupation, will be least able to claim that they have derived their strength from a concern for the welfare of mankind. They really are unable to say why it is that they do physics, must do physics – why they keep returning to the laboratory or go on searching for mathematical theories. They just do it; that is, in the end, the only reason they can offer with some credibility.

For all that, this honest explanation of the reason for working in physics seems to me a better justification of its value than the ideally constructed one. Physics may or may not be useful or valuable for humanity – the most important point to be made is that it is a need, that it grows in men just as does the urge to live or to play or to build a society with other people. It is the desire to *know* to which it is ultimately to be traced and which can become so strong that this most sublimated form of curiosity dominates a person's life, enabling him to overcome all material or economic hindrances. It is the yen to "discover what makes nature tick", as a great physicist once explained to me. To be sure, this wish can display itself in various forms. It drives one person to more and more experiments, to the gathering of facts; it drives another to the construction of theories and to concentrating upon the explanation of a single fact, comprehension of which appears more important to him than the gathering of new material. It compels one person to be careful and precise, to select ideas critically; it stimulates another to more fluent thinking and to imaginative creation. The one comes near to fulfillment in the mental exercise of planning and designing; the other finds satisfaction only in secure results. But the force behind all these activities is always a focus upon knowing, the desire to learn something of the secrets of nature. None can say from whence this desire comes; it is itself one of the ultimate facts in which the physicist must believe.

This fact alone constitutes a reply to the question as to the ultimate value of physical science. All theories about the value of science for civilized culture, when asked about the value of civilization, are eventually faced with the necessity of appealing to some fact of this kind. We want to do such and such a thing: this is the fact from which every value theory must begin. To be sure, we draw distinctions, setting aside certain desires as inessential. But the very deepest desires force themselves upon us with such a degree of self-evidence that we have no choice but to regard this obvious self-evidence as an expression of the highest value.

If we were to accept a general value for civilization as the basis of the value of physical science, we would be obliged to derive the goal of physics from this general value, the goal being the realization of this value. Goal and value are in general so related, the evaluation prescribing the goal. For instance, if we were to make technical utility the basis for the value of physics, the goal of physical research would have to consist in the discovery of technically useful effects; anything else would be permissible, at best, from the standpoint of the possibility of future utility. But it is quite apparent that technology can never represent the philosophical justification of physics, for the achievements of technology, too, only serve to satisfy human needs, and precisely those that, unlike the thirst for knowledge, do not seem essential in any profound sense. Surely technological value can be no more than a byproduct of physical research, not to be despised with the arrogance of the theoretician, yet not to be permitted to determine the goal of research. To take another example, if we were to see the value of physics in the establishment of a world-view, then the goal of physics would have to be regarded as the acquisition of knowledge of a kind susceptible of philosophical interpretation, e.g., of cosmological hypotheses. But even this would give a one-sided determination of the goal, conceiving the essence of physics far too narrowly. We are spared such restrictive determinations of a goal – and even the most tolerant determinations of this kind will impose limitations, judged from the breadth of the drive to gain a knowledge of nature – if we refuse to formulate a supreme value. If we view the fact of the will to know as the ultimate ground of value, we can regard as the goal nothing other than to do everything that serves this will.

The contents of this goal may be gathered from the science in question and its historical development, which shows us the goals along with the facts of scientific knowledge. An inductive determination of this sort must appear far more fruitful, far closer to reality, than deduction from general principles. To be sure, physics has not always been conscious of its own goals, and at least of all the individual physicist himself, whose work may have made these goals clear in the first place. It is one of the most curious features of the human mind that it is able to pursue the right course without clearly discerning its end, that, indeed, the man who follows his course with easy confidence rather than reflecting upon it attains his destination better and sooner than the excessively introspective thinker. An observation regarding the goals of science exceeds the boundaries of science itself. It belongs instead to philosophy, and it cannot be said that philosophical insight is a necessary condition of research in the individual sciences. It is, rather, a later stage, presupposing a certain degree of maturity in the individual sciences. Hence the relation between physics and philosophy is asymmetrical: philosophy can learn a great deal from physics, but physics very much less from philosophy. In spite of this, the physicist cannot avoid occasionally waxing philosophical and asking after the goals of physics. This question confronts him when, pausing in his work, he reflects upon his own activity, when he scans the general outlines of the course followed by his science, and he will soon notice that these philosophical questions are able to establish themselves with the same persistence as questions regarding the secrets of nature. The question as to the course and the end of human *knowledge* is, as a question of a higher logical order, rooted in human nature in much the same way as questions concerning the course and the structure of nature, and the physicist who refused to recognize the philosophically oriented question as justified would be removing the rational underpinnings from his own activity. The only form of philosophy that must be rejected is that which attempts to prescribe for the physicist the results or methods of his science. But a philosophy that draws its facts from science, that is able to shed light upon the mysteries of scientific research and clarify for the investigator by means of his own accomplishments the ends and the methods of his work, can only serve as a welcome ally along the path to knowledge.

These introductory remarks are intended both to illuminate the valuation of scientific activity and to indicate the direction to be taken by the investigation that follows. It does not aim to be a critique of physical research, scouting out or inserting values from some biased standpoint. It wishes to do justice to the wealth of scientific activities while at the same time revealing the general outlooks lying behind them that can be logically derived from the particular contents. Within the framework of an encyclopedic handbook of physics, such a picture can only be sketched in its rough features, omitting details of justification. Yet it is to be hoped that this framework can show that investigation in the philosophy of natural science should be carried out in the closest conjunction with physical science itself.

2. DEMARCATION BETWEEN PHYSICS AND THE OTHER NATURAL SCIENCES

We have described the aim of physics as the knowledge of nature, yet we know that there are other sciences that pursue the same goals. Thus there arises the task of drawing a line of demarcation between physics and the other natural sciences.

The demarcation can be carried out in one of two ways. The first separates two sciences according to the difference between their *objects*, the second according to the difference in their *methods*. Let us take a look at the first approach. Inanimate nature is differentiated from animate, and the former assigned to physics, the latter to biology. The division thus gained is in some ways pertinent; above all, it corresponds to the historical development of both sciences, the distinctive character of which has grown precisely out of their concentration upon certain subject matters. It appears, nonetheless, that this kind of delineation cannot be strictly carried out if logical standpoints are put before historical. For animate nature also displays phenomena governed by physical laws; organisms, including man, follow the laws of gravity, the law of energy, laws of electrical conductivity, and so forth. In setting up these laws, physics always claims that they apply equally to organisms. They leave the testing to the biologists simply because these laws are often cloaked by complex phenomena – think, for instance, of the flow of electric current in the body of an animal – and their clarification presupposes a precise knowledge of the organism. Conversely, the materials of inanimate nature play a role in biology; consider, for instance, the process whereby animals receive nourishment, in which even inorganic materials such as salts are of great importance. Thus we cannot draw the distinction between physics and biology by means of a line of demarcation running through the whole of nature.

For this reason we select the second means of demarcation, which employs the difference in methods. Such a division has the advantage of remaining valid even when the subject matters of the two sciences partially overlap or even coincide. For purposes of clarification, let us consider the subdivision of physics into mechanics, thermodynamics, statistics [statistical physics – Ed.], and so on. Thermodynamics and molecular statistics treat identical subject matter throughout, but using different methods[1]. The former begins with a few basic concepts, such as energy, entropy, and phase, combines them into a few very simple laws, and deduces from these a large number of phenomena. Statistics, on the other hand, does not initially recognize the basic concepts of thermodynamics; it employs only mechanical concepts, and by applying them to a large number of similar particles in a state of motion, in conjunction with the axioms of probability theory, it eventually reaches complex concepts, such as the probability of a constellation class, which it then coordinates with the thermodynamic concepts. Thus it offers explanations of the same subject matter, but with a higher degree of precision. For statistics, thermodynamics seems to offer a mere approximation, yet an admissible one, and in view of this limitation on precision the two disciplines can never contradict one another. In a certain sense, indeed, statistics has rendered thermodynamics unnecessary, for everything asserted by thermodynamics can basically also be accounted for statistically. However, matters are quite different from a practical viewpoint. There are many occasions on which statistics is of no use, precisely because it operates with such exactitude, whereas general thermodynamic methods can be implemented without difficulty. For instance, it seems hopeless to try to comprehend the occurrences in a steam engine by statistical methods, which give out when confronted with the complexity of the phenomena of currents in a steam cylinder. But thermodynamic methods can be successfully applied to the steam engine and deliver just exactly what we want to know. A method is sometimes too fine an instrument for a particular subject matter and is therefore unsuitable for revealing the gross features in the structure of that subject.

The situation is similar with respect to the difference between physics and biology. Both can treat the same subject matter, and the division of labor that assigns living organisms to the one and inanimate objects to the other is not of a fundamental nature. It is comparable to a division of labor that assigns the steam engine to thermodynamics and rarefied gases to statistics. The difference lies far more in the method, finding its clearest expression in the fact that biology employs concepts as elementary concepts that appear extraordinarily complex to physics. Concepts such as life, organism, organ, nourishment, reproduction, and development are accepted by biology as elementary concepts, in the same way that thermodynamics speaks of a quantity of heat or an aggregate state, and it attempts to elucidate by means of these concepts phenomena of a complexity impervious to the precise instrument of physics. The concept of life stands in the foreground, occupying much the same position as the concept of energy in thermodynamics; biology clarifies the concept of life by pursuing its implications through the whole of the field, just as thermodynamics is forced ultimately to define energy as that entity the consequences of which possess such and such properties. Thus biology gains a characterization of the phenomenon of life by means of methods that physics would regard as macroscopic or phenomenological.

Any development extending beyond the method of biology would have to consist in providing the concept of life with a content in some other way. Life would have to be shown to be, say, a characteristic of protein constructed in a particular way or a stationary condition of certain chemical reactions. But up to now such attempts have failed. And here we come to the great *difference* between biology and physics on the one hand and thermodynamics and statistics on the other. It has been shown that thermodynamics is merely a macroscopic form of statistics; thermal energy can be completely explained as the mechanical energy of all the molecules taken together. But that biology is a macroscopic form of physics – a sort of physics of protein – that the concept of life can be reduced to physical concepts, has *not* been demonstrated. The question whether biology will ever become a part of physics in the way that thermodynamics has become a part of mechanics must be put aside for the time being. The concept of life in itself may still contain problems which have nothing to do with physics. *Vitalism* perhaps is correct in asserting that a living being is not merely a particularly complicated physical apparatus, like a machine, but something fundamentally different, the comprehension of which calls for new, non-physical assumptions. It seems idle to make predictions concerning the course of research. In any case, the theory of evolution, important as it has become through combining the various biological disciplines together into a single great science – it achieved in biology something similar to what is to be observed in the combining of optics, the theory of electricity, and mechanics into general physics – has not been able to solve the problem of life itself and has not realized the hopes pinned upon it in this respect. It is important to emphasize that this problem has not been resolved. The representative of a specialized subject has a tendency to see in his own science the exclusive form of all knowledge, and hence it is that the physicist likes to believe that biology will one day be absorbed into physics. This possibility certainly cannot be rejected out of hand, but whether or not it will come about will depend upon the way in which the problem of life comes to be solved. Should it turn out that certain physical laws are absolutely inapplicable to life processes, the absorption of biology into physics will never take place, and physics will instead have to refrain from touching upon areas in nature to which it would not apply[2].

We find, then, that the question of whether two sciences will coalesce cannot be answered until a train of scientific development has reached its end. The answer is itself a scientific *result* and cannot be deduced from general scientific principles. The combining of mechanics and the theory of heat was such a physical result, not predictable *a priori.* The answer to this question will determine the extent to which the difference in *methods* is related to a difference in *epistemological objects*. It is undoubtedly true that the methodological difference between biology and physics has grown out of the characteristics of the material with which these sciences are preoccupied. But for the time being, as long as the problem of vitalism is unresolved, we can establish only the difference in methods as a certain result. In any case it is possible that, as with thermodynamics and statistics, it does not correspond to a fundamental difference in the subject matter. But if it does, there will be at the most only a few areas where physics is inapplicable to animate nature; in general it has been clearly established that the laws of physics apply in biology also. These are the reasons that cause us to choose the second means of demarcation; in demarcation according to method we have a solution that is independent of the future development of science.

But we cannot carry out here a comparative characterization of the methods. Such an investigation would require a very precise analysis of conceptual formation within both sciences, and this belongs in the field of comparative studies of science, once again the object of intense research. For instance, Kurt Lewin has carried out a comparative investigation of the concept of genesis in physics, biology and the history of evolution (in his *Begriff der Genese in Physik, Biologie und Entwicklungsgeschichte*)[3] demonstrating that peculiar logical differences exist between the structure of the world-lines in these two sciences, i.e., in the identity series of individuals. He has further indicated that the comparing of two sciences is extraordinarily difficult due to the fact that all sciences undergo an historical development and are only commensurable at stages of equivalent historical maturity. Studies of another kind have been made by E. Becher[4], Paul Oppenheim[5], and even earlier by Heinrich Rickert[6]. Yet we must content ourselves here with giving a very general characterization of the methodological difference, adequate for laying down a line of demarcation for practical purposes.

We find just such a distinguishing mark, adequate in practice, in the fact that physics employs *mathematical concepts* in great numbers. Indeed, it is its mathematical character that constitutes the peculiarity of the physical knowledge of nature; it is the quantitative mastery of natural events with the help of equations. Physics has long been seen as the *mathematical science* among the natural sciences, the model of a *strict* natural science, and therefore we can briefly describe physics as an *exact science.* In establishing laws, physics makes quite different demands regarding precision from those made by biology. This has even led to the physicist holding a degree of precision such that he demands it to be the first requirement of all knowledge of nature, and he is inclined to accuse biology of inexactitude. But here he is forgetting that this degree of precision is possible only for physics. For the time being, at least, biology cannot entertain the notion of emulating the strictness of physics. Endeavouring to attain the degree of precision found in physics would only serve to blur its problems; it would be like an engineer wanting to take account of the phenomena of oscillation in gases in order to make calculations about steam-engines. This is just why we cannot consider precision to be a special mark of the method of physics. Physics is able to be precise only because it excludes the problem of life from its arena for the time being. It seeks out in every phenomenon that which can be mastered by mathematical methods, paying for its precision by renouncing those problems that resist this method. Therefore, precision is a true mark of the physical method.

To be sure, there are some exact laws in biology, and even mathematical concepts occasionally find application. For instance, the Mendelian law of heredity is quite certainly established, and even a law such as the principle that every individual has two parents, applicable to the higher classes of animals, may be called exact. Yet is it only a question of a few integral laws which, while having been discovered, cannot be *established* with the same precision. It is significant that the exact biological laws contain only whole numbers; they are really only an expression of qualitative knowledge that has been developed to the point of strict classification. Thus the above-mentioned law of reproduction is explained in biology by saying that a foreign nucleus must penetrate the female egg in order for fertilization to take place, but this is a qualitative piece of information, comparable, say, to the physical proposition that there are two forms of electricity. The whole conceptual apparatus of biology is tailored with far less precision than that of physics. Above all, it lacks the concept of a mathematical function, which makes changes in one quantity dependent in a regular fashion upon changes in another quantity. Related to this is the fact that *geometry* plays no significant role in biology. The few geometric-mechanical considerations that are used, for instance, in the anatomy of bone structure are not comparable to the extensive application of geometric considerations in physics. For this reason even the concepts of space and time have quite a different significance in physics than in biology. Biology needs scarcely more than a workaday knowledge of these concepts. Physics, on the other hand, has come to view all processes as laws concerning spatio-temporal changes. The presentation of world-events within the framework of a spatio-temporal coordinate system is specifically physical, and possesses no analogue in biology.

When we speak of the precision of physical method, then, we mean precision in a broad sense: a thorough-going application of the mental apparatus of mathematics, not the mere detection of some statistical relations. We may therefore refer to the *exact knowledge of nature* as the goal of physics and designate physics as an *exact natural science.*

It will perhaps be objected that this difference is merely temporary, that biology lags behind physics in terms of historical development and may ultimately be able to attain a similar degree of precision. This possibility certainly cannot be ruled out, but if biology were ever to arrive at this condition – if, for instance, it became able to represent a living being as the solution of various differential equations concerning albuminous matter – it would no longer be different from physics. Should this occur, that coalescense of biology and physics would have been completed. But the question is precisely whether this can ever succeed, whether or not living creatures by nature contain some element that prevents such formulations. As long as this question remains undecided, we will regard only physics as an exact science.

The demarcation of physics as against biology is the most important distinction that we have to carry out. Differentiating it from the other sciences appears much easier. Most important, the problem of physics and chemistry appears finally to have been resolved: today it is possible to say that chemistry is a part of physics, just as much as thermodynamics or the theory of electricity. Here we find already completed that development as to which we can only express conjectures concerning biology. The great achievement of Bohr's theoretical model of the atom has been its ability to interpret chemical laws as physical. Even if this cannot be carried out yet with complete satisfaction, its success up to the present leaves no doubt that, basically, the reduction of chemistry to physics has been achieved. Today, the distinction between physics and chemistry exists only for technical working purposes, with no fundamental significance.

The situation is similar with the sciences of astronomy, geodesy, and geology. They, too, are natural sciences; we may not believe that, say, the former two, as 'applied mathematics', occupy a position midway between physics and mathematics. Physics, too, is applied mathematics; in the application of mathematics to natural objects we find expressed precisely what is characteristic of the exact sciences. The characterization as applied mathematics is called forth by the fact that these sciences get along with relatively simple physical presuppositions and yet lead to very complicated mathematical problems, especially of a geometrical nature. But since astronomy has of late ceased to be a science so overwhelmingly concerned with measurement and has inclined once again more toward physical hypotheses, and since the theory of relativity has revealed the physical character of geometry as a science of real space, we can no longer doubt that these, too, are basically divisions of physics and that only the division of labor justifies the existing separation of specialties. We may, then, say that there are only two natural sciences today, physics and biology; these have absorbed all the individual disciplines, and the only question that remains unanswered is whether the development will continue in the same direction and ultimately also combine these two sciences into a single science.

3. PHYSICS AND TECHNOLOGY

Technology deals with the same object: nature. How is it to be marked off from physics?

Technology today makes extensive use of the methods of physics. It is not only in our time that physics and technology have progressed hand in hand; the same phenomenon is to be observed even in ancient times. Nonetheless, we may not conclude on this basis that they are parallel sciences. Technology is not a science at all, but the application of a science to practical use. The more general concept, under which technology in the narrower sense is subsumed, is itself to be designated as technology and placed over against the concept of science. The technological specialties include such subjects as mechanical engineering and technical architecture; but also agriculture and medicine, for their ultimate aim, too, is practical application, not theoretical research itself, and they bear approximately the same relation to biology that mechanical and electrical engineering bear to physics; they are, so to speak, biological technology.

For technology, science is never anything more than a resource – a powerful resource, to be sure, yet never an end. Technology can be forced to undertake scientific investigations in order to solve its problems of application, but it will always rest content once practical utilization has succeeded, and will abandon theoretical research to the scientists. On the other hand, technology can learn methods of practical application from science and, indeed, makes extensive use of these. In many cases technology is indebted for its practical achievements to the discoveries that scientists have made with purely theoretical aims. A famous illustration of this kind is Hertz' discovery of electric waves, which came about as a consequence of Maxwell's theory and later became the basis of the technology of the wireless telegraph. Conversely, scientific discoveries have arisen out of technical problems; an example is the law of the conservation of energy, the discovery of which is due to those unfortunate inventors who tried to construct perpetual motion machines. Nonetheless, technology treats its problems in a fundamentally different way from science. It circumvents genuinely theoretical problems by means of its 'phenomenological' methods: through these, it seeks to ascertain the functional relation of the variables, not by appealing to universal laws of nature but by making direct measurements of objects, thereby obtaining a whole series of 'material constants' that it simply accepts as given. Examples of this procedure are found in the technological theory of elasticity and in the acceptance of characteristics of, e.g., electron tubes. In this way technology finds a function that has been empirically established to be quite adequate for purposes of practical application; indeed, it finds it more useful than the theoretical formula because it is more precise. But for the physicist, such a phenomenological formula can serve only as a starting point. In his eyes, it is not a theoretical result, as it is for the technologist, but just empirical material which he uses only to test out his theories. Phenomenological measurements carried out by technology can be extremely useful to the physicist in that they relieve him of the necessity of making experimental tests. Conversely, the theoretical investigations of the physicist may prove useful to technology when it is a question of finding a new material or a new procedure that frames the empirical constants in a more advantageous manner. With the help of the theoretical connection discovered by physics, the technologist is then able to establish the direction his modifications must take: what substances he must use, say, to alloy steel, or how he should alter the electrode dimensions in a vacuum tube. He would never reach his goal by means of experimentation alone, but requires an anchor for the direction of his alterations. To be sure, he does not usually employ the physical theory in its quantitative form for such purposes; he uses it more as a qualitative guide, adding, at the most, differentiations of degree. In making the ultimate choice of the new procedure, he will revert to phenomenological methods, taking the various characteristics one by one and testing them out. Thus the very cases where the technologist makes use of physics display his specifically technological orientation.

The difference between physics and technology is not a difference between types of science but one of a more profound sort, involving the ultimate valuation that determines the aim of the activity. That unconditionally given and ineradicable interest that the scientist has in knowledge of nature is not the overriding influence for the technologist. He is concerned with action, not knowledge, with the mastery of nature, not observation. A glance at the fundamental orientation of typical representativies of science and technology makes this difference immediately evident. It can be so extreme that the one type completely fails to understand the other, i.e. finds his goals incomprehensible. In the average person, to be sure, the two tendencies will be more or less combined. Thus the engineer will repeatedly experience moments when a theoretical interest catches him by surprise at his work, leading him into what are technologically totally unfruitful investigations that only benefit science. And on the other hand, the scientist will from time to time see the practical utility emerging from a new discovery and pursue it even though it offers nothing new to the understanding of the phenomenon at hand. It is never to be expected that the theoretically distinct human type will be found in a pure form in individual persons; and it will in general be advantageous for the particular work of any one person if he combines various tendencies within him. With technology in its present state of complexity, an engineer who does not possess a goodly quantity of theoretical interest will not be well suited to the discovery of new technological paths. Conversely, the physicist sensitive to technology will excel in his experimental methods.

These remarks apply to the discipline of technical physics, a concept we hear mentioned today and which we find already existing in the academic curricula. It stems from the fact that technology, having had a glimpse of new discoveries in the theoretical sciences that beckoned it on to technological application, at the same time faces the necessity of improving its past methods with the help of theoretical insights. Both facts encourage a thorough knowledge of physical science, yet they equally encourage the establishment of a goal directed toward technology. The former demands the 'preparation' of the new discovery for technology, i.e., its phenomenological examination for the purpose of increasing its achievements. The latter calls for 'theoretical instinct', in order to predict the direction to be taken by new experiments. Important, then, as technical physics is, and develop as it may from sociological necessity – for it requires a new form of union between scientific knowledge and technological goals – it is, after all, by nature technological and not scientific. This comment is not intended to be demeaning. Quite the contrary; the achievements of technology will fill the physicist in particular with admiration, for he can see in it a completely different way of thinking and a different kind of use of the scientific tools from that to which he is accustomed. Yet clarity of conceptual definition demands a strict separation, and we will be unable to pay further regard to technical physics in the remainder of this investigation.

4. PHYSICS AND MATHEMATICS

There is one last problem of demarcation, that of drawing the boundary between physics and mathematics. Clearly mathematics is not a branch of technology, but a science, for it aims at knowledge. When we observe the close intertwining of these two sciences, the separation of mathematics from physics may appear dubious. Historically, mathematics has developed through tackling problems of geodesy and astronomy, that is, physical problems, and again and again the posing of such problems has proven immensely fruitful for mathematics; we need think only of potential theory. Conversely, physics has been greatly advanced by means of mathematical discoveries, as is demonstrated by the introduction of the infinitesimal calculus, among other examples. Are we faced here with a mere working distinction; is it possible that physics will some day be absorbed as a subdiscipline into mathematics, just as physics itself has absorbed the other sciences?

In this connection we are likely to think of the development of the theory of relativity into a world geometry, yet it would be quite erroneous to interpret this development as signifying a fusion of physics and mathematics. The general theory of relativity by no means turns physics into mathematics. Quite the opposite: it brings about the recognition of a physical problem of geometry. This interpretation will be more precisely worked out in Section 15 of this essay. For the present, we are merely investigating the relation between the two sciences in general. Further on, we will discover in the relativistic solution of the problem of space a confirmation of the strict separation that we are simply asserting here.

Although both mathematics and physics are sciences, the difference between them is fundamental, and we must put it down as quite impossible that it will ever disappear. For mathematics is not a natural science at all.

It undertakes investigations quite independently of whether they bear any relation to reality; it is a purely logical science [*Denkwissenschaft*], in which the concept of experience plays no role. There are no experiments in mathematics; all diagrams and models are merely visualizations, and, unlike observations, they are not the factor determining acceptance or rejection of a suspected law, but simply serve as aids to logical thinking, helping it to make the decision with its own faculties. Thus the problem of perception plays no role in mathematics, nor does mathematics face a 'problem of the external world'. Mathematics does not impart to us any substantive information about things that have reality outside ourselves.

For this reason, mathematics is faced with quite a different problem concerning unification or fusion: the question of whether it will be fused with logic. Today, after the publication of Russell and Whitehead's *Principia Mathematica*[7] and after Hilbert's investigations of axiomatics, we can regard this question as decisively answered: mathematics has become a part of logic – that part, in fact, requiring particularly broad development for use in our knowledge of nature. Logic, however, is the normative science of thought. It teaches us what is correct for thinking, but nothing about the world of experience.

This is true, with but a single qualification. The world of experience, too, must make use of the laws of logic; nothing can be admitted as correct in this world that contravenes logic. To this extent, then, logic teaches the *formal* structure of experience, but it teaches nothing as to the *content* of experience. Logic teaches what *can* and what *cannot* be, but it does not teach us what *is*. Therefore logic, and with it mathematics, is the science of *possibility*, whereas physics is the science of *actuality*. Here lies the profound and unbridgeable difference between physics and mathematics – and herein, too, the great significance of mathematics for physics.

For here we uncover the reason that mathematics has become the great intellectual machine of physics: physics needs, for purposes of method, a science of formal laws. The achievements of mathematics in this respect go in two directions. On the one hand, it shows physics to what extent the observed case is only a special case; it teaches the physicist how to work out more general possibilities and consequently to look for extensions of his experiments that are free of the limitations of the special case. Thus physics, by observation, found the law of the conservation of areas to hold for forces that decrease according to the square of the distance, but mathematics teaches that this principle holds quite generally for central forces. As a result, certain problems of the atomic model can now be treated with the assistance of the law of areas, although only the more general presupposition concerning central forces applies to it. Take another example, although for a more precise proof we must again refer to Section 15. Mathematics teaches us the possibility of non-Euclidean geometries, and physics derives from this the consequence that Euclidean space is really to be found only under certain restricted conditions, while non-Euclidean geometry holds everywhere else. In this instance, the mathematician shows the general possibility, but this is not always the case. Usually the physicist himself commences the investigation with the assistance of mathematical considerations. Ordinarily, the mathematician is not sufficiently familiar with the problem of physics to spot the *direction* in which the generalization should be made with a view to physical success. The other direction, taken by mathematical work, consists in finding out what is compatible with certain given presuppositions. This is the true process of constructing physical theories. An hypothesis is posed: what sort of facts are to be deduced from it? In this way mathematics establishes a logical relation among facts: on the one hand, it teaches what sort of facts are to be deduced from given facts, and on the other, what sort of facts cannot be hypothetically accepted, given a certain system of observations. Mathematics teaches all this quite independent of the facts; this is precisely its function as a science of possibility. Nonetheless, it is most efficient for such deliberations that mathematical and physical work be united in one person, for it requires a special instinct to guess what facts are capable of being brought into a mathematical relation. Purely mathematical testing will not be helpful. This explains the current development of the profession of the theoretical physicist, which illustrates once again that professional activity can require the simultaneous occurrence of two theoretically separate interests in a single person. The ultimate aim of the theoretical physicist is, to be sure, the ultimate aim of physics, i.e., knowledge of nature. The separation of experimental from theoretical physics is no more than a working arrangement within physics. The combining of mathematical and physical work in a single person is no proof that the two sciences are not logically separate. Mathematics is the intellectual tool of physics; it teaches what is *permissible* and what is *forbidden*, but never what is *physically correct*.

The intimate relation between mathematics and physics may in the end be regarded as a proof that these two forms of science are permanently separate. There is no other science with which physics is so closely intertwined as it is with mathematics; biology, chemistry, and astronomy have never had for physics as a whole the universal significance possessed by mathematics. Nonetheless, these two sciences have not, up to now, grown together. For this form of entanglement is possible precisely because mathematics is not a natural science, because it makes no material assertions regarding the objects of physics, teaching only the rational relations to which our knowledge of nature must conform. It is in the interplay between possibility and actuality that a knowledge of nature is acquired. In discovering possible processes, we learn to understand the peculiarities of the actual processes, and in excluding impossible processes, we learn to infer others from given facts. That is why mathematics is the universal instrument of physics, and also why the description of physics as the *mathematical* natural science constitutes its most precise characterization.

With this we conclude our observations regarding the demarcation of physics from the other sciences. The following investigations will concern physics only, analyzing more precisely the process of physical research.

5. PERCEPTION

The characteristic quality of physical knowledge shows up most clearly in its delineation from mathematics: physics treats of objects that have their particular form of existence independently of the knowing person. The means whereby we come to have knowledge of these object is *perception.* Sense organs are our key to the external world, and all physical knowledge begins, therefore, with perception.

This point appears as complicated and puzzling in the light of philosophical scrutiny as it appears simple to naive thought. In what follows, we will go into the questions related to the problem of perception[8].

If we subject perception, as used in science or in daily life, to a more intense scrutiny, we soon notice that it is never 'pure' perception. We say: "The ammeter indicates a current of 2.4 amperes" or, more simply, "There is a house", and call this a fact established by perception. But a little thought shows that this statement asserts far more than perception actually teaches. It is not in the least true that perception reveals a current of 2.4 amperes: it can at the most be asserted that the pointer on the instrument stands at the number 2.4. If we assert something about the current above and beyond that, the assertion contains a *theory* extending beyond perception. It is basically the same with the assertion about the house. If we had not already often had this perceptual image and learned that it can be taken as evidence of the presence of the thing we call a house, we could not in any way arrive at the assertion, 'There is a house.' This assertion, too, already contains a theory, goes beyond what is given in perception. And this consideration leads us, in turn, to make a second correction in the first example. We said the only fact we could assert on the basis of perception was that the pointer of the instrument stood at 2.4 – but we cannot even regard this as a fact of perception. Here, too, all that we directly possess is a single perceptual image, which we can interpret as an image of such things as a pointer, instrument, or number 2.4 only because certain experiences have justified us in so doing. This assertion, too, then, embraces a theory extending beyond perception. To be sure, it contains a significantly lesser degree of theory than the assertion regarding current, being roughly comparable to the assertion about the house, i.e., it contains no scientific theory, but only the 'theory of daily life'. This certainly is only a difference of degree. Thus even the very elementary facts of perception that are at the basis of reading the measuring instrument and contain no genuinely physical theory are nonetheless not 'pure' facts.

We can demonstrate these relations by ranking the facts. First-level facts are the facts of daily life; to them belong our examples, 'There is a house' and 'The pointer stands at the number 2.4.' Second-level facts are the data given by the measuring instruments, for instance, 'The ammeter indicates a current of 2.4 amperes.' We can easily progress to higher and higher levels. If, for instance, in order to ascertain a certain quantity, we use a complicated correction formula that itself is a correction of a datum, obtained from a measuring instrument, that has been interpreted as a second-level fact, we then have a third-level fact. An example would be the weight of a body as indicated by scales, where the correction is made with respect to a vacuum. A fact of a still higher level is, e.g., the discovery of a spectrum. The so-called facts of experimental physics are all of a higher level. At the most, they descend as low as the third or second level, but the first level is scarcely mentioned.

Yet we found that not even the facts of the first level are 'pure', and therefore we must search for *zero-level facts,* as we would like to designate pure facts. These zero-level facts are obviously the immediate perceptual experience themselves; they take the form, e.g., 'Now [there is] blue', 'Now [there is] a bang', or 'Now [there is] a triangle'. The contents of perception described in these sentences are called *sensation.* This word usually suggests at the same time the experience of self, an ego-experience, and we formulate the expression correspondingly, 'I hear a bang', etc. However, this second formulation goes somewhat further. To be sure, it is always correct if the first one corresponding to it is correct, but it asserts something different, for it contains the term 'I', and therefore an assertion about the 'I'. The mere report of sensation does not contain this, and that is why we choose the first formulation; essentially it is the appearance of the term 'now' that cannot be omitted in reporting the experience of a sensation.

One vital mark of perception is that it is not subject to our will. I can conjure up a mere image of a house by means of my own will, but not the perception of a house[9]. In the act of perception I am only an observer, accepting as given the colour or figure that is seen. That we distinguish our active from our passive posture is a fundamental fact that we will discuss further in Section 6; it is essentially connected with our belief in the existence of things independent of ourselves.

Although we omit the term 'I' from our formulation of perceptual experience, the aforementioned examples are to be regarded solely as a description of sensation, not as an assertion about the objective cause of sensation. The statement, 'Now [there is] a bang', does not signify that a bang is taking place in the air, for the assertion would also be admissible if the auditory nerve were properly stimulated by some other means, e.g., by means of an electric current. It certainly is not false in that case. But if, by a bang, we mean an occurrence in the air, then the given assertion becomes a first-level fact; as a zero-level fact it applies equally to the case in which a vibration takes place in the air and that of electrical stimulation of the nerves. It is just here that we can see how a process of inference can lead from a zero-level to a first-level fact, and how even this step can be false. The so-called sensory illusions all find their roots here. For instance, the same water is judged, upon insertion of the hand, to be cold or warm, depending upon whether the hand has been dipped in warm or in cold water beforehand. In this situation, the zero-level fact is the correct assertion that the one sensation signifies 'cold', the other 'warm'; what is false is only the inferred fact of the first level that these differences of sensation correspond to a difference in the objective temperature. The zero-level facts are completely certain simply because they do no more than report sensations.

But at the same time we recognize that this certainty is of no use to us, for neither in science nor in ordinary life can we get along with nothing but zero-level facts; we must always make the transition to higher-level facts. Since, as a result, we have very little to do with zero-level facts, it require some practise to draw these facts out from our complex perceptual experience. This experience itself usually skips over the chain of inference and accepts the first-level facts as completely certain. This is what makes sensory illusions so striking. Also related to this is the fact that the contents of the experience of sensation can actually be altered by means of additional knowledge concerning the objective state of affairs. For instance, we judge the brightness of an illuminated wall differently when we are able to estimate its depth. Psychologists have very precisely investigated these relations, but it is clear that such investigations prove nothing opposed to our epistemological reflections, and serve only to uncover the psychological complexities of the process of judgment.

Our deliberations show that knowledge of nature, with the help of perception, is a process that cannot come into being solely by means of perception. Perception is but the key to knowledge of nature; this knowledge itself is a path branching out in many directions that begins behind the unlocked gate. We are led along this path by *theoretical thinking*. This will become still clearer when we consider the whole system of scientific thinking with its convoluted mental constructions. Science can even be described as a methodical working up of perceptual contents into theories. In the process it progresses to facts of ever higher levels, as, for instance, to the assertion that matter consists of atoms, or that for every genuinely mechanical motion a particular function of the parameter becomes a minimum, compared with certain other motions. Such assertions are usually called theoretical knowledge and are contrasted with perceptual knowledge. But our presentation demonstrates clearly that the difference is only one of degree and that it would be completely wrong to designate such assertions as 'fictions' or 'mere theories'. They are basically gained in the same way as the so-called perceptual facts. Not only higher knowledge, but *all* knowledge of nature, without exception, is founded upon the interaction of perception and reasoning.

6. THE PROBLEM OF REALITY

Our investigation of perceptual knowledge led to the isolation of zero-level facts as experiences of sensation. With this separation the division into an 'internal world' and an 'external world' is given, and the question arises what right we have to infer the existence of things outside ourselves from sensations.

We begin by establishing a point that is independent of any metaphysical assumption. All physical knowledge is found by means of a mental construction that is connected with perceptions; and its experimental test, too, consists in the last analysis in the experiencing of certain perceptions. Thus it establishes a sequential relation between perceptions, the assertion of which can basically be written in the following form: 'If a certain perception a'_1 occurs, then a certain perception a'_2 will also occur.' If we write this implication[10] with the symbol '$\rightarrow\!\!\!\ni$', the proposition will then take the form

$$a'_1 \rightarrow\!\!\!\ni a'_2 \, .$$

Generally this form of notation is possible only when the a'_i are already made up of combinations of temporally successive or simultaneous perceptions; this possibility must therefore be included in our notation. We must further bear in mind that, even with this extension, the individual implication does not exhaust the given physical knowledge; rather, a whole series of such implications must be enumerated.

When we search in an example for the a'_i that belong to a physical proposition a, we set out from a. Nonetheless, from the epistemological standpoint, the relation is just the other way round, for the a'_i are all that is given, and a has to be constructed out of them. For this very reason it must be basically possible to exhaust a by means of a series of implications between perceptions. To begin with we shall disregard the fact that a conclusion is involved; in any case, we may establish the more modest claim that the cognition a is unambiguously *coordinated* with a series of such implications. Coordination will be symbolized by a double arrow. We can then symbolize the relation as follows:

$$a \longleftrightarrow \left\{ \begin{matrix} a'_1 \rightarrow\!\!\!\ni a'_2 \\ \cdots\cdots\cdots\cdots \\ a'_{2n-1} \rightarrow\!\!\!\ni a'_{2n} \end{matrix} \right\} \qquad \text{(1a)}$$

If we abbreviate the propositional system on the right by using the symbol "a'", we can also symbolize it as:

$$a \longleftrightarrow a' \, . \qquad \text{(1b)}$$

Here we can think of a as being made up of elements a_i, the things, just as a' is made up of elements a'_i. On the left there is an 'external' assertion; on the right, the a'_i contain only perceptions, that is, zero-level facts, so that the right-hand side is of an 'internal' nature. *With the help of this relation, every proposition concerning objective things can be transcribed into an assertion about perceptual experiences.*

If, for example, we take as a the statement, 'Matter is made up of atoms', then the a'_i are the perceptual experiences of the chemist when he checks out the chemical combining weights, of the physicist when he makes radio-active experiments, etc. Of course, these perceptual experiences have nothing to do with the atom directly; they are not, as it were, 'pictures' of the atom. Rather, they directly present only measuring instruments, spectral lines, luminous streaks of mist, etc., and yet these things, too, are not the a_i', but only the first-level objects that correspond to them. The atom, on the other hand, is a higher-level object relating to the same perceptions. In this example the proposition a is not itself an implication; but this fact is immaterial, for it could as well be an implication, such as 'If an electric current is flowing, a magnetic field is created.' On the other hand, the right-hand side must always contain implications if a proposition stands on the left.

There can remain no doubt that this coordination is universally applicable. For if a proposition a existed that could not be coordinated in this way with a system of propositions a', then this proposition could assert the existence of a state of affairs with no consequences susceptible to experience; but natural science makes no such assertions. Whether or not philosophical systems have always accepted this basic principle can be left an open question; natural science, at least, has always employed it, using it at critical moments as the very definition of what is to be regarded as a 'meaningful assertion about nature'. Einstein's development of relativistic thought rests upon this fundamental principle; and more recently, Heisenberg has demonstrated a similar point of view regarding the determination of the concept of an electron (cf. Section 24). We will, therefore, accept as confirmed the feasibility of coordination (1).

The significance of these considerations is that, in opening up for discussion the interpretation of the double arrow in (1), they make possible a sharper formulation of the problem of reality. Two conceptions emerge as a result:

Positivism interprets relation (1) as equivalence [identity], replacing, the double arrow by the identity symbol '≡'.

According to this reading, the meaning of every 'external world proposition' a is given exhaustively by a system a' of 'internal world propositions', which rids us of the problem of a special thing-like [objective] existence. Natural objects can then be obtained by means of 'constitution', by which is understood the process of defining a 'higher-order construct' [*Gebilde*] by means of propositions about 'elements' (which may in turn be 'lower-order constructs') which are such that every proposition concerning the construct is capable of being transformed into a system of propositions about the elements. Positivism simply defines the term *construct* by this process of coordination. In order to understand this concept, which grew out of mathematical logic, we must note that, e.g., the classes or sets in Russell's system of logic are constructs in this sense. Nonetheless, they are, of course, only special constructs, and many complex constructs can also be defined. This process of constitution can be applied to the elements a_i of a, i.e., for natural objects; they are defined as constructs from a_i' by means of the coordination (1), conceived as an equivalence relation. *The natural object is, then, a 'perceptual construct'.* This conception of the object originated principally with Mach[11]. More recently it has been substantially sharpened by Russell[12], who believes the concept of a class to be the only one required, and by Carnap[13], who uses the term 'logical complex' in place of 'construct'.

Realism, on the other hand, does not permit the relation (1) to be treated as an identity. On this view, the left-hand side has an additional meaning, which consists in the assertion of existence. The proposition that objects exist independently of us is held not to be exhausted by a proposition about perceptual experiences, but to contain, instead, an independent and indefinable basic concept, that of existence. This idea can be interpreted, in a somewhat abbreviated form, as the claim that perceptibility [the ability to be perceived] is the *criterion* of existence, but not its *definition.* This interpretation of the concept of existence is contained in many philosophical systems, even though they may largely contradict one another in other respects; for instance, the Kantian idea of the 'thing in itself' can be interpreted in this way, but so, too, can Schlick's realism[14].

It must be admitted that, with the instruments of science alone, we cannot decide between the two conceptions. We cannot, for instance, show by means of experiments that objects exist in the sense claimed by realism, for the ultimate data offered by experiments are themselves only perceptions, and the problem of interpretation exists for these perceptions in the same sense as for all other perceptions. Nor is it possible to prove realism by appealing to past experience and showing that, heretofore, the existence of real things has been assumed without encountering contradictions, for the same experiences can just as easily be interpreted to favor the positivist interpretation. The question here is not at all *whether the objects of the external world really exist,* but rather *what we really mean when we assert their existence.* Things exist just as much for the positivist as they do for the realist; yet the former is of the opinion that the concept of existence is a reducible concept that can be traced back to perception and its lawfulness.

If, then, it is an error to present realism as a necessary foundation of natural science, it is, in turn, just as mistaken to construct a special concept of existence, applicable only in science and different from the workaday concept of existence. Positivist reasoning has sometimes been construed as giving the 'more abstract objects' of science, such as the atom or the electric field, a special kind of existence different from the existence of the 'concrete things' of ordinary life, such as a house or a telescope. The atom has been called a 'conceptual construct' that science has introduced simply as an abbreviation of its mode of expression and without ever being able to prove its existence. The error in this conception is quite apparent: either we interpret the relation (1) as an identity or equivalence, in the sense of positivism – and then the house and the telescope are 'conceptual constructs', 'logical complexes', in the same sense as the atom – or, following realism, we interpret the concept of existence as an unanalyzable basic concept – but then the atom is just as real as the house and the telescope. That these two kinds of existence are *of the same kind* is epistemologically provable. This idea has profound consequences; it shows that natural science is justified in assuming the existence of things that are not as accessible to our senses as the objects of daily life. The philosophical schools that dispute this idea err less in their interpretation of scientific inference from perception to object than in their conception of the existence of 'concrete things'; for they overlook the fact that these objects presuppose the same process of inference[15].

The objection of the realist to positivism rests upon the complication of the concept of existence that is caused by the constitution theory of positivism. The positivist can formulate the existence of constructs only by means of a peculiar twist in the interpretation of the concept of existence which he uses. The existence of constructs can be interpreted in two different ways. Let us take, as an example of the first kind, a wall that is constructed out of bricks. It is a construct, for every proposition concerning the wall can be replaced by propositions concerning the bricks. The wall is, of course, not the mere 'sum' of the bricks, just as its monetary worth is not given simply by the sum of the monetary value of the bricks. Rather, it bears to the bricks the complex relation of propositional coordination. Yet in one respect the wall is 'similar' to the bricks; its spatio-temporal position is given exhaustively by that of the bricks. To be sure, there are small gaps between the bricks, so that they do not solidly fill up the whole of the spatial area assigned to the wall. Nevertheless, this spatial area is determined by the spatial area of the individual bricks; the wall is 'in the same place' as the bricks. The same thing is true of the temporal duration of its existence. Of course, the wall can be dismantled in such a way that the wall no longer exists, whereas the individual bricks do; conversely, the bricks can be individually replaced and destroyed, so that the wall exists longer than any of the bricks. Yet it is true to say that as long as a brick is an element of the wall, its temporal situation is identical with that of the wall. Let us call constructs the spatio-temporal position of which is directly determined in this sense by the spatio-temporal positions of their elements *coincident [artgleich] with their elements.*

But the existence of objects as constructs from perceptions, as taught by positivism, is of quite another sort. Perceptions exist for only a few moments; like the perceptions from which we infer phenomena lying in the historical past, they may even occupy a completely different temporal position from the object that they construct. That is why the object, as a perceptual construct, is not coincident [*artgleich*] with its elements. For instance, if we accept as the elements of the construct 'house' the perceptions of this particular house that I have experienced during my lifetime, we cannot assign to this 'house' the temporal position of its elements. If, perhaps, we wished to say that the construct exists so long as at least one element of it exists, then the construct – that is, the house – would not exist during the intervals between the individual perceptions of the house, and would also not exist any more after my death. The proposition that the construct 'house' exists even when I do not see it or when I am no longer living takes on quite a different form in positivism. It is not formulated as a proposition about the *temporal position* of the constituent elements, but is transposed into propositions about the *material [inhaltliche] qualities* of the elements and their *regular [gesetzliche] order.* Thus the proposition that the house exists even when I am not looking at it is expressed by means of a regular connection in the form of an implication: *if* I have these particular perceptual experiences (of the garden, say) and this particular sensation of the position of my head, then there will occur a perceptual experience of the house. And the proposition that the house already existed one hundred years ago is expressed in the form: my perceptual experiences of the house stand in a certain regular connection with other perceptual experiences, for instance, those I have while reading the land register. This regular connection between present experiences defines what is meant by 'existence one hundred years ago'. In this way, the positivist must transcribe the proposition that the house existed at a certain time. The house does not exist at the same time as its elements; the object as a construct is *not coincident* with its elements [but is of a different type (*artverschieden*)].

This transcription can certainly be carried through without contradiction, but it has the effect of completely transforming the meaning of the concept of existence. The realist can thus rightly object that the positivist's concept of existence for natural objects does not possess the direct meaning that we associate with existence. While the positivist employs this direct concept of existence for *elements,* he rejects it for *constructs [Gebilde]:* the positivist's concept of existence is a *construct even for natural objects,* not a basic concept. We cannot prove that there is only one type of existence – that everything which exists, exists in same sense and that differences in 'levels of existence' are possible only in the sense of element and coincident construct. If we were to prove this, the conception of objects as constructs from perceptual experiences would be refuted, as it presupposes the acceptance of a construct not coincident with its elements; yet without the presupposition of the identical nature of all existence no such proof can be given. Thus the positivist is preserved from contradiction so long as he questions the claim that every construct must be coincident [*artgleich*] with its elements.

But at this point, the realist appeals to immediate evidence. He holds it to be absolutely necessary to grant to the existence of physical objects the same immediate and direct meaning as the existence of experiences. He considers existence to be an unanalyzable basic concept that, like certain fundamental concepts of logic, must be discerned as meaningful, and that possesses the same meaning for every existing thing, and is not capable of being constructed. He believes the statement, 'Objects exist in the same sense that I and my experiences exist', to be a meaningful proposition that it is absolutely impossible to reject, while the positivist must declare this statement to be meaningless.

The establishment of this concept of reality rests principally upon the distinction between perceiving and imagining [*Vorstellung*] mentioned in Section 5. The passivity of perception is in the end a merely qualitatively characterized experience that any normal person undergoes intuitively; it is through just this experience that perception takes on for him the meaning of an 'announcement' of objects existing independently of him. To be sure, some thinkers have disputed this, attempting instead to establish the difference between perceiving and dreaming or hallucinating solely by means of differences in their sequential regularity. According to this conception, the equating of the two would lead to contradictions in the relation between experiences. For instance, if the hallucinating wanderer in the desert takes the dream-like experience of a water-hole for a perception, he will recognize his error when he discovers that the implication, 'The experience of water and drinking is followed by the experience of quenched thirst,' fails to apply. Yet we can imagine a 'consistent hallucinator' who always produces hallucinations that do fulfill implicative connections – whose thirst, in our example, is really quenched. Such a person will process his experience in the correct logical order; if we nonetheless consider him sick and reject his dream world as being *unreal,* our only justification can be the qualitatively experienced difference between perceiving and imagining that the healthy person accomplishes intuitively.

For the hallucination or the dream image is to be differentiated from perception because, in them, the will is the determinant, albeit usually unconsciously so, because the knower plays not a passive but an active role. If the applicability of a perceptual implication is to be interpreted as confirmation of a suspected regularity, it is essential to presuppose that *the expectation of the future perception will not in any way influence its coming into being.* That it is difficult to free ourselves totally from such an influence is well known. Every experimental physicist or astronomer knows, for example, that it is advisable to write down numbers read off from observations before he has an opportunity to realize how the calculations based upon them will affect the expected result. Hallucinations and the wishful fantasies of the sick are extreme cases of the dependence of experience upon our expectations, and this is precisely why they cannot be accounted as perceptions, as reports from the external world. The influence of expectation is not always readily apparent, and modern psychology has shown that even normal perception is influenced throughout by expectation and desire to a far greater extent than we ordinarily believe. If, then, we require psychological methods, i.e. scientific knowledge, to obtain 'pure perception', we are nonetheless dealing with a difference that can in the end be only qualitatively characterized, that requires an insight. It is upon this insight that the distinction between the world and the self [*Ich*] is grounded.

We therefore cannot accept the frequently presented view that the distinction between perceiving and imagining can be carried out simply with the assistance of the lawful connections that have been constructed for them. It is only occasionally that the healthy person can make use of the regular laws, as a standard against which to check an uncertainty in this distinction. We sometimes ask ourselves, for instance, whether the bell has really rung and then go to the door and decide, upon failing to perceive anyone standing there, that our experience was not a genuine perception. In general, however, we must assume this distinction as given, for otherwise, the problem of ordering our experiences would be indeterminate. If we observe the procedure actually employed in acquiring knowledge, it is difficult to maintain the possibility that, along with ascertaining the regularity *within* perceptions, we can also accomplish the *separation* of perceptions from other experiences without a greater store of 'initial distinctions'.

We do not propose to lay down here all the details involved in the decision between positivism and realism. Let it simply be noted that a number of other weighty arguments against positivism can be enumerated and that the author has decided in favor of realism. On the other hand, positivism and realism can to a considerable extent run a parallel course. In particular, the realist can adopt the idea of the process of constitution, i.e., the definition of constructs by means of coordinating propositions. Yet this idea does not lead him to a theory of *objects*, but to a theory of *concepts*.

What is a concept? That concept and object are to be distinguished from one another has been clear for a long time. Concepts are coordinated with but are essentially different from objects. But if the problem of the existence of objects is shrouded in darkness, the problem of the existence of concepts is very much more so: If a concept is not an object, how can it exist? Ever since the time of Plato, this difficulty has given rise to the idea of a special sphere of existence for concepts; concepts, it is said, exist in the 'ideal' or 'conceptual' sphere: The danger contained in this conception is that it attempts to solve the problem of concepts by means of an analogy, a blurred image. It is, in fact, impossible to say what is meant by the spatial image of this sphere or in what sense the term 'existence' is used in this context. We wish to show that the constitution theory is able to provide a better answer.

It is quite certain that concept and object have no internal similarity. The concept is merely coordinated with the object, is a *sign* [*Zeichen*]. This symbolic interpretation of concepts has been emphasized of late, especially by Schlick[16] and by Hilbert[17], who in the course of his logical investigations coined the motto, 'In the beginning was the sign.' But what is a sign? If we were to introduce a special form of being for signs, we would simply be reintroducing the murky notion of the idealistic theory of concepts.

We therefore put forward another interpretation, viz., that, to begin with, signs are also nothing but [physical] things. In fact, every sign that we use is a thing. The written sign is a material construct composed of particles of pencil lead, the spoken word is a material occurrence made up of vibrations of sound, and similarly, a banner is a sign of certain attitudes; such an object acquires its meaning as a sign only by being coordinated with another object. There is, then, nothing to prevent concepts from being things even though they are signs.

But what sort of thing? It is at this point that the theory of constitution comes into play. With the right side of (1b) we can coordinate an equivalent assertion a^*:

$$a^* \equiv a' \tag{2}$$

a^* is, in turn, coordinated with a in such a way that, to every element of a_i of a, there corresponds uniquely an element of a_i^* of a^*, so that

$$a \longleftrightarrow a^*. \tag{3}$$

The difference between this formula and (1b) consists in the univocal nature of the coordination, which is attained by means of the fact that (2) carries out a transformation of a'. While the elements a_i' of a' are experiences, the elements a_i^* of a^* are experiential *constructs,* that is, summaries of a_i', and because of the univocal character of the coordination (3), the a_i^* are precisely what we call concepts. Here the equivalence of type [*Artgleichheit*] between experiential constructs and elementary experiences is admissible, for the concept exists at the time of the experiences, not at the time of the objects.

For example: Let a be the assertion, 'The table is rectangular'; then a' is the system of assertions of the form, 'When I see "a plane with legs", I also see an "oblique parallelogram" ', 'When I sense gliding around the edge, I sense "corner" four times', and so forth. In contrast, a^* is the assertion, 'Whenever I find the experiential complex "table" to be realized, I also find the experiential complex "rectangular" to be realized.'

We have now reached a theory of concepts: concepts are constructs out of perceptions that are defined by means of the propositional coordination (2). Concepts have existence, just as do all complex psychic experiences, and thus are objects; they exist as long as the person thinking about them exists, independent of the temporal position of the object coordinated with them. They receive their significance as signs by means of coordination (3), which establishes a mediation between the totality of all objects and these special objects. As clarification, we could consider the image of a map that is spread on the ground *somewhere in the country that map depicts;* it coordinates all the points in a large spatial area with a small selection of them.

If the way in which we sharply separate concept and object makes us appear closer to traditional philosophical views than to positivism, we nonetheless adopt, as does positivism, the idea of constitution that has grown out of mathematical logic. The superiority over older theories of the theory of concepts thus attained consists in its taking propositions as primary and concepts, in contrast, as something secondary that is defined only by means of propositions. Concepts have no meaning by themselves, but only in conjunction with propositions. In this change in the relation between concept and proposition, we can observe one of the most important results of mathematical logic. Its relation to 'Gestalt theory' is unmistakable. Propositions are of a 'gestalt character'; they are not constructed by joining elements having an isolated existence and lacking any gestalt character, i.e., concepts, but the other way around. Concepts are defined through their relation to gestalt-like elements, i.e., propositions[18].

On the other hand, even if positivism shares the logical benefits of the method of construction with our form of critical realism, we must nevertheless describe its conception of the equivalence of the propositions a and a^* as an identification of object and concept[19]. This is perhaps the most obvious reason for our rejection of positivism. For only by means of this sharp distinction, it appears to us, can the fact that the objects of nature have an existence independent of our experiences be exhaustively formulated. We cannot abandon the primitive concept of existence employed in daily life; indeed, it seems quite impossible to call this primitive concept of existence into serious doubt. It is, rather, a question of implementing this concept consistently. Admittedly, some work remains to be done in this regard. If existence represents a coordination with concepts, then it must be possible to coordinate some object with every concept that is legitimately introduced. Yet a feeling for language sometimes calls for distinctions that are not always tenable. We dislike calling a light ray a genuine visual image, an object [*Ding*] – still less, giving this name to a temperature or an electric current. If we examine the reason for this, we find that our sense of language attempts to do justice to the logical distinction between element and relation, attributing existence to the first but not to the second. Here the philosophical concept of substance has its roots; presumably it can be traced back to the concept of the logical element. Language has quite often coined nouns for relations, and this gives rise to the doubt about the applicability of the concept of existence with respect to these relations. For instance, because they are elements in the logical sense, we more readily attribute existence to atoms than to the strength of an electric current. This latter is a 'substantivized' relation, the elements of which are electric currents. (That is, a proposition concerning the 'magnitude' of an electric current can be transformed into propositions about relations of 'greater or lesser magnitude' between various electric currents.) The electric current, on the other hand, is of the character of a substance. To date, no investigation of scientific concepts has been carried out from this standpoint. In particular, there remains open the question whether the distinction between element and relation is really an ultimate distinction; perhaps each element can, in turn, be resolved into relations between other elements. In order to circumvent these difficulties, which rest upon the logical structure of the concept, it is expedient to introduce the concept of 'reality' and give it a wider range than the concept of existence, which remains confined to substantial (elemental) objects. The magnitude of an electric current and the temperature will, then, possess reality, but so, too, will, e.g., the relation, 'to the north of', in which, say, a house stands to a mountain. Furthermore, we will have to ascribe reality to a 'state of affairs'; every proposition will then be coordinated with a state of affairs in the same way in which each concept is coordinated with an object.

7. PROBABILITY INFERENCE

In the preceding section we explored the significance of the double arrow used in (1); we must now investigate the meaning of the other sign in (1), the sign '$\Rightarrow$'.

This sign asserts an *implication,* for it lays down a connection of the form, 'If a_1, then a_2.' But it does not wish to present this connection as being absolutely necessary, but only as probable. Hence we speak of *probability implication.* Strict implication would correspond to the limiting case in which the probability equals 1. However, we must note even this limiting case is not identical to logical implication, but only corresponds to it, for logical implication signifies a connection between *propositions,* while probability implication represents a connection between *things* and *events.*

The procedure whereby the assertions on the right-hand side of (1) are obtained gives the basis to speak here only of probability, not of certainty. There is, basically, only one possible procedure, as follows:

We establish that a regularity holds for previously experienced perceptions and then assert that future perceptions will display the same regularity. Here we are dealing with an inference of the character of a leap; it is known as an *inductive inference.* It is by means of induction that the concept of probability is introduced into the knowledge of nature, for assertions of this sort concerning future perceptions can be pronounced only with probability. Thus inductive inference is also known as *probability inference.*

Along with it, a most extraordinary metaphysical assumption is made with respect to knowledge. For this inference means a prediction of perceptions; yet, because perceptions are not subject to our will, we are in principle unable to know whether they will occur. Thus the inference is ultimately of the form, 'If I have seen green connected with blue seven times and I see green for the eighth time, blue will be there, too'. To be sure, the prediction does not generally rest upon such a simple regularity as the enumeration of similar cases given in this example. Rather, the whole of the scientific theory construction enters into the prediction. In the language of the right-hand side of (1), this assertion would run: Given a selection of perceptions, we demand the fulfillment of a whole series of conditions before we connect to them the perception that is to be observed. We are schematizing when, in the above example, we replace these conditions with the simple colour description 'green', yet the characterization of the a_i' is basically of just this sort. Therefore, every probability inference is just as unsatisfactory as the example given; we cannot justify it, but can merely summarize our knowledge about it in the following three propositions.

1. Probability inference cannot be established logically, for it goes from observed cases to unobserved cases, about which absolutely nothing can be known.

2. Probability inference cannot be justified empirically, for it is impossible to infer its validity from the fact that it has so frequently worked in the past; this inference is itself a probability inference.

3. Probability inference is indispensible in science.

Hume was the first to give a precise formulation to the first two characteristics of probability inference[20]. He also mentioned the third characteristic; nevertheless, he did not recognize the central position held by probability inference in our system of knowledge. He believes that this form of inference can only be explained as an unjustifiable habit; but this view cannot help toward a solution of the problem.

We therefore proceed in another fashion, by introducing the concept of probability as a basic logical concept, the meaning of which we accept as given axiomatically. We make, that is, the following two axiomatic presuppositions.

1. It is meaningful and permissible to infer with probability from a finite number of cases to all cases.

2. It is meaningful and permissible to infer from the probability of a phenomenon the probability of its respective occurrence in a finite number of cases.

The justification for setting up assumption 1 and 2 will not be examined here. It would scarcely be possible to give a satisfactory answer to this question, for it embraces what we suppose is the most obscure problem in epistemology. On the other hand, we can establish that we constantly make use of the concept of probability, in the sense of assumptions 1 and 2, in daily life and in science, and that we are absolutely unable to dispense with it.

It should be pointed out that the problem of probability arises not only for realism but also for positivism. For even if positivism can get along with assertions of the form on the right-hand side of (1), it needs the concept of probability implication and probability inference, since the concept of probability occurs even on this side of (1). The positivist, too, cannot avoid asserting the *lawfulness* of the connections between perceptions. Thereby even positivism contains a metaphysical hypothesis, namely, that of the probability inference, and is therefore just as unable to set forth a completely rational account of the problem of knowledge.

On the other hand, it is clear from our presentation that, according to realism, the occurrence of the concept of probability does not result from an inference from perceptions to transcendental objects. It is not the assertion of existence that introduces the concept of probability into knowledge, but the assertion of a lawful connection in general, regardless of whether the connection is applied to objects or experiences. Thus it would be meaningless to say that we can speak of transcendental objects with a certain probability. The concept of probability is not applicable to the existential axiom itself, since this axiom is not inferred inductively. The relation between the concept of probability and the concept of existence is quite different. If, with positivism, we construe the double arrow in (1) as an identity, then the assertion of existence becomes identical to the assertion of probability; i.e., the belief in the existence of objects is identical to the belief in the axiom of probability[21]. Thus two metaphysical hypotheses are reduced to one, and this is perhaps the strongest source of support for positivism. According to realism, this identity cannot be carried out; for realism, the probability axiom is merely one component part of the assertion of existence, which as a whole, however, possesses surplus content.

8. THE PHYSICAL CONCEPT OF TRUTH

We can now proceed to one of the most important questions in epistemology: When is a physical proposition true?

First, a preliminary comment. People often speak of a *definition* of truth and designate, accordingly, the conceptions to be described in what follows as various definitions of the concept of truth. This designation is nonetheless misleading. No doubt we can define the concept of truth in any way we please, but this is not the significance of the problem of truth. The question is, rather, what is the concept that we think of when we speak of truth? Or better: What is the concept of truth that modern science actually employs? This question cannot be answered by means of a definition; the reply will itself constitute substantive information. In what follows, then, we will not speak of the *definition* of truth, but of the *characterization* of truth.

The most ancient characterization runs: Truth consists in the correspondence between an idea and its object[22]. This answer proceeds from the assumption that ideas depict objects in much the same way in which, say, photographs depict objects. But our earlier investigations show that this is an untenable assumption. To get from the perception to the object, we require a logical chain that first coordinates the perceptions to a concept (as in (2), Section 6) and then, in turn, coordinates this concept to an object (as in (3), Section 6). Thus the perception is coordinated with the object in a very complicated manner, and it is only in a few simple cases, with first-level facts, that the coordination is of such a simple character that anyone could come up with the idea of taking the perception for a picture of the object. With higher-level objects it is certainly no longer possible. And as the transition is continuous, it cannot make sense to regard perceptions as pictures even with objects of the first level. The type of coordination that pertains between concept and object is comparable only to mapping in the mathematical sense, not to depicting in the sense of 'creating a similar picture'. Consequently the concept that we form of an object has no further similarity to it. It is simply coordinated with it, and it is impossible to characterize truth by means of a comparison of the concept with the object.

For all that, this most ancient characterization contains one correct idea, for it proceeds from the insight that the assertion of truth is related to the coordination between concept and object or between proposition and state of affairs, as the case may be. We can give this idea a precise formulation on the basis of our earlier presentation. From the standpoint of realism, we, too, had to coordinate an object with the concept. But this coordination is valid precisely for a true proposition only, not for a false one, so that we may say: A proposition a is true when objects a_i are coordinated to its concepts a_i^*. It turns out, then, that the concept of truth is connected with the concept of existence; the proposition 'a is true' is identical with the proposition 'The objects a_i, that are coordinated with the concepts a_i^*, exist'.

This conception of the problem of truth coincides with the characterization given by Moritz Schlick[23]. According to Schlick, knowledge consists in the coordination of concepts with objects, and truth demands simply that, going from concept to object, this coordination must be *univocal.* (The coordination need not be univocal when proceeding in the opposite direction). For our presentation, too, of course, the univocal nature of the coordination is essential; otherwise, it does not make sense, and any objects could be arbitrarily coordinated to the concepts a_i^*. The formulation we offered must, then, be put more precisely, as: 'The objects a_i, that are univocally coordinated with the concepts a_i^*, exist.' Thus the formulation, 'Truth consists in univocality of coordination', is equivalent to the formulation, 'Truth consists in the existence of univocally coordinated objects.' We prefer the latter formulation because it allows for a clear presentation of the connection between the problem of truth and the axiom of existence.

The analysis of the concept of truth by means of the concept of univocal coordination stems principally from Schlick, who recognized the formation of this characterization of truth from the older one through the transformation of the naive concept of depiction into the mathematical concept and who supplied its philosophical formulation. Helmholtz had already developed this conception: "Every law of nature asserts that upon preconditions alike in a certain respect, there always follow consequences which are alike in a certain other respect. Since like things are indicated in our world of sensations by like signs, an equally regular sequence will also correspond in the domain of our sensations to the sequence of like effects by law of nature upon like causes."[24] Heinrich Hertz, too, takes up this idea in the introduction to his work on mechanics: "We form for ourselves images or symbols of external objects; and the form which we give them is such that the necessary consequents of the images in thought are always the images of the necessary consequents in nature of the things pictured."[25] This is also tantamount to the assertion that the conceptual system we create for ourselves is univocally coordinated with what happens. Hertz had already rejected the notion of the similarity of the images, in speaking expressly of "images or symbols": ". . . for their purposes, they need not in any further way resemble the objects." And Helmholtz writes: "But a sign need not have any kind of similarity at all with what it is a sign of"[26].

Nevertheless, we cannot yet declare ourselves satisfied with the characterization of truth that has been presented. We have two objections to it.

First, the characterization in question pertains solely to the ultimate goal of knowledge and can consequently never be attained. We will never, for example, be able to determine the concept earth so precisely that it will correspond completely to the object earth. We have to include in the concept, as a characteristic, some sort of spatial figure, say, that of a rotating ellipsoid, and yet we know very well that this is a schematization that cannot do full justice to the real earth. It will not do to reply that we can offer, as a characteristic of the concept earth, the spatial form that the earth really possesses, that we can determine the concept earth by means of the object earth by coordinating the thing to the concept. This process simply gives a *definition* of the concept earth, in the sense of a coordinating definition (cf. Section 10), it does not amount to *knowledge* of it. It is only knowledge, not a definition, that possesses the character of truth, and if what we mean by truth is a characteristic coordinating concept and object, the concept cannot be determined by means of this coordination but solely through its connection with other concepts. Hence the univocality of coordination is unattainable precisely where it would constitute knowledge.

Second, this characterization does not offer any means whereby the truth of a given physical proposition can be tested. For, as we established earlier, the coordination with real objects is exactly what cannot be shown; for purposes of testing scientific propositions, all that we possess are perceptions and theoretical connections. We must therefore seek a different characterization of truth that says nothing about a correspondence with objects, but concerns only perceptions and theories and is, for that reason, suitable for application to science.

The gist of these two objections is the same. The first asserts that the given formulation characterizes truth solely from the standpoint of the *limit*, the second, that in our characterization we should employ the method of approximation that is used in the acquisition of knowledge.

But we have already, in the right-hand side of (1), given expression to this method of approximation by introducing the concept of probability implication. We have purposely avoided bypassing the method of approximation, as is commonly done in epistemology, introducing, instead, the ideal limiting case into (1), in which the exhaustive characterization of a_i' is achieved and the probability is equal to 1, for we believe it to be unjustifiable to formulate propositions about this limiting case that are not equally capable of being formulated as propositions about the method of approximation. However, this forces us to replace the concept of truth with the concept of probability. We will no longer be able to speak strictly of the truth of a proposition, but only of its degree of probability. The broader concept of probability, as an undefined basic concept, now replaces the concept of truth in the sphere of knowledge; absolute truth is only the limiting case, in which the probability equals 1. Thus we say: a is probable if a' is constructed in accordance with the rules of probability. We regard the rules themselves as self-evidently given.

At the same time, the uncertainty that has been introduced extends to the relation of correspondence between concept and object. We may no longer speak of univocality of correspondence, but must say instead: the correspondence conforms with probability if a' is constructed according to the laws of probability.

It is useful to introduce the concept 'correct', taken in the sense of 'highly probable'. This enables us to say that we are no longer offering a characterization of truth, but a characterization of correctness instead. To be correct is to be connected with perceptions in accordance with the rules of probability. Truth is merely the limiting case that correctness approaches, though without ever reaching it. By following the rules of probability in constructing a', then, we can achieve a situation in which the proposition is correct, or in which its concepts are correctly coordinated with objects, but we can never achieve univocality and truth. The designation varies, however; the limiting case is often referred to as 'strict' or 'absolute truth', while the term 'truth', without any such modifiers, is used solely to mean correctness. In any case, many claims about the truth of physical propositions can be maintained if and only if truth is understood to be correctness. In accordance with this uncertainty in usage, we, too, will use the expression 'truth' to mean 'correctness' from time to time in later sections.

Thus the new formulation presents the earlier one as the limiting case. It signifies, at once, both less and more. It asserts less in that it is applicable even where the limiting case does not exist at all; it need not presuppose that there is some ultimate conceptual formulation than can be coordinated with reality. On the other hand, it asserts more in that it stipulates the manner in which the approximation is to be executed; the formulation of approximation can not be deduced from the formulation of the limiting case.

In contrast to its predecessor, this characterization of truth does not employ the conception of a correspondence between concept and objects. Thus it is a characterization stemming from the system of scientific knowledge and not giving any direct attention to the relation between this system and reality. Once we note that every item of scientific knowledge depends upon every other, that all scientific propositions are interconnected, we can say that truth, according to this characterization, is a characteristic of the system; truth pertains only to the system as a whole, and only from this standpoint can truth be transferred to the individual assertions. This characterization of truth as the coherence of the system has been carried out principally by the Neo-Kantians (Cohen, Natorp, Cassirer, Görland). To be sure, they have paid inadequate attention to the basic position of the concept of probability, without which the system could never be called coherent because a precise agreement between all observations can never be attained. Also, it must not be forgotten that the *logical* coherence of the system is not sufficient for the characterization of truth. Perceptions must be included in the system, and the truth of the system demands that no *contradictory perceptions* occur. Perception is, as it were, the point at which the system is connected to reality: *Truth does not only signify freedom from logical contradiction of the system within itself, but also agreement at the connecting points.* This is just why we can ascribe to this system a meaning for reality over and above its mere logical meaning, just why we can regard it as a description of reality. The concept of truth thus gained also makes a claim regarding a correspondence of the system to reality, even though it only characterizes the system from the inside, so to speak; that the system is true means also that corresponding objects and states of affairs exist.

The characterization of truth in the system theory, too, implies at the same time a criterion for existence. *Whatever is inferred from perceptions in accordance with the laws of scientific concept formation exists.* This is the notion underlying the slogan coined by Max Planck: "Whatever can be measured also exists." For to measure means nothing other than to carry out a theoretical calculation in accordance with the rules of scientific concept formation, this calculation being connected to certain perceptions (observations of measurements). It is a narrower concept, to be sure, and therefore we pronounce the converse only for our general formula: *Only that exists which is deduced from perceptions in accordance with the laws of scientific concept construction.*

Noting at the same time that, according to the above, only the system as a whole can be characterized as true and that, therefore, the inference that an individual object or state of affairs exists is always grounded upon the system as a whole, we must recognize that statements about the existence of an individual object can only be made in the context of the existence of reality as a whole. *In what manner reality is divided into elements is not given as such; this process is accomplished only by the conceptual system in that it defines elements by isolating them from the complex of reality.* Thus the coordination in (3), Section 6, contains a definition; it defines first what is to be designated an *individual* state of affairs or object and, at the same time, what is to be accounted the *same* state of affairs or object. For, strictly speaking, we never encounter the same states of affairs or objects, since slight differences will invariably be revealed by precise analysis.

Nevertheless, coordination (3) indicates in what cases we speak of the same state of affairs in spite of this. It is able to do so because it rests upon laws of probability. The coordination between reality and concepts that we carry out in knowledge is, then, of a very peculiar nature. One side contains no elements that are defined in themselves; they are defined instead by means of the coordination.

But we must distinguish between the definition of an object and the definition of existence. The coordination can certainly define what an individual object is, but it does not thereby define what constitutes the existence of the object. In accordance with the realistic conception developed in Section 6, our characterization of truth offers only a criterion of existence, not a definition.

9. PHYSICAL FACT[27]

Our analysis of the problem of epistemology showed that the more general concept of probability takes the place of the concept of truth in the study of nature, the former containing the latter as the limiting case. This is to say that physical facts cannot be judged to be absolutely true or false, but only more or less probable.

We now recognize the deeper significance contained in the rank-ordering of facts presented earlier (Section 5). According to that presentation, the ordering of facts proceeds according to degrees of probability; there are no discrete levels, the facts being ordered instead solely according to continuous degrees of probability. Only zero-level facts possess probability 1, but it is precisely they, as the limiting case, that do not constitute *objective* facts and are therefore not included among physical facts. First-level facts have a probability somewhat less than 1, which can nonetheless scarcely be distinguished from certainty. It is only in the facts of higher levels that the probability deviates noticeably from 1, and it is at this point that the field of hypotheses begins. The aim of research is to increase the probability of the hypotheses as much as possible by uncovering new theoretical connections with the facts of lower levels from which it is possible to infer an increase in probability.

Research in physics deals only in higher-level facts; daily life has instructed us so clearly regarding lower-level facts that we no longer notice the transition from perception to fact, carrying it out automatically. So it is that the physicist believes himself to be dealing at all times with facts alone, not with perceptions. He contents himself with reducing complex reality to facts of a lower level. In publications, for instance, he may still give the current strengths and temperatures, without describing his perceptual experiences when noting them down: the probability of the inference employed in making the transition to these facts may, for all practical purposes, be set at 1.

For the same reason, the probability inference is generally not carried out between perceptions, but between objective facts. For instance, we carry out a series of measurements, enter them as points in a coordinate system, and draw the *simplest curve* through them. The measurement data are lower-level facts, being themselves inferred from perceptions with such a high degree of probability that they can for all practical purposes be regarded as certain. A probability relation between perceptions is now replaced by a probability relation between the measurement data corresponding to them. This case represents the very prototype of a probability inference as employed in physics: an inductive inference is made from observed facts to facts that have not been observed. For in drawing the continuous curve through the points of measurement, we are asserting that the curve reproduces, in the intervals between two measured points as well as at them, the objective relations that would be revealed by measurement if the appropriate experimental conditions were established.

We can now also grasp the significance of the concepts of *theory, hypothesis*, and *experiment*. Experiment reveals lower-level facts, which are also inferred from perceptions with the assistance of theories, yet with such a high degree of probability that we can speak of 'experimental fact' with a certainty adequate for practical purposes. The high degree of certainty of lower-level facts stems from the fact that they are relatively invariant in the face of extensive alterations in theory. We learn from experiment, say, of the appearance of a new spectral series: even this 'experimental fact' contains certain theoretical presuppositions, such as the law of dispersion for prisms or the theory of diffraction by a grating. If the 'observed factual state' is used to alter ideas about the structure of matter, then, ultimately, the theory of dispersion or of diffraction must also be revised, since the interaction between matter and light plays an essential role in this theory. Strictly speaking, the 'observed factual state' would also have to change as a consequence. Yet what may be a very extensive change in the theory from a conceptual standpoint results in only very minute changes in the occurrences in a spectrometer, so that it is not necessary to supply a new interpretation of the bright streaks observed in it. To take another example, the confirmation of Einstein's theory of gravitation by means of astronomical measurements (through the deflection of light near the sun, say) is carried out with data from observations that could not have been acquired from the telescope readings without the assistance of the Newtonian theory of gravitation (with the sun's mass, say); yet this involves no internal inconsistency, since interpreting the telescope readings in accordance with Einstein's theory would make no noticeable quantitative difference in the case of these particular observational data. Thus only the *ranking of facts according to their degree of probability* enables experiment to be the test of theory. Such ranking is the basis for the *process of approximation,* without which scientific research would be, for all practical purposes, impossible.

The transition involved is gradual; if less firmly established theories are used to interpret a state of affairs, we speak of an hypothesis. An example is the penetrating radiation which is thought to come from the vicinity of the Milky Way through to the earth. The observations now at our disposal do not support an inference regarding this radiation with so much certainty that it can be called an experimental fact, yet the manner of establishing its existence is fundamentally no different from that in which, for instance, the existence of radiation from radium was established. The difference between experimental fact and hypothesis is only gradual. The difference between hypothesis and theory is also fluid, hypothesis being understood more as an individual assertion, theory as a combining of many individual assertions into a system. Thus we speak of atomic theory, meaning thereby all the relations connected with the assertion of the existence of atoms, but we call the existence of the atom itself an hypothesis. To be sure, the hypothesis in this example has already gained such a high degree of certainty that it could just as well be called an experimental fact.

The theoretical construction of physical knowledge can take two different directions. The first approach sets out from lower-level facts and proceeds to higher-level facts, displaying the degree of certainty of the individual assertions by setting the facts out according to their ranking. This *inductive* method corresponds to the epistemological construction, which proceeds from what is more certain to what is less certain. The second, *deductive* approach places a fact of a higher level at the apex and derives from it deductively facts of a lower level. This may, in certain situations, lead to new lower-level facts which can then be directly tested. Examples of a deductive theory are Maxwell's theory of electricity and the theory of mechanics founded on Hamilton's principle. Textbooks of experimental physics usually proceed inductively. Historically, the method of science is an interplay between the two directions: often the unknown is inferred from the known, but sometimes, for intuitive reasons, a higher principle is placed at the apex, new lower-level facts then being derived from it. By direct experimental testing of these facts the more basic principle can, in turn, be confirmed.

10. PHYSICAL DEFINITION

The coordination of concepts with objects that is carried out in natural science is not always such that it yields knowledge. While the matter of which concept should be coordinated is, in general, established as a result of research, the coordination may in particular instances consist of an arbitrary stipulation. These cases constitute physical definition. Their mark of distinction is that, instead of coordinating a concept with a particular combination of other concepts, as is done in definitions of concepts, they coordinate a concept with an *existing object.* In the end this coordination can only be given ostensively [*durch Hinweis*]: 'that thing there' is to correspond to such and such a concept. As the coordination between object and concept is peculiar to this type of definition, we shall refer to it as *coordinative definition*. It is also commonly known as *real definition*.

The necessity of such coordinative definitions becomes particularly apparent wherever measurements are involved. As the unit of linear measurement, the meter is not defined by concepts but by reference to the original meter, which is housed in Paris. The earlier definition of the meter by means of the circumference of the earth is of the same sort. The only difference is that a series of concepts is interposed; the circumference of the earth does not correspond directly to the concept of a unit but to a logical function of this concept, viz., the concept 'forty million units'. This process of interposition is used very frequently.

If we have a class of similar objects at our disposal, each element in the class can be used for purposes of definition. For instance, the linear unit may be defined by means of the wave length of the red cadmium line. The similarity of all cadmium atoms is used for this purpose, and there is no need to store a special unit at a definite location. In this instance the coordinative definition is simplified, because certain *facts* permit it. For the fact that the objects in question are similar is, of course, not established by definition but is a fact that must be discovered.

Coordinative definitions are used at many points in the study of physics. It is not always easy to recognize them as such and to distinguish them from assertions of fact, and some well-known scientific disputes stem from seeking empirical knowledge where definitions belong. Most notably, the significance of the theory of relativity arises from its having shown definitions to be necessary at certain points where, earlier, scientists sought for facts; examples are the congruence of distant spatial segments and simultaneity. We have a certain freedom in employing definitions and facts. It is only when a definition is given in one place that another assertion becomes an assertion of fact; conversely, the second may be regarded as a definition, which makes the first into an assertion of fact. A general systematic study of physics from the standpoint of the separation between definition and fact has yet to be carried out – hence there are still a number of ambiguities in the foundations of physics.

11. THE CRITERION OF SIMPLICITY

We discovered that we cannot choose among the possible explanations of an experimental finding by comparing them with the actual state of affairs, as this can only be inferred through theories, and that the selection can therefore be made only by means of an internal comparison of the theories. We have declared probability to be the criterion [*Gesichtspunkt*] according to which this comparison is made. This assertion needs to be laid out in greater detail, for a different criterion is usually employed in this connection: the *simplicity* of a theory.

It must, in fact, be admitted that the criterion of simplicity has played a major role in the construction of physical theory. Again and again physicists have espoused a particular theory because it gave a simpler explanation of the observed phenomena, and in philosophical studies simplicity has been granted a decisive position in the description of nature. In the famous words of Kirchhoff, mechanics has the task of "describing movements in nature completely and in the simplest manner possible"[28]. This goal, which can likewise be extended from mechanics to the whole of physics, includes other conditions in addition to simplicity. Kirchhoff's remark is clearly intended as criterion of truth, for the concept of truth must surely be contained in any such goal, and there is no other element in his formula which could represent it. Kirchhoff might more properly, then, have formulated the task as being to "describe completely and truly". His insertion "in the simplest possible manner" in the place of "truly" can only signify that he intends at the same time to give a characterization of truth: truth here is equated with simplicity.

This interpretation of Kirchhoff's formulation is confirmed by the passage following the above citation, in which Kirchhoff discusses how the simplest description can only be ascertained in the course of the gradual evolution of science. It is only a step from this characterization of truth to a *renunciation* of the concept of truth, and Kirchhoff's much-quoted formulation has, in fact, often been interpreted as meaning that there is actually no truth in natural science and that truth is to be replaced with the simplest description. Mach[29] had an even greater influence in this direction; he proposed the *principle of economy* for scientific research and taught that there is no such thing as a true description of nature, but only a *most economical* one.

This theory of truth, also known as the pragmatic theory, cannot, however, withstand a deeper criticism. For it rests upon a confusion of two meanings that converge in the concept of simplicity.

The first meaning of the concept of simplicity has no connection with the concept of truth. There are numerous cases in which several *equivalent* descriptions can be given of one and the same state of affairs; one of these may be the simplest, without its in any sense being truer than the others. For example, it is simpler to present mechanics by using vectors than by using coordinates, but not therefore truer. Again, the metric system is simpler than a system in which the factors for the transformation of one measure into another are not equal to 1, but it is not 'the true' system of measurement. To be sure, physics will choose the simplest presentation in each of these instances, but it makes no claim as to its truth, operating really only from motives of economy. The object is to spare physicists unnecessary efforts. This simplicity is a characteristic solely of the description, saying nothing about the state of affairs. We therefore call it *descriptive simplicity*.

The second meaning of the concept of simplicity, however, has an express connection with the concept of truth. For instance – to use a previous example – we choose the simplest curve to join a series of measurements. And in doing so, we are making an important assertion regarding reality, viz., that future measurements will lie along this curve rather than along a variety of curves waving back and forth between each pair of measurements which might have been drawn between the points instead. The preference given to the simplest description is thus related to a *truth claim*. This claim is based on a probability principle, that *the simplest description appears to us to be the most probable.* The principles of probability may appear in various guises, but they are always characterized by a certain evidence that cannot be logically justified and from which we are nonetheless unable to free ourselves. Thus the selection of the arithmetic average of various measurements seems to us the simplest assumption, yet it, too, is of course a probability assumption. The special significance of the concept of simplicity deserves a more detailed investigation within the context of the theory of probability. But even here we can establish that giving preference in this way to the simplest description is to be interpreted as a probability assumption and is, therefore, an application of the criterion of truth that we offered. Since the simplicity of inductive inference is involved, we speak of *inductive simplicity*.

The confusion of these two meanings of the concept of simplicity is the source of the pragmatic theory of truth. The purely economic character of the property of simplicity is taken from the first meaning, its significance as a criterion of truth from the second; the two are then combined into the assertion that there is no such thing as the truth of an investigation, but only of its economy. This confusion has given rise to numerous misinterpretations in physics. It is to be dispelled by a thorough investigation of each case to determine which variety of simplicity pertains to it. Thus the simplicity of the Copernican system as opposed to the Ptolemaic is of a purely descriptive nature, as we have known ever since Einstein: it is therefore pointless to introduce the simplicity of the Copernican description in order to combat the relativistic view. Relativity theory in general offers numerous examples of descriptive simplicity, and the significance of this theory consists in large measure in its having revealed the purely descriptive character of the simplicity of particular modes of describing the universe. An example of inductive simplicity, on the other hand, is Bohr's approach to the explanation of the spectral series, which is not only more economical than older theories but also asserts the existence of a different state of affairs. Only *one* of two theories differing with respect to inductive simplicity can be true.

Thus the difference between descriptive and inductive simplicity is related to the difference between definition and fact. By means of a skillful selection of definitions, descriptive simplicity can be attained, but never a truer description. In contrast, the selection of propositions asserted to be facts can only be carried out by means of inductive simplicity. The difference between these two forms of the property of simplicity is not usually made sufficiently clear in physics. This clarification of physical knowledge will be attained only when the separation of definition and fact in general has been carried out.

12. THE GOAL OF PHYSICAL KNOWLEDGE

According to the above considerations we may present as the goal of physics the gaining of true propositions about reality. Yet this statement is not quite adequate for the specification of this goal. It lacks a more precise characterization of the tendency of physical research; for this we would have to add, for instance, that it aims at acquiring the *greatest possible number* of true assertions. And even then, a yet more precise characterization is necessary.

For the mere piling up of propositions is not the aim of physics, even though it is greatly interested in the dissemination of knowledge. Rather, we distinguish between a *breadth* and a *depth* of knowledge, assigning greater value to knowledge that penetrates deeper. It is not difficult to say what we mean by depth: we mean the combination of a number of propositions into a single proposition, the explanation of a series of phenomena by a single law. For instance, Newton's law of gravity did not, in his day, constitute an extension of knowledge; the laws of freely falling bodies had been established by Galileo, the laws of planetary motion by Kepler. Nonetheless, Newton's law signified a step forward in knowledge in that it combined these various laws into a single law. This is a typical case of knowledge in depth. Of course, the Newtonian law also increases the sum of our knowledge by one new true proposition, but it is the depth of this proposition that gives it its particular value.

This direction of research is what we call *explanation.* A new fact adding to the breadth of knowledge does not *explain,* but rather *requires* explanation; that is, we require that it be incorporated into the system. In this there is more than the mere search for truth. The same inclination to construct a system is shown in the coalescence of the various subdivisions of physics into the one science of physics. This is the tendency described by Max Planck in his lecture, 'Die Einheit des physikalischen Weltbildes' (The Unity of the Physical View of the World)[30]; this coalescence signifies an increase in explanation and therefore a progressive step of knowledge in the direction of depth.

The tendency holding sway here can be clarified by the example of mathematics, where we also find research in depth; this is called *axiomatics.* The search for axioms is a search for a system of true propositions that combines the numerous mathematical propositions into a few general propositions from which all the rest can be logically deduced. A *minimal principle* is at work here; axiomatization means reducing the number of presuppositions to a minimum. It is impossible to say, of course, that precisely one single presupposition must be sufficient for the deduction of the rest of the system. This will generally not be the case, and the best we can do is search for a minimum. We regard the establishment of the minimum as having logical value.

The situation is much the same in physics. Physical explanation follows the same direction in its investigations as axiomatics, and the most general laws of physics, from which it derives all others, are its axioms. If a physicist occasionally feels that axiomatization is 'unphysical', he is displaying a failure to understand the principle of depth in physical research, which is directed entirely upon axioms. Newton's law, the principles of mechanics, the fundamental laws of thermodynamics, Bohr's determination of frequency and his correspondence principle are all axioms. Apart from the fact that physical axioms lay claim to hold for reality, the difference between mathematical and physical axiomatics is superficial: mathematical axiomatization is concerned with maximum elegance in its inferences and the proper construction of its logical connections, while physics, mindful of the fact that its systems of axioms have only an interim character in the face of experience, is frequently content with an approximate investigation of the connection. To be sure, in a more advanced state of physical science this provisional system of axioms would no longer suffice. It would then be completely in the interests of physics to increase the precision of its deductive system; thus axiomatic construction is ultimately as much the goal of physics as it is of mathematics. Examples of more advanced axiomatic construction are already to be found in thermodynamics, mechanics, classical electrodynamics, and relativity theory. Quantum theory, on the other hand, is still in an interim state, a way out of which has at last been opened up by the quantum mechanics of Heisenberg and Schrödinger. (Cf. Section 24.)

The burden of physical axiomatization is substantially lightened by the fact that it need not make a particular investigation of the logical fine structure of its method of inference, but can adopt it from mathematical axiomatization. It is therefore able to combine these axioms into a single axiom that asserts the applicability of mathematical modes of inference quite generally and is placed alongside the physical axioms. Usually this procedure is presupposed, without explicit expression as an axiom.

The minimal principle at work in axiomatization takes its place beside the principle of truth; the goal of physical research is not determined until the two are put together. We seek true propositions with a minimum of concepts. The tendency toward a system that finds its expression here has always been of particular interest to philosophy, and especially emphasized by the neo-Kantian school. While this school of philosophy considers the concept of truth to be exhausted by its incorporation into a system, Moritz Schlick set forth with great clarity the independence of the two concepts[31]. We must now investigate this question in greater detail. If characterized by the uniqueness of coordination, the concept of truth is undoubtedly independent, but the situation changes if we make our theory of probability the basis for truth.

Let us now examine the transition in mathematics from one system of axioms to another, deeper system. We can differentiate among three forms of transition. First, the second axiom system can be equivalent to the first, i.e., each one can be derived from the other; in this case, they have the same propositional domain. Second, the second axiom system may have a greater domain than the first; in that case, the propositional domain of the first system is contained in it as a subset[32]. Finally, a third situation is also possible; the second axiom system may be more general than the first and contain it as a special case. This situation must be distinguished from the one preceding it. Here the propositional domain of the first system is not contained in that of the second at all, but is derived from the other only through the addition of new 'specialized' axioms. This case is profoundly significant for mathematics. An example of an entire ordered series of such axiom groups, progressing from the universal to the particular, is the construction of geometry from topology via projective geometry to metrical geometry. With the transition from a special to a universal system of axioms, the group of axioms in question becomes recognized as one special case alongside others. Here the advance in knowledge is due to the transitional process.

This particular form of axiomatics is important for physics. For the progress of physical knowledge consists in establishing more general laws, for which the previous law is a special case. To take our initial example of progress in the direction of depth, the Newtonian law of gravitation contains Galileo's earlier law of falling bodies only in the sense of a special case. The constant, $g = 981$, in Galileo's law occurs as a variable in Newton's law, and alongside Newton's universal law there are placed particular supplementary conditions (namely, limitation to a small area on the surface of the earth) which have the effect of making this variable precisely equal to 981. Bohr's quantum principle will stand in this relation to classical electrodynamics once the two have been joined without contradiction; classical electrodynamics will then become a special case for large quantum numbers. This process illustrates precisely the meaning of the term 'explain'; to explain is to incorporate a state of affairs as a special case within a more general state of affairs.

We are principally indebted to Ernst Cassirer[33] for recognizing the fundamental significance of this procedure.

This species of axiomatic construction has yet a further significance for physics that does not apply in mathematics. For the transition to a more general system entails a *sharpening in precision.* The coordination of the special system of axioms with the state of affairs at hand falls short of precision, for the particularizing supplementary axioms are not strictly fulfilled. The introduction of the general system of axioms by means of such supplementary axioms, which do greater justice to the state of affairs, signifies, therefore, an improvement in the precise degree of validity[34]. Thus the institution of the axiomatic principle is connected with a progress toward truth whenever it is headed from the particular toward the more general; it is only then that physical *progress in the true sense* occurs. Compared to it, the two other species of axiomatic reduction play merely subordinate roles. The second kind that was mentioned occurs but rarely, the first more often, and both signify merely a *formal progress.*

This distinction offers us a vital insight into the meaning of the minimal principle. Once again we are faced with a principle which might be, and has been, regarded as a principle of economy[35]. But here, too, we find that this involves a misinterpretation. Only axiom systems of the same degree of generality are distinguished from one another by their descriptive simplicity, and, therefore, in accordance with economy of thought; but such systems play no essential role in physics, serving principally as an arena in which physicists compete for mathematical elegance. Formal progress of this nature can only be secondary in the increase in knowledge, if, for instance, it makes the system logically more intelligible or teaches mathematical methods that can be fruitfully employed; as, e.g., the four-dimensional geometry introduced into relativity theory by Minkowski. But generally the minimal principal is effective not only in simplifying the state of our knowledge but also in formulating the truth more precisely and thus effecting a transition in accordance with inductive simplicity. The probability nature of all knowledge entails that research in depth results in greater precision at the superficial level as well, and ultimately in a better correspondence in the implications on the right-hand side of formula (1) in Section 6. The drive in physics toward the strictest possible truth forces it at the same time toward axiomatization and systematization.

To be sure, we cannot conclude from this that striving for generality in knowledge derives its value solely from the simultaneous increase in truth. This principle also possesses its own logical value, and there are instances in which we are far more interested in the incorporation into a general law than in the improvement in precision connected with it. The process of comprehending the particular by incorporating it into the general is simply the way of comprehension for human reason; it is the only form whereby the seemingly impossible task of gaining knowledge of complex reality on the basis of a finite number of perceptions and within the finite confines of human thought becomes possible. Yet the fact that this process at the same time brings about a heightening of truth signifies a fruitfulness extending beyond its original intent and reveals *de facto* coupling of two logically distinctive approaches of the will toward knowledge. Thus the probability theory of knowledge, which sets out from the process of learning rather than from an ideal and perfect world-picture, teaches us to understand the relation between the different approaches pursued by physical research. In the constructive development of science into a system, two tendencies become combined in such a manner that the one always supports the other; the striving toward the most general truth possible goes hand in hand with the striving toward the most precise truth possible.

Part (b) Empiricism and Theory in the Individual Principles of Physics

13. THE PROBLEM OF THE *A PRIORI*

Our investigation of scientific knowledge has led to our conceding far more territory to theoretical thinking than the professional scientist normally supposes to fall under its sway; even the most elementary physical fact contains theoretical thought. Along with this, a series of more or less conscious presuppositions enter into science, and philosophy, particularly since Kant, has regarded the discernment of these presuppositions as a vital task.

We have already declared the probability principle to be the most important of these presuppositions. Others listed by philosophy are the principles of time and space, the principle of causality, and the principle of the conservation of matter. Logic, also, is to be included, of course, since scientific thought assumes the logical laws to be correct. We must now enter into a more precise discussion of these principles. But before investigating the individual principles, we shall look at a general issue, common to all principles. This is the problem of the *a priori,* which is revealed along with the epistemological presuppositions of principles.

We must first of all exclude logic from the complex of questions. We are unable to escape the rigorous compulsion of its laws; yet its conclusions appear empty to us, for they signify nothing new, but simply express the self-evident. We certainly are not taking the view that the epistemological problem of logic is exhausted with this statement. But these questions have as yet undergone so little study that we will exclude logic from epistemological discussions, as it has been customary to do since the time of Kant.

And in fact, the problem that is far more urgent for natural science first arises with the other principles we have listed. In contrast to logic, they are characterized by not being necessary for thought; we can easily imagine, for instance, the occurrence of an event without a cause, or the total quantity of matter increasing of itself. Nonetheless, we use these principles, and are even deeply convinced of their validity, which gives rise to the question from whence we derive our right to this belief. We might suppose that these principles are thought to be trustworthy solely because they are constantly being tested in science, that they are, then, empirical principles. However, Kant, and many other philosophers along with him, takes an opposing view. He begins by making the aforementioned distinction and calls logic an analytic principle; the other principles he calls synthetic, for they lack logical force. But he goes further to say that these synthetic principles are nonetheless *a priori,* that their validity does not stem from experience; they are instead, he says, true independently of all experience.

Kant's proof of this conception can be roughly formulated as follows: the presuppositions in question are used in the construction of propositions from perceptions; therefore no proposition can ever contradict them, for any such proposition must itself have come into being with the help of these presuppositions. Thus the truth of these principles is completely certain; furthermore, they apply to the external world, for they are contained in every proposition about the external world. They are the 'constitutive' principles of the external world.

This is not the place for a detailed presentation of Kantian philosophy. We will have to make do with the above brief summary, and our critique must be correspondingly brief[36]. To begin with, we must dispute the proof of irrefutability. Kant assumes [*fingiert*] the following logical sequence: a series of propositions a_i, all containing the presupposition A, is given; it is concluded from them that A is false. Kant then pronounces this to be a *false proof;* we cannot, he says, infer not-A, since the a_i presuppose A. But he is wrong, for the proof is *valid.* It shows that the presupposition A leads to contradiction and thereby proves that A is false. This is the *deductio ad absurdum* of the ancient logicians, but, it is, at the same time, a mode of inference very familiar in physics. A series of phenomena is explained by a theory, which is then seen to lead to contradictions; instead of concluding that the method of confirmation is false, the scientist concludes that the proposed theory is false.

This procedure can be used to judge any theory that is not directly used in establishing the contradiction. The existence of the contradiction must be firmly established independent of the presupposition, in essentially the following way. We anticipate certain perceptions a'_i, but experience different perceptions b'_i. We will not, for the present, go into how we can judge perceptions to be the same or different, but will assume throughout that we are able to do so. It is nonetheless clear that the *a priori* principles do not enter into establishing the existence of the contradiction. The only question here is whether the perceptual experience takes place or not, and no theory is required to answer it. Thus it is in principle possible to decide whether the *a priori* principles lead to a contradiction.

Yet this assertion must be qualified. The contradiction will generally not be established according to the strict method we described. We cannot expect the perceptions a'_i to be precisely those we counted upon; we will have to concede certain deviations, in accordance with probability, if we are ever to reach a positive result at all. Making the decision will consequently follow a somewhat different route. We will test to see whether the observed a'_i are compatible with the anticipated perceptions, taking into account the laws of probability. We cannot reach any correspondence unless we presuppose these laws of probability. Conversely, a negative proof is not carried out by interpreting negatively every departure of the a'_i from what was expected; we consider the contradiction to be proven only if the variations contravene the laws of probability. That is, the laws of probability are not drawn into the establishment of the contradiction, but they must always be included in the construction of theories if accord with experience is to be attained at all. This is the reason for their special position.

This has the following immediate consequence. Of all the principles used in the construction of theories, the probability principle is the one that we will maintain to the very last. We will, for instance, invariably try to modify the principles of time and space before we call into doubt the validity of the laws of probability. Conversely, a refutation of these other principles can be carried out only if we hold to probability, for otherwise the disparities may always be interpreted as failures on the part of the theories, which have not yet reached a sufficiently precise command of reality. Owing to the limited precision of our knowledge, every assertion we make regarding nature will be indefinite unless we assume the laws of probability. There will then be no true or false, for concepts that are defined only for the limit are of no use in dealing with approximations. Paradoxical as it may sound, the laws of probability are our most secure possession.

Must we then regard the laws of probability themselves as irrefutable? This would claim too much, even though they are the ultimate presuppositions. All we can say is that without the laws of probability we would certainly run into contradictions; but we may not assert that contradictions can invariably be avoided *with* them. We will invariably maintain the laws of probability for as long as possible, for all other principles must fall with them; yet it is in principle possible for a theory presupposing only the laws of probability to lead to contradictions. In this case, we would have to state that knowledge of nature is impossible.

In theory, at least, we must take account of this possibility. It will not do to prove the validity of certain principles on the grounds of the possibility of knowledge, for this possibility is itself merely experienced and perhaps itself exists only within limits. Of course, this is not to say that the occurrence of occasional difficulties in the explanation of physical phenomena justifies this conclusion. And another distinction must be borne in mind. Only the negative formulation, "Up to now we have been unable to attain any knowledge free of contradiction", does not assume probability. On the other hand, the positive formulation, "Knowledge of nature is impossible", does presuppose the laws of probability and should be stated more precisely as, "Knowledge of nature is probably impossible." Closer investigation will show that this, too, is a meaningful assertion.

Combining these various considerations, we may assert that physics contains no principles that are *a priori* in the sense of 'independent of experience'. We hold to even the most general principles of knowledge solely because they prove themselves in experience. These principles signify something about reality; they are not empty; rather, they formulate characteristics of the world.

14. THE PLACE OF REASON IN KNOWLEDGE

The philosophy of the *a priori* started from the idea that knowledge is the result of a combination of thought and perception, and it was the belief of its proponents that they had discovered in the universal principles of knowledge an expression of rational thinking, the *rational components* of knowledge, as it were. We must now regard this as an error, for these principles have not only a rational characteristic but also a *reality component.* On the other hand, it is one of the profound ideas of philosophy that the laws of thought play a role in knowledge and that our system of knowing is determined not only by reality but also by the nature of our thinking. If we reject the characterization of the rational component of experience by means of the most general principles of knowledge, we must then find another way to show the place of reason.

The greater portion of our world-picture is certainly determined through characteristics of rational thinking. The nature of thought expresses itself in the whole structure of science: in the division into concepts and propositions, in the progression from one stage of knowledge to another, in the application of mathematical and logical concepts, in the distinctively net-like character of the relations among the elements of our knowledge. Apart from perceptions, what we meet in thought is never anything but thought. We experience only a net-like schema, which represents reality to us in a way no different from the way in which the motion of the strings in a puppet theatre represents the motion of the figures to the actor behind the stage. It is therefore more difficult to identify in this schema the elements determined by reality than the elements determined by thought, and yet this is precisely the task faced by physical science.

It reaches this goal by using the difference between possibility and reality. Thought allows more schemata than are compatible with reality; the properties of reality are characterized by the selection of admissible schemata from among the possible. On the other hand, it emerges that this procedure does not lead unequivocally to a single schema. We encountered this indeterminacy in the concept of descriptive simplicity, which rests upon the possibility of equivalent descriptions, among which none is superior with regard to truth. Thus a narrower class of admissible schemata are selected from the broader class of possible schemata; it is in this that the process of physical knowledge consists.

There are two ways in which we can obtain a list of those characteristics of the schemata that stem solely from reason and that therefore signify features not of reality, but of reason. First, we can indicate within the narrower class of admissible systems those characteristics that are not valid for all these systems; they cannot signify characteristics of reality, for reality does not necessarily require them. Thus Euclidean geometry is not a characteristic of reality, for it cannot be maintained in the face of admissible modifications of the schema (cf. Section 16); nonetheless, it is actually as such a characteristic of reality to be *geometrically comprehensible.* Second we can seek out within the broader class of possible systems those characteristics that are valid for *all* systems, for they cannot signify any specific difference for the *narrower* class and, therefore, cannot designate any characteristic of reality. However, it is more difficult to pursue the second course for the task of formulating propositions about all possible systems of knowledge appears hopeless if we bear in mind the preceding discussions of the refutability of presuppositions and are aware of the fact that no limitations are set on any possible perceptions. Yet we may entertain the supposition that a sweeping proposition of similar scope can be formulated respecting logic.

Both methods taken together signify the execution of the idea that *the characteristics specific to the narrower class, and only these characteristics,* constitute a characterization of reality. They are therefore able to isolate those characteristics of our system of knowledge that stem from reason, and yet it cannot be claimed that these characteristics are thereby exhausted. For the formulation of the characteristics specific to reality is also carried out with the help of reason. Therefore we find in these propositions an interaction of both components.

It is nonetheless important to stress the subjective component, especially in connection with the first method. For physics usually does not work over the entire general sphere of the narrower class, but selects from among the admissible schemata that one that appears most suitable with respect to descriptive simplicity. This is the reason that at times we erroneously attribute an objective significance to some of those characteristics of the schema that are in truth purely subjective in nature. We can distinguish characteristics of this sort by transforming to some other admissible schema whatever is altered in the process possesses no objective significance. It is not always easy to achieve this transformation. The notion that certain elements are of an objective nature has been so firmly implanted that it frequently takes generations of work before they are seen to be subjective. We need only recall the theory of relativity, the epistemological significance of which consists precisely in its having revealed the subjectivity of certain elements formerly believed to be objective. H. Thirring has listed other examples of elements the subjective nature of which, although actually known, is easily forgotten[37].

In such investigations, the human mental faculty proves to be more susceptible to variation than we ordinarily assume. Caution is advisable in following the second method. The structure of logical thinking perhaps may not be fixed to the extent that we are inclined to believe. When we consider the evolutionary origin of human thought, it seems plausible that even the most general forms of thought have arisen through a special adaptation of man to his environment and therefore do have, after all, in certain respects the character of reality. It is quite possible that there are no absolutely universal characteristics in the possibility class, that its characteristics never possess more than relative generality over against the narrower class in the temporary state of knowledge at a particular time – that logic, too, will ultimately be shown to be subject to evolution. To be sure, we cannot tell with the logical instruments at our disposal what the significance of this may be. All we can do is to refrain from formulating *any* proposition about the possibility class, either positive or negative.

Having completed these general considerations, we proceed to the more precise analysis of the most important principles of physical science. In so doing, we will pursue the idea of separating the rational portion of these principles from their empirical content. The arguments carried out in Section 13 have shown us that they must have an empirical content. In order to locate it, we must show what part of the relevant principles originates in subjective arbitrariness and what part is independent of it. Our earlier concept of a coordinative definition will prove fruitful in this process, and we shall discover that our inquiry concerning the relation between empiricism and theory coincides with the distinction between fact and definition, a necessity we have already indicated.

15. SPACE

The heterogeneous variety in the judgment of the problem of space has its root in the failure to distinguish adequately the mathematical problem of space from the physical. As long as there was no non-Euclidean geometry, there was only *one* model of space for mathematics and physics alike. It seemed, therefore, that physics could simply take over from mathematics the concept of space and its laws, while mathematics, in turn, had the task of examining these laws by means of its own methods, thought to be independent of all experience. Thus arose the subjective conception of the problem of space, for space appeared to be a subjective element, a member of the rational component of knowledge, corresponding to nothing in reality. This is also the conception of Kant. The situation changed, however, when the discovery of non-Euclidean geometries brought to light the possibility of several quite different types of space. From that time on, the mathematical problem of space parted company with the physical. It was recognized that mathematics dealt only with *space as a type [Raumtypus]*, with possible spaces, whereas physics was obliged to decide which of these possible types accords with reality. The peculiar position of mathematics as the science of possibility in relation to physics as the science of reality was once again made clear, and the objective conception of physical space took root once more, despite Kant. To be sure, the full development of this conception took nearly a century. The prophetic words of Riemann[38], from the beginning of the movement, were first brought to realization through Einstein's relativity theory.

The peculiar nature of the mathematical problem of space becomes clear if we consider the logical position of the geometrical axioms. Within the mathematical sphere, the judgment true or false is not applicable to the geometrical axioms. Mathematical axioms are *definitions,* comparable to the rules of chess, and they merely construct the relevant type of manifold. Therefore it is permissible to transform individual axioms into their converse and then combine them with other axioms. As long as the axioms are mutually independent, no fallacies can result from this process, and instead, a new type of manifold emerges. The objects or elements of geometry are defined by means of the axioms, in the sense of implicit definitions[39]; anything that satisfies the relations set forth in the axioms is itself an admissible element of geometry. It turns out to be not at all only intuitive elements such as a point, or a straight line, that fulfil the conditions; the same axioms apply among quite different elements, such as numbers. We need only recall the various models of non-Euclidean geometry that have been constructed since the time of Klein.

In physics, however, the problem is fundamentally different. Physics must decide which axioms are *true;* its axioms are not definitions, but facts. Physicists must test whether the Euclidean axiom of parallels applies to physical space or not. By carrying out these investigations, they decide which of the mathematical spatial models conforms to reality. This conception has admittedly come under attack. It has been said that physics could not possibly make such a judgment, that in making measurements physicists can only *presuppose* geometrical axioms, not obtain them as a result of the measurements. In connection with this objection we must refer to Section 13, where we showed that at least the falsity of presuppositions can be judged on a physical basis. Moreover, this objection overlooks the fact that, if Euclidean geometry is presupposed in the small, a method of approximation can be used to prove the validity of a non-Euclidean geometry for large realms of real space. It is, for instance, a logically acceptable procedure to use astronomical measurements to conclude that the universe is non-Euclidean, even though the Euclidean character of astronomical instruments is presupposed.

To be sure, we must make one concession to the subjective conception of the problem of space, which employs these very ideas in its defense. The results of spatial measurement are not free in the sense of subjective arbitrariness as is often believed by the proponents of the objective conception of space. At particular points, definitions must be laid down before any spatial measurements can be made at all; and which geometry will appear as the result of measurement will depend upon the choice of definitions. These definitions are coordinative definitions in the sense given in Section 10. Thus the establishment of the definitions of spatial order constitutes an important epistemological task, for only by means of this analysis can we gain a definitive clarification of the epistemological problem of space. This task has been carried out in connection with the theory of relativity[40].

The first coordinative definition for space is the setting of the unit of length. It is also the most obvious type of a coordinative definition, and has long been recognized as such. Yet what was not recognized for a long time is that a second coordinative definition is needed to establish a comparison between lengths at various points in space. We may visualize schematically that a line segment is represented at every point in space that will serve as a unit[41]. Whether or not these segments are equal in the ordinary sense is immaterial, for they can be made equal by definition. The necessity of a coordinative definition to stipulate the congruence of line segments at a distance is understood since it is in principle impossible to compare remote segments with respect to congruence. Transporting rigid rods cannot help, for we have no way of knowing whether or not the rod changes during the journey. Quite the contrary: the comparison of lengths during the transport of rigid rods is rather to be conceived of as a definition.

A further coordinative definition concerns motion and rest. Which system is to be designated as at rest is established by a coordinative definition, which subsequently determines the motion of all other bodies. Historically, this idea was the starting point of relativity theory. It also applies to motion of rotation, which is why the Copernican and the Ptolemaic system are equivalent. The former is certainly simpler, but this is merely a matter of descriptive simplicity, which signifies nothing as to its truth.

That this question has always been connected with the theory of gravitation is well known. As long as the Newtonian theory of gravitation was upheld, the Copernican system was regarded as the only true system; the relativistic conception has been recognized only since the establishment of Einstein's gravitation theory. However, this conception is only partially correct. Motion requires a coordinative definition, quite independently of any physical theory, and every physical theory, including Newton's, can be transformed in such a way that it does not designate any state of motion in an absolute sense. The Mach-Einstein idea has a physical import over and above this epistemological relativity, viz., that inertia is to be found not only in instances of motion relative to all the masses of the universe but also, in a correspondingly smaller degree, in the relative motions of individual smaller masses in any spatial direction. Thus Friedländer came to suspect that the rotating fly-wheel in a steam engine generates a centrifugal field in its interior. The Einstein equations also contain this Machian principle, as Thirring has demonstrated[42]. Through the appearance of controllable effects, we are able to recognize that we are dealing here with a more sweeping proposition than the epistemological relativity of a coordinative definition, even though such a definition may have furnished the logical occasion for it. Conversely, the epistemological relativity of motion is for this very reason independent of these effects[43].

Finally, yet another coordinative definition is found in the problem of motion, one that is frequently overlooked. Not only the state of rest of a whole system, but also the state of rest of each point must be laid down specifically. We are relieved of the necessity of giving this further definition only if the physical rigidity of the system is presupposed; in this instance, the rigidity itself realizes this further definition. The relation between this definition and those preceding it is similar to that between the definition of the congruence of distant segments to the definition of the unit; in each case, the latter definition can be joined up with the former with the assistance of the rigid body.

The definition in question may also be described as a definition of the *relative rest* of points. It is therefore identical with the definition of the spatial coordinates system. A system of points that swirl about every which way in relation to rigid bodies can be defined as a system of reference 'at rest in space and in itself'. The import of this idea is that relative motion, too, is not an absolute. Mach's view that the relative motion of the earth and the system of fixed stars, at least, existed objectively must be modified to point out that here, also, there is present 'only a relative motion with reference to the rigid body'. Thus we may speak of a relativization of relative motion, introduced along with the general theory.

Along with the possibility of establishing the coordinative definition of space by means of rigid bodies, there also exists a means of establishing it solely through the use of light signals, which leads to the construction of a special *light geometry*[44]. Imagine observers located upon particular mass points that are swirling about in space who can only communicate with one another by means of light signals. With the help of these signals they are able to define the relative rest of points, the equivalence of line segments in space, and so forth, using light rays as the definition of straight lines. To be sure, the univocal designation of systems that are rigid within themselves in the sense in which a body is rigid can be carried out with the assistance of light geometry only if we regard the absence of singularities as an admissible mark of distinction; otherwise we will have to augment the definitions by material entities such as rigid rods and natural clocks[45].

The coordinative definitions that we have listed contain the *subjective components* of spatial order. The *facts* of spatial order – that is, its *objective components* – can be formulated independent of them. That two rigid rods which are found to be of equal length when placed side by side somewhere in space will turn out to be of equal length when compared side by side anywhere else in space is a characteristic essential to the designation of the rigid body. In connection with light geometry, further objective characteristics have been formulated as *light axioms*, to which *physical axioms*, such as the one just presented, are then added.

Spatial order results from the combining of these facts with the definitions. If we say, with Einstein, that physical space is of a spherical nature, we are asserting a complex fact that presupposes elementary facts and definitions. If we were to select definitions otherwise, not drawing upon the rigid body, the geometry would be different. The same is true of the shape of bodies in space. Even the conclusion that the earth is spherical is related to the definition of congruence by means of the rigid body and is significant only as a relative result.

The solution to the problem of space described here is to be attributed principally to the work of Riemann, Helmholtz, Poincaré, and Einstein. Helmholtz, the first to acknowledge the significance of Riemann's idea for physics, indisputably deserves the major credit for the recognition of the definitional character of congruence in physical space[46]. Poincaré coined the term '*conventionalism*', which refers to the definitional character of the congruence of line segments and designated the definition in question as a *convention*. At the time he introduced this idea, Poincaré still believed that the convention of the rigid body led to Euclidean geometry, not knowing that Einstein was soon to take up the idea of conventionalism in all seriousness and apply non-Euclidean geometry to physics. Final clarification came about with the philosophical discussion of Einstein's general theory of relativity[47]. Unfortunately, the viewpoint of simplicity played the greater role in this discussion. The opponents of the theory of relativity wanted to designate Euclidean geometry as simpler, whereas its proponents declared Riemannian geometry to be simpler by appealing to the fact that the physics which came into being along with geometry would have to be employed in judging between the two. But this observation only confounded the problem. It is *descriptive* simplicity that is in question here. It makes no difference which geometry leads to simpler relations; the one is not in any sense truer than the other. Rather, the relativity of geometry occurs along with the relativity of motion, for geometry, too, rests upon coordinative definitions[48].

If physics nevertheless gives preference to one particular congruence definition, it does so because it must choose *one.* It will select the one which formulates the relations in the most lucid manner, without thereby binding the selection to any truth claim. It is solely on such grounds that physics chooses the definition of congruence through the rigid body.

16. THE IDEALISTIC AND REALISTIC CONCEPTIONS OF SPACE

Some take the view that the structure of space does not contain any objective determination of reality, that it is a conceptual element, introduced into the explanation of nature by human reason, to which nothing objective corresponds. Indeed, the relativity of geometry is employed as an argument for this interpretation, and it is usually connected with the more sweeping claim that Euclidean geometry can be established by definition, for if it is in any case a matter of something subjective, nothing hinders us from selecting this distinctive form of geometry. Proponents of this choice appeal principally to the intuitively vivid nature of Euclidean geometry, denying to non-Euclidean geometry any possibility of being visualized.

Yet this conception, which has emerged as the latest offshoot of the Kantian philosophy of space, is nevertheless not viable. It is certainly true that geometry deals with a conceptual system, but not in some special sense different from that of natural science as a whole. This conceptual system may be determined by reason, yet the problem of the *coordination* between the conceptual system and reality does not depend upon reason; its solution can only be given by experience and constitutes, therefore, a characterization of reality. It is only the selected *definitions* of spatial order that are of a subjective character; the *facts* that stand alongside them are objective, which is why only the realistic conception offers an adequate explanation of our findings.

And for this reason, too, the so-called intuitive character of Euclidean geometry provides no claim to a superior position. Starting from the assumption that we totally disagree with assigning it this mark of distinction, we can demonstrate that, on the one hand, Euclidean geometry can be considered intuitive only to a very limited extent and that, on the other, non-Euclidean geometry is equally susceptible to intuitive visualization[49]. However, the following point can be made quite independent of this viewpoint. Even if the assertion about the intuitive quality of Euclidean geometry were correct, it would have no decisive bearing upon the epistemological value of the theory of absolute space and time. It would merely signify something about the preferability of one of the logically equivalent conceptual systems from the standpoint of psychological facilitation – something, that is, about the subjective abilities of human beings. But the question of which conceptual system fits reality is to be answered quite independently, for we should not presume that reality orders itself according to the psychological abilities of man. In particular, the question of physical space must be decided independent of intuitive qualities [*Anschaulichkeit*], for it is a question of the relation of a conceptional system to reality, not of its relation to the human capacity for intuitive visualization.

Those who press the point further next seek to defend their position by appealing to the relativity of geometry. In Section 15 we showed that reality does not unambiguously prescribe one geometry and that, in choosing the definition of congruence, we have it in our power to determine the nature of the geometry that will subsequently emerge. This being this case – so runs the argument – there can be no objection to employing in physics that geometry that is superior in intuitive qualities. We need only set up the definition of congruence in such a way that the geometry is Euclidean.

Such a definition of congruence is frequently possible, but we must note that, in doing so, we introduce a peculiar indeterminacy into physics. If measurements with rigid rods – think, for example, of the measurement of the circumference and diameter of a circle and the calculation of the relation between the two – indicate a deviation from Euclidean geometry, then the preceding argument means that we must interpret this deviation as action of a *force* that deforms the measuring rods. But further experimentation would demonstrate that this force has the same effect upon everything, so that its existence cannot be proven in any other way through the deviations from

Euclidean geometry that are found in practical measurement. But to admit the existence of such universal forces in physics would be to introduce uncertainty into all practical measurements. For instance, it would become meaningless to talk about the spherical form of the earth's surface, for if these forces exist, we might just as well conceive of the earth as an egg-shaped body or a cube and consider earlier observations to have been distorted by universal forces. It therefore seems more to the point to exclude universal forces by definition from the outset, stipulating that physics will acknowledge as physical only such forces as result in *differential effects;* e.g., heat would be an admissible force because of the variation in the coefficients of expansion. Given this stipulation, the question of the geometry of space is completely determined. This idea has also been formulated as the proposition that only physics and geometry taken together as a whole is subject to the test of experience[50]; but this assertion certainly is not intended to deny the possibility of indicating in physics the exact places containing the definitional prescriptions that make the question about geometry an empirical problem.

Yet another argument for the special position of Euclidean geometry has been put forth in that a rigid body cannot be defined in itself, but already presupposes Euclidean geometry in its definition[51]. But this idea is refuted by the preceding arguments. A rigid body is defined as a system having only a slight, elastic reaction to physical forces of the above-mentioned sort. If we add the stipulation that the influences of physical forces are to be eliminated by means of corrections, while universal forces are inadmissible, we have a complete definition of the transmitted lengths of rigid bodies carried over during transport. Thus the exclusion of universal forces determines the rigid body, too, and can take the place of a rule stipulating the form of the rigid body by means of Euclidean geometry.

However, we must take up the idea of the relativity of geometry once again and investigate the extent to which the relativity of geometry, which in itself certainly exists, permits Euclidean geometry at all. Mathematically, the relativity of geometry rests upon the possibility of transforming geometrically different spaces into one another. But clearly this transformation can be carried out *uniquely and continuously* throughout only when the spaces are topologically equivalent. If measurement of a space carried out in accordance with the definition of congruence by means of a rigid body reveals a form that deviates topologically from Euclidean geometry, then the introduction of Euclidean geometry by definition entails a violation of the principle of causality. If, for instance, non-Euclidean space turns out to be spherical and closed, then transforming it into Euclidean geometry by means of a definition will have the consequence, among others, that a light ray can traverse the infinity of Euclidean space in a finite time and come back from the opposite side. To admit such possibilities seems physically meaningless, especially for the philosophy of the *a priori,* and therefore their exclusion confines the relativity of geometry within certain limits, which in some circumstances prevents the introduction of Euclidean geometry. This might be the most important argument against giving a special position to Euclidean space[52].

As a consequence, the topology of space is to be regarded as a more fundamental determination than the metric; the topology is prescribed by reality once certain elementary presuppositions about causality have been made. This also leads to a clarification of the physical significance of the number *three* as the number of spatial dimensions which itself signifies a topological determination. It is known from mathematics that spaces possessing different numbers of dimensions can be transformed into one another, yet never uniquely and continuously throughout. According to the fundamental principle of the relativity of geometry, we can transform three-dimensional space into four-dimensional space and thereby take from the number of dimensions all objective meaning, but then the continuous nature of the causal connection would be lost. Consequently, we can establish the following objective meaning of three as the number of dimensions: only for three dimensions is a continuous causal ordering of reality possible. This is not to be conceived of as a characteristic having an *a priori* connection with the number three, but solely as a result learned from experience. Even if the continuity of the causal order, and particularly the implementation of the principle of action by contact (cf. Section 18), is to be viewed as a requirement of a definitional character, it is nonetheless a fact of experience that there is a particular number of dimensions that fulfills this requirement and that this is precisely the number three. In this sense, then, the number of dimensions is prescribed by reality and is, conversely, an objective characterization of reality. It is entirely incorrect to suppose that the number of dimensions of space is determined by the nature of man – say, his makeup in the sense of the psychology of sensation. The number of dimensions is not subjectively determined; on the contrary, evolutionary adaptation had developed in man precisely that disposition that singles out *three* dimensions[53].

17. TIME

Time also contains coordinative definitions; only when they are revealed can the philosophical problem of time be clarified.

The first coordinative definition concerns the *unit of time*, and it is, of course, well known as such. However, we find the problem of congruence with respect to time as we did before for space. The *congruence of consecutive segments of time* requires a coordinative definition. It is impossible in an absolute sense to compare two consecutive units of a clock; if we nonetheless wish to call them equal, this assertion has the nature of a definition. If two natural clocks standing side by side have equal first units, then their second units will also be equal – this is a fact, and it is taught to us by experience. But if we assert that the first unit is equal to the second for every clock, this is not what is taught by experience, but a definition instead. Thus the metric of time, time's uniformity, rests upon a definition. This definition is, again, arbitrary, and there could be nothing to prevent us from defining the metric of time in such a way that, e.g., a freely falling body falls with uniform speed. However, we possess a number of distinctive ways of defining uniformity, their distinction consisting in the fact that they all lead to the same metric. They are as follows:

1. Definition by means of natural clocks, e.g., the rotating earth. The natural clock is defined for this purpose as a closed periodic system. This concept contains problems similar to those of the rigid body, and other difficulties as well. For the best clocks, such as atoms with their rotating electrons, do not themselves directly give their unit of revolution as a frequency, but only define a period by the number of oscillations of the light emitted, which is related in turn in a complicated manner (according to Bohr) to the unit of revolution. If, finally, with modern quantum mechanics, the conception of revolving electrons is to be given up altogether, it must remain an open question how the atom is to be fitted into the concept of a clock.

2. Definition through the law of inertia. A force-free moving mass-point covers equal paths in equal times. Thus the metric of time is given through measurement of spatial segments; i.e., it is traced back to the metric of space via the law of inertia. This definition has the disadvantage of placing the end points of the equal segments of time at different space-points, so that simultaneity is already included in the definitions.

3. Definition through the motion of light. A light ray also covers equal paths in equal times and, like the moving mass-point, it can be used to define

uniformity. In order to eliminate simultaneity, we can imagine the light ray as being reflected back and forth between two mirrors connected by a rigid rod; it will then separate into equal time segments at each end point (Einstein's light clock). Yet spatial measurement cannot thus be eliminated, for the rigid connection between the mirrors is essential. Only with the help of light geometry is it possible to give a definition of the metric of time by the movement of light alone, yet the rigid rod is unavoidable if the definition is to be made unique.

All these definitions are arbitrary, but the fact that they are univocal is independent of our will, and this is their mark of distinction.

The third coordinative definition concerns *simultaneity*, i.e., the congruence of parallel time segments at different points of space. It is essential that the space-points be different; the juncture of two events at the same space-point belongs to another group of problems and is called a coincidence rather than simultaneity. Under the name 'relativity of simultaneity', the third coordinative definition has become the foundation of the special theory of relativity. To obtain a philosophical understanding of this definition, we must look more deeply into the concept of time.

The three coordinating definitions listed above concern *metric* characteristics of time, but in addition to these we must consider its *topological* characteristics. The most elementary characteristic of time is the *time sequence,* the purely topological ordering of time-points in the sense of one after another. All events taking place at a given space-point P are ordered in such a way that of any two, one is later. This relation is transitive, so that the time sequence, which we assume to be a linear continuum, develops at a single point. The fundamental topological relation, then, is 'temporally later' along with its inversion, 'temporally earlier'. Is it possible to reduce this relation to another relation; is it possible, that is, to give a coordinative definition of the relation of time sequence?

With the assistance of the concept of causality, such a definition can, in fact, be formulated. If a certain event E_1 is the cause of an event E_2, then E_1 is earlier than E_2. We can, then, trace the concept of time sequence back to the concept of a cause if we distinguish the order of E_1 and E_2 *without* using the concept of time. We must search in the events E_1 and E_2 for some discrimination whereby E_1 can be recognized as the cause of E_2. This can be accomplished in the following manner.

If we attach a *mark* to E_1 – i.e., make a small variation in the parameters – this mark will also be observable in E_2. If, on the other hand, we attach such a mark to E_2, it is *not* observed in E_1. This is the distinctive characteristic of the causal relation: the effects propagate only temporally forwards, not backwards. Therefore the nature of the propagation of effects can, in turn, be used in defining the sequence of time (cf. Section 21).

We find the same idea contained in the concept *signal.* By signal we understand a physical process that is propagated from one space-point P to another space-point P'. This 'propagation of the same process' can only be characterized with the help of a mark; a signal is a process whereby the mark moves from point to point. The signal, then, is the model of a transfer of action and can therefore also be referred to as a *causal chain*. The term 'signal' suits precisely the property of propagating the mark, for it signifies a transmission of marks.

Only such a signalling process is a *real sequence,* which must be sharply distinguished from an *unreal sequence.* If, for instance, we hold two rulers across one another and move one diagonally, the point of intersection will move rapidly along the edge; the velocity can be increased indefinitely by holding the rulers so that they are practically parallel. However, no mark is transmitted in this case. If one of the rulers should happen to have a small projection somewhere along its edge, the sequence will be interrupted at that point; yet this has no influence upon the subsequent process, which continues unaltered. Such unreal sequences offer no mark of direction and, therefore, no definition of the direction of time. In emphasizing the mark as the means of characterizing real sequences, von Laue was probably the first to call attention to the difference between real and unreal sequences[54].

We can, then, give the following *topological coordinative definition of temporal sequence:* an event Q_2 is said to be *later* than an event Q_1 if Q_2 can be reached from Q_1 by means of a signal.

Next we formulate a second topological concept of temporal order, which follows by analogy from the concept of temporal sequence: Two events Q_1 and Q_2 are called *indeterminate in respect to temporal sequence* if a signal can be sent neither from Q_1 to Q_2 nor from Q_2 to Q_1.

It will be noted that this definition captures precisely the essential property of the concept *simultaneous.* For the intuitive image of simultaneity means just exactly that simultaneous events cannot stand in a cause-effect relation to one another: an event that is occurring simultaneously with my present action is temporally situated in such a way that I cannot influence it and it cannot influence my action. Moreover, in speaking of simultaneity, we wish to assert something further; we wish to attribute to the events in question the same time values. But clearly this is admissible only if the stated condition is fulfilled.

How are we to accommodate this additional condition? There is no way to make a meaningful choice. We must admit that the condition is not only necessary, but also sufficient for simultaneity. Any two events that are indeterminate as to time sequence may be called simultaneous, for this assertion can never contradict the definition of temporal sequence, given our conceptual specification of *indeterminate in respect to temporal sequence.*

It would be quite possible that this topological specification of simultaneity will nonetheless lead to a univocal simultaneity; this would be the case if the velocity of the causal transference could be increased indefinitely. Then, for every event Q_1, there would be only one event Q_2, indeterminate as to temporal sequence, in another causal chain. The concept, indeterminate as to temporal sequence, would then be identical with the concept of simultaneity of the classical theory of time. Yet this is not necessary, for there may be a finite upper limit to all cause-effect transference. If so, the topological concept, indeterminate as to temporal sequence, will not lead to a univocal [absolute] simultaneity.

And this is the actual state of the case; the speed of light is the upper limit of all cause-effect transference. This is to be viewed as an experimental result; a decision between the two possibilities cannot be taken *a priori.* However, we still must show how it is possible to formulate this assertion without using the concept of simultaneity. Since it is to be used for the purpose of judging the univocality of simultaneity, it cannot presuppose the concept of simultaneity.

We therefore formulate the existence of an upper limit to all cause-effect transference as follows: If, at time t_1, a signal is sent from a space-point P to another space-point P' and back again to P, then there is at P a time-point $t > t_1$ such that, for all physically possible signals, the return to P does not occur earlier than t. This formulation assumes only the concept of temporal sequence, not that of simultaneity.

The upper limit t_2 for all eligible t corresponds to the return of the light signal. Light is therefore the *first signal,* for it comes back first of all. If we designate by t' the moment at which light arrives at P', we thereby coordinate the interval from t_1 to t_2 at P with the time-point t' in P'. Every point within this interval can be called simultaneous with t', so that, within these limits, simultaneity is arbitrary.

This is the strict presentation of the relativity of simultaneity; it rests upon the fact that, as a result of the existence of an upper limit of cause-effect transfer, the topological concept *indeterminate as to temporal sequence* does not lead to a univocal stipulation of simultaneity. Thus simultaneity can be established only by adding a metric coordinative definition to the topological specification. In the special theory of relativity, Einstein stipulates that $t' = t_1 + 1/2\,(t_2 - t_1)$, but this is clearly only a definition and cannot be said to be true or false. Every definition of the form $t' = t_1 + \epsilon\,(t_2 - t_1)$ is admissable, where $0 < \epsilon < 1$. Choosing $\epsilon = 1/2$ merely leads to certain advantages in the sense of descriptive simplicity.

The relativity of simultaneity has often been established by introducing two observers and justifying the difference in time by their difference in positions or states of motion. But this is a misleading presentation. As we have shown, the relativity of simultaneity is connected with the *logical* problem of temporal order, not the *psychological.* The problem of defining time exists in the same way for every observer, and the individual observer by himself has available to him all the possibilities of temporal definition. We simply get a better *intuitive image* of the situation if two different definitions of time are assigned to two different observers. Yet the observer who is at rest is just as able to define time as the observer in motion, although the speed of light will not be constant in his system, but will be different for the two directions along a segment. This is not a matter of a difference in positions, but of a difference in the *logical presuppositions of measurement;* before any measurement can take place at all, these must first be arbitrarily laid down.

For this reason, it is impossible to circumvent the relativity of simultaneity by constructing some suitable mechanism. Some have tried to define absolute simultaneity by means of an electrical mechanism in which a current is closed at two separate points: but closer analysis shows this to be untenable because of the finite speed of propagation of electrical effects. Others have attempted to establish absolute simultaneity by means of the transport of clocks, failing to notice that this device, too, could only produce a *definition,* not any *knowledge.* Furthermore, in order to be univocal, this definition requires a presupposition that can only be tested empirically; this is the independence of the transported simultaneity from the speed of transport. Relativity theory denies this very supposition, and consequently this procedure is not suitable even for a *definition* of simultaneity, let alone the establishment of absolute simultaneity.

18. THE CONNECTION BETWEEN TIME AND SPACE

The relativity of simultaneity makes measurement of space dependent on measurement of time, with respect to moving objects.

A segment at rest – at rest, that is, within the relevant system of co-ordinates – is measured by laying a rigid rod alongside it. A moving segment cannot be measured in this way; that is, we can lay off against it a rod that is moving along with it, but this will only give us the length of the segment relative to its own system at rest, not relative to the original system of co-ordinates. We must therefore give a new definition establishing what is to be meant by the length of a moving segment.

According to Einstein, this definition consists of the following stipulation; the length of a moving segment is the distance between the simultaneous positions of its end points. The segment, then, is projected into the system at rest and the resulting distance measured according to the procedures for measuring segments at rest.

This, too, is a matter of definition, and therefore this conceptual construct is arbitrary. That some such stipulation must be made, however, is not subject to our discretion. Pre-relativity kinematics did not include such a definition because scientists failed to notice that it was needed. From a logical stand-point it was just as necessary then, for without it, it is impossible to see what could be understood by the length of a moving segment.

In relativity theory it is demonstrated that, as a result of the Lorentz transformation, the length of the moving segment becomes shorter than its length when at rest. Once it is clear that two completely different concepts are at work here, the difference in the measured quantities ceases to be puzzling. This contraction of moving segments in the direction of motion has been suitably designated as the *Einstein contraction.* It must be distinguished from the *Lorentz contraction,* which has nothing to do with the relativity of simultaneity and relates to other quantities[55].

The dependence of measurement of space upon simultaneity is expressed clearly in the combining of space and time into four-dimensional geometry, as created by Minkowski. It is there expressed as the dependence of the positions of the directions of the axes of space and time, although the Minkowskian presentation is not to be interpreted in such a way that time becomes a dimension of space. On the contrary, time maintains its characteristic properties in the Minkowski world, its peculiarity finding its most visible expression in the negative sign of the temporal term in the line element.

More precise study shows, indeed, that time is the deeper concept, from which space is to be derived. This becomes evident in light geometry (cf. Section 15), which begins with the measurement of time and moves onward from there to the measurement of space. Spatial distance is measured by means of time, which is necessary for cause-effect transference. Spatial neighborhood, then, means the possibility of rapid causal transference, while spatial contiguity means causal connection. Epistemological analysis doubtless teaches that there is a connection between space and time, but in a different sense from that of the graphic representations generally used to clarify the connection between time and space in relativity theory.

In Section 16, we gave a thorough justification of the realistic conception of space. The preceding considerations demonstrate clearly that time similarly requires a realistic interpretation. There are, to be sure, subjective elements in the coordination – given through the definitions – but, independently of these, the temporal order also contains facts ultimately signifying properties of the causal chains. Time, therefore, is not in some special sense a schema imposed upon nature by the human mind, different in kind from any other conceptual formulation of our knowledge of nature. *The objective significance of time consists in its formulating the type of order of causal chains.* It is, then, a physical theory of a very general nature, but not in any way the product of a special intuitive human faculty.

Causality is the deeper concept, to which time is to be reduced. But since, as we have already established, space can be reduced to time, it follows that spatial order also reduces to the concept of causality. The deepest formulation of the realistic conception of space and time is to be found in the assertion that space and time are nothing other than the expression of the *causal structure of the world.*

19. SUBSTANCE

Along with the concepts of space and time, the concept of substance is also interwoven with philosophical ideas, for it is the most general expression of the concept of existence in physics. By substance we mean what exists as such, that which constitutes the foundation of all happenings: properties and laws are merely relations between elements to which the nature of substance pertains.

For this very reason there is a certain obscurity about this concept. Substance is a philosophical concept, and physics has a healthy tendency to avoid traditional philosophical concepts, replacing them with its own. For it often turns out that the basic concepts of philosophy are nothing more than the physical concepts of past generations which have not participated in the development of physics but which are considered by philosophers, who are far too historically minded, to be eternal necessities. Thus physics prefers to speak of matter, mass, energy, and not of substance, and physics has added certain derivative notions to these concepts, such as force, effect, and momentum, so that only the sum total of this conceptual system expresses everything that is meant in philosophy by the concept of substance. Deep physical discoveries such as those concerning the identity of mass and energy or the combination of momentum and energy in a tensor would not have been possible had a clean separation between these concepts not been made beforehand, and had the vague concept of substance not been abandoned altogether and the new concepts precisely defined without philosophical considerations.

The handling of the problem of the ether provides the most notable instance in which this procedure proved necessary. For here, in particular, it became apparent that the attempt to maintain in all circumstances the concept of substance used in classical physics and in daily life leads to contradictions that cannot be removed by any theory, no matter how complicated. The sort of substance which is presupposed by an elastic ether is indeed nothing other than the macroscopic entities presented to us in solid bodies, liquids, and gases, which we have long recognized as an aggregate of elementary particles of quite a different kind. Yet it is extremely improbable that these elementary particles display, in turn, those properties that, in macroscopic matter, are only properties of the aggregate; and physics has, therefore, been quite right in refusing to regard the macroscopic concept of substance as *a priori* necessary and to make it the basis of the theory of matter.

It has been replaced, in physics, by the concept of a field. Since Einstein, a field has meant a substance which does not have the property of having its particles naturally defined. Atomic matter is quite different; each atom is an individual that can be pursued through time as the identical object. In the language of Minkowski, matter dissolves into a bundle of world-lines, the position of which in the space-time world is determined by matter. We can, to be sure, also imagine a field-like substance being split up into particles (leaving open the question whether the particles touch each other like the volume elements in a continuum, or are separated like atoms, the answer being significant only in another context); but *how* the division into world-lines is to be made is not determined by nature. Within the restrictions of their timelike character, the direction of the world-lines can be selected. While one division results in the particles *a, b, c* . . . , another yields the particles *a*′, *b*′, *c*′ . . . , the world-lines of which intersect obliquely with those of *a, b, c* . . . , so that the particle *a*′ is identical for a time with *a*, then with *b*, with *c*, and so on. The system at rest which is marked out by *a*′, *b*′, *c*′ . . . thus has a different state when in motion from the system marked out by *a*, *b*, *c* . . . This situation constitutes the significance of Einstein's assertion that the extended entity made up of such particles does not possess a defined state of motion. "There may be supposed to be extended physical objects to which the idea of motion cannot be applied. They may not be thought of as consisting of particles which allow themselves to be separately tracked through time"[56]. Such a development of fundamental physical concepts is, of course, only possible when it can participate in the progress of natural science, free from the constraint of stipulations laid down *a priori*[57].

The situation is analogous for the law of conservation. Kant postulated the law of the conservation of matter to be *a priori* necessary, and most philosophers followed him in this point. Physics, on the other hand, thought it more proper to submit this principle to experimental test. The conservation of matter during chemical reactions had already been tested by Lavoisier, who sealed substances in a glass flask and weighed them before and after the reaction. Later, Landolt[58] carried out more precise experiments and found the principle to be confirmed within the margin of error. Furthermore, physics did not hesitate to universalize the principle of the conservation of substance. According to Einstein, only the sum of mass and energy is constant, and we must note that Lavoisier and Landolt, had they been able to make more precise measurements (as is possible today), would have discovered, not a constant mass, but a decrease in weight, for a portion of the energy escapes through the glass in the form of heat radiation.

Philosophy, for its part, has tried to justify the principle of conservation as being *a priori* in another way. According to this interpretation, the principle of conservation is strictly correct, but what the physical thing is that constitutes substance is to be learned only through the gradual progress of knowledge[59]. Thus in modern theory, it would no longer be matter alone, but the sum of matter and energy that constitutes true substance, which takes two phenomenal guises. This conception requires yet further amendment today, for scalar energy has been replaced by the Einsteinian energy tensor; accordingly, substance can only be characterized by means of an entity having the complex nature of a tensor. Admittedly, this appears at first glance to be a possible extension of the older philosophical principle of conservation. But we must not concede that this method of extension will always remain possible *even if,* e.g., the principle of conservation turned out to be a statistical law not applicable to an individual, as is assumed in a theory proposed by Bohr, Kramers, and Slater[60]. It is irrelevant to the philosophical problem that, for physical reasons, this theory is no longer maintained by its authors. Thus it is senseless to want to maintain the principle of conservation as an *a priori* principle; it must be viewed as an empirical result and therefore as subject to all the modifications that may be made in empirical knowledge.

On the other hand, that aspect of the problem of substance concerned with differentiating object and relation does appear to us to be of philosophical significance, as was indicated at the end of Section 6.

20. CAUSALITY

The cluster of problems connected with causality has also not been sufficiently studied for us to be able to give a conclusive presentation. We must therefore content ourselves with sketching out its general features. To begin with, it seems most important to formulate the content of the causal assertion. We must realize that it is not at all a matter of a single assertion but of a large number of assertions, all of which are included in the complex principle of causality.

1. The most general point that can be made about causality is that it asserts the existence of a functional dependence of physical quantities. If *A,* then *B,* is the general form of the causal assertion; thus it represents an implication. A dispute has been carried on as to whether the concept of a cause is exhausted by the concept of a function. While some like to imbue causality with an element of mystery, a metaphysical creation or connection, others have objected strongly to this metaphysical interpretation, claiming that causality is purely descriptive, signifying a connection in no other sense than that signified by the concept of a function[61]. The latter conception is correct insofar as it holds the causal concept to be capable of exhaustive analysis by purely logical means, yet incorrect insofar as it identifies the concept of a cause with the concept of a function. For not every form of function corresponds to causality, which is, rather, a functional connection of a very specific character. It is the job of epistemology to spell out more precisely the determinants of this function. In carrying out this task, we will find that such logical analysis achieves exactly what the metaphysical

conception only dimly sought: it captures just that characteristic of causality that we intuitively sense as 'becoming' and 'connection'. We will attempt to give these more precise specification in what follows.

2. The implication connecting A and B is an *asymmetrical* relation. In order to indicate the order of the two members, we must write it in the form

$$A \rightarrow B\,.$$

We explained in Section 17 the means whereby the direction of the relation is to be recognized. Further, this implication is *transitive* and *connected*[62], and the resulting series has the character of a *linear open continuum.* The points (point-events) of a causal series comprise, therefore, the *field* of a relation among these properties; only then is the causal relation characterized in such a way that a *causal structure* of the world emerges.

3. We assert further that the relation between A and B is *continuous,* i.e., that small modifications in A also produce small modifications in B, if the modifications in A are assumed to be sufficiently small. This is an extraordinarily important property. Without it, the causal relation would be practically useless, since we never encounter objects or events that are strictly speaking the same.

4. We assert further that the elements between which the causal relation holds, that is, the parameters of the events, are *not* the space-time coordinates. To formulate it another way: the coordinates do not appear *explicitly* in the physical equations. This is an assertion about space and time rather than about causality, and can be regarded as a *definition* of the space-time coordinates. We continue looking for factors determining an event until we find those of which we can say that they would produce the same effect in a different spatial position at a different time. As long as we have not found them, we simply say that a field of force, or something of the sort, exists at the pertinent spatial position. However, closer study is required to determine to what extent we have here a mere definition and to what extent an assertion of a restrictive nature about reality.

5. Also included in causality is the idea of an event that spreads continuously so that all intervening points are traversed. This is the *principle of action by contact,* which may be formulated as follows: "After a time Δt lying beyond a particular distance r_1, an effect that begins in a space-point P cannot be noticed in any ray r that goes through P after a time Δt lying beyond a particular distance r_1, where r_1 increases continuously with Δt, and $r_1 = 0$ for $\Delta t = 0$". This is not yet to say that there is a finite upper limit to the speed of an effect, but only to exclude infinite speed. Thus the assertion

that the speed of light is the highest signal speed (cf. Section 17) is not a consequence of the principle of action by contact, but rather an assertion going beyond it.

6. We assert further that the effect decreases with distance: as r_1 increases, the intensity of the effect decreases. This is an extension of the principle of action by contact, which is generally valid for the direction of time also. The decrease of the effect with distance is evident in field theory, since effects spread over ever-expanding spherical surfaces. Yet the situation is different for transported material particles or for needle radiation. Closer investigation is required to determine to what extent this principle can be maintained in these circumstances.

7. Finally, there is one last claim connected with causality that represents an extrapolation beyond its character as an implication, namely, that the course of the universe is strictly determined, that, given a precise knowledge of the state of the world at a time-point t, it is possible to calculate univocally the past and the future. This assertion is known as *determinism.* It is one of those risky consequences that are drawn when the probability nature of knowledge is not given sufficient attention. We therefore replace it with a more modest assertion relating to the method of approximation employed in science: *through more exact analysis of the effective parameters, the probability of the prediction can be increased arbitrarily close to probability* 1. We refrain from any assertions as to the limit and leave open the question whether the probability 1 corresponds to any defined state of the universe. At this point, determinism collapses. Pursuit of this idea leads to a theory of the probability connection of the universe in which there are only degrees of probability: at the same time this sheds new light on the problem of time, if considered side by side with the observations set forth in Section 17. Results of these investigations are to be found in other works by the author[63].

We have listed seven separate assertions that are included in the concept of causality, and we cannot exclude the possibility that there are more. We will now turn to the question whether causality can correctly be called an *a priori* relation.

As a logical relation, it certainly is something stemming from the rational component of knowledge, as do all theoretical components of knowledge. We cannot claim that this relation occurs in reality, for this makes no sense. Instead, we can only ask whether it is possible to apply this relation to reality, whether it can be justified. The idea of the philosophy of the *a priori* must then be formulated so that the causal relation can be maintained under all circumstances, despite any experiential material whatsoever.

We cannot look into this question thoroughly here. In order to confirm the claim, we would have to investigate separately each of the component assertions about causality. We shall simply indicate that this question can only be handled in connection with probability theory, and that causality can be shown to become an empirical assertion once the concept of probability is acknowledged to be a fundamental concept. It is possible to imagine experiences that would force us to give up the principle of causality, at any rate for individual phenomena or in the atomic sphere. Only the future can tell whether physics will take this direction (cf. Section 24).

21. THE ASYMMETRY OF CAUSALITY

A more thorough examination of this fundamental question is required in order to respond to the objections that can be made to the repeatedly established asymmetry of causality. They are related to the problem of the second law of thermodynamics.

According to a widely-held conception, the elementary processes of all physical events are of a *reversible* nature. This is most clearly expressed in the differential equations of mechanics, which are of the second order and which therefore admit, along with every solution for a possible motion, another one that is obtained from the first by the transformation $t = -t'$. According to these, the laws of nature do not indicate any direction of time, and the events could just as well run their course heading from future to past as the other way around. Boltzmann endeavoured to show that this interpretation does not contradict the occurrence of irreversible processes in thermodynamics by giving the irreversibility a *statistical* basis. According to his theory, the asymmetry of the direction of time arises only in connection with *macroscopic* occurrences resulting from mixing processes; for instance, the probability is very great that the course of the diffusion of two gases will take place in the direction of mixing, while a separation, though it cannot be declared impossible, is highly improbable.

This argument has been adopted by a number of philosophers, and the view has been expressed that every indication of temporal direction must conform to the Boltzmann schema[64]. For instance, M. Schlick indicates that the concept of a 'trace' is to be interpreted accordingly; supposing we find imprints of a human foot in the sand, we are justified in inferring the temporally prior presence of a human being only in the sense that it is overwhelmingly probable, for it is merely improbable, not impossible, that the impression in the sand results by chance from the sand's swirling about in the wind. In the same way he interprets the principle of the mark employed by the present author. For instance, if we make a chalk mark on a stone, this process – Schlick would say – is, strictly speaking, reversible; it could happen that, just as we touch the stone, which already has some chalk marks, with the piece of chalk, the particles of chalk on the stone coincidentally become detached from the stone and attach themselves to the piece of chalk. In fact, such an occurrence cannot be regarded as impossible, but only as highly improbable.

The criticism of this conception must take two directions. First we must study the question whether, if elementary processes are reversible, Boltzmann's inference about the irreversibility of macroscopic processes is viable. This, in turn, brings us to the second question: What justification is there for establishing the thesis that the elementary event is reversible?

Let us take up the first question. The author is of the view that Boltzmann's attempt to derive irreversibility at the macroscopic level from reversible elementary processes is to be regarded as unsuccessful. For Gibbs' *Reversibility Objection (Umkehreinwand)*[65], which was published during Boltzmann's lifetime, is quite correct, so long as we confine ourselves to closed systems. If B is an improbable (unmixed) state of a gas and C is a mixed state succeeding it, Boltzmann would say that the probability of the occurrence $B \rightarrow C$ would be very much greater than the probability of the occurrence $C \rightarrow B$. The reversibility objection, on the other hand, runs as follows. If A is a state defined by the fact that we imagine the direction of motion of all particles to be the exact opposite in C, then B must develop out of A; but according to Boltzmann's principles, A is just as probable as C, because the probability of a state is, in Boltzmann's view, independent of the sign of the velocity; therefore $A \rightarrow B$ will occur as often as $B \rightarrow C$. This conclusion contradicts Boltzmann's basic idea.

Clarification of the contradiction requires the use of the concept of *relative probability*[66]. Then we can express Boltzmann's idea as follows. The relative probability of $B \rightarrow C$ is high (i.e. the probability of attaining the state C, *given* the state B), the relative probability of $C \rightarrow B$ low; likewise, the relative probability of $A \rightarrow B$ is low, that of $B \rightarrow A$ high. The reversibility objection, on the other hand, runs: the absolute probability for the occurrence of a segment $B \rightarrow C$ is just as high as for the occurrence $A \rightarrow B$ (or $C \rightarrow B$). Both assertions are correct, and they do not contradict one another.

But if the contradiction is thereby resolved, the designation of the direction of time has been lost at the same time. This is best seen in the following way. Given a state B, we may conclude it to be overwhelmingly probable that, at a later time, the system will be in a state of higher probability, for instance, C. But the reverse is not true: given two states B and C, we may not conclude that B occurs earlier, for it is just as probable that C will occur earlier. Thus only the inference from the direction of time to the event is justified, not the inference from the event to the direction of time. But it is only the second inference that would indicate a direction of time: that the first one offers no indication of direction is best seen in the fact that, if we take state B as a starting point, we must conclude with overwhelming probability that the system was *previously* in a state possessing greater probability.

This will change only if we inspect closely a system in an improbable initial state and concede the origin of the initial state to be due to outside causes, i.e., cease to regard the system as closed[67]. If we take, e.g., a gas system in state B, in which nitrogen and oxygen exist side by side, not mixed, it is far more probable that the system arose by means of separate production of the two gases, for instance, by means of chemical transformations, than that it decomposed on its own through a closed process. Given two states B and C (C being in a well-mixed state), we may infer that B is the earlier state, for the origin of B is not connected to the improbable case $A \rightarrow B$, but may be traced back to outside causes. In this instance the second inference, the one from the states to the direction of time, is justified, and in fact it becomes possible to indicate the direction of time.

The outside causes can be included in the system, in which case the arising in outside causes is synonymous with the increase in probability in the more comprehensive system. If we proceed in this way to the world-system as a whole, we can say that the probability of the world-system increases steadily, thereby indicating a direction of time. If we may regard the total evolution of the world-system as a steadily climbing curve, a directional indicator is indeed present.

A difficulty arises only when we conceive the totality of the universe as a closed system made up of a finite number of reversible elementary processes. In that case, our previous comments apply to the probability curve of the universe, and once again *no* direction is indicated. Thinkers have occasionally tried to extricate themselves from this situation by carrying out the assumption that, at least in the area of a sharp decline of the probability curve, the directional indicator does exist after all, so that, for periods of (admittedly) cosmic duration, the direction of time may be defined as climbing upwards. According to this theory, for instance, a different race of men living later than ourselves as seen from our vantage point and following a branch of the curve headed in the diametrically opposite direction would define the reversed direction of time as positive[68]. Nonetheless, the present author cannot embrace this solution. If it is possible for the direction of time to change during the course of events, it no longer makes sense to speak of events, of something evolving, in the sense of a development; the concept of becoming is forfeited, and the temporal course of events takes on a statistical character.

This answers our first question. If what happens in the world consists of a finite number of reversible elementary processes, then the indication of a direction of time does not hold. We could, to be sure, still ascribe to causality a 'relative probability asymmetry' (given by means of the difference between the relative probabilities of $B \to C$ and $C \to B$), but this is not sufficient to designate a direction of time. The second question, concerning the justification of the presupposition of reversibility, still remains to be answered.

We could dispute the assumption that the number of elementary processes is finite, but it would be easier to cast doubt upon the reversible character of elementary events. For it is an idea that arose essentially through the influence of the mechanical view of the world, which can only be maintained in conjunction with the idea of determinism. There can be no strong reason for maintaining this in the age of quantum mechanics[69]. Epistemologically speaking, it would be sounder to conceive events in the universe through a schema of laws that do not contain any idealizations, such as the idea of determinism, that extend beyond the course of what can be experienced. The present author has developed just such a schema on the basis of the concept of probability[70]. The indication of the direction of time by means of causality appears to be a very elementary fact, and we will do better to order our conceptions of the elementary event in such a way that this fact is preserved.

We should point out that, in substituting the concept of probability for the concept of causality, we do not have to use the *idea of mixing processes* in particular to establish irreversibility. It suffices that there exist probability chains such that the absolute probability that the chain $B \to C$ will occur is very much higher than the absolute probability that the chain $C \to B$ will occur; we may then assume with probability that B is the earlier event. Although the concept of mixing processes emphasized by Boltzmann establishes similar relations, it is nonetheless not *necessarily* presupposed by these relations. Its preponderate usefulness probably stems principally from the fact that the macroscopic world happens to be an aggregate of many minute particles in which, therefore, mixing processes play a large role. A simple mechanical model for indicating direction by means of probability without appealing to mixtures can be set up in the following way. If we take a Galton board (a board covered with nails standing upright), set it aslant, like a writing desk, and shoot a marble sideways into it from the left-hand side, the marble will describe a curve similar to the probability curve of a gas system. (It may occasionally roll backwards, too, but that is irrelevant.) The curve thus described by a single marble defines the direction of time just as well as the gas system. It yields differences in relative probability for the sequence of higher and lower positions of the marble, just as the gas system does, and differences in absolute probability only where the curve as a whole reaches the lower positions, for instance, when we take note of modifications caused by friction. Another way of indicating temporal direction by means of the concept of probability, making use of three events, was presented by the author in an earlier publication[71].

The asymmetry of the causal relation laid out in Sections 17 and 20 is to be understood in this sense and can therefore be strictly justified only by appealing to the concept of probability[72]; we might call it an 'absolute probability asymmetry'. But *if* we are going to take the purely causal approach at all, eliminating the concept of probability by setting small probabilities equal to zero, it is justifiable, and even necessary, to represent it by means of an asymmetrical schema such as we have presented with the assistance of the mark principle[73].

22. PROBABILITY

We have already made extensive use of the concept of probability in our presentation. For what we have here is a basic concept of knowledge, the form of the concept of truth that is employed by physics, and it is therefore impossible to treat it simply as a special problem. Nonetheless, we will discuss here the particular application that has been found for the concept of probability in the statistical laws of physics.

The concept of probability has entered physics at two points. On the one hand, it was introduced along with the theory of error, which was constructed principally in connection with astronomical problems. On the other hand, it entered in conjunction with the kinetic theory of gases in thermodynamics, celebrating its greatest triumph in Boltzmann's exposition of the principle of entropy. From that point it was taken up into the theory of matter and today constitutes an indispensible component part of physics. *Statistical* regularity has taken its place alongside causal, or *dynamic* regularity[74].

In addition, however, there is a mathematical theory of probability, the probability calculus. It was developed in connection with games of chance – that is, through practical application and, therefore, through physical problems – but it then came to treat these physical applications more as schematic examples and took on the dimensions of an independent mathematical discipline. This mathematical theory of probability has no special epistemological problems. It sets up certain initial axioms, which, like the axioms in geometry, are viewed as being in the nature of definitions, and nothing more. It develops the consequences from these axioms in accordance with the usual logical-mathematical methods and is interested solely in this relation, the comprehension of which, of course, requires a high degree of mathematical acumen. Mathematical probability calculus is, then, a logically strict science, the results of which are certain in just the same way as the results of geometry; its concept of truth is that of strict logic and not, for instance, the concept of probability. Thus the genuinely epistemological problem begins only after the mathematical questions. Are these axioms applicable to reality? – that is the epistemological question, which also relates to the application of the probability calculus in physics. If this question is answered in the affirmative, then physics can, of course, adopt the entire logical structure of the probability calculus.

The connection is just the reverse in the actual practice of science. Physics does not start out by asking for the axioms; it takes on the probability calculus as a whole and works with it. Thus it is convinced of the validity of the axioms. Acknowledging the fact that the laws of probability are applied in physics, we must go into the question of what justifies this procedure: in using it, what claim does a scientist make about reality?

Some have answered that using the laws of probability is never more than an expedient – albeit a necessary one – to which we are driven by ignorance and which, under ideal circumstances, physics would be able to avoid. This is the *subjective theory of probability.* According to this view, for instance, Boltzmann's principle is only a makeshift aid; we would have no need of the concept of probability if we could follow precisely the motion of the molecules. The subjective theory of probability was founded principally by C. Stumpf[75]. It bases the judgment of *equally probable* cases (usually referred to by the less happy phrase *equally possible* cases) upon the *principle of insufficient reason,* according to which we call the six sides of a die equally probable because we have no reason to prefer the one side. An opposing view, the *objective theory of probability*, has been put forward by J. von Kries[76], E. Zilsel[77], and the present author[78]. The proponents of this theory grant, of course, that we are frequently unable to carry out a precise calculation, but they refuse to infer from this that the concept of probability must be interpreted subjectively. For even if we ourselves were able to follow exactly the elementary processes in, e.g., a game of dice, we would not make our calculations for the overall processes on any basis other than the equal probability of each side of the dice[79]. Similarly, if we had a precise knowledge of the path of every gas molecule, we would still base our calculations on the fact that the gas passes from a state of lower entropy into a state of higher entropy. For experience shows that the probability calculation is *confirmed* by reality; thus Boltzmann's principle asserts not only that we know nothing more exact, but also that its own contents are nonetheless correct. That we are able, by using probability laws, to succeed in making accurate assertions about nature is proof that we are dealing here with something more than a lack of knowledge, and that we possess instead, in the concept of probability, a very positive form of knowledge. Why do we lay down as equally probable precisely those conditions that correspond roughly to the ergodic hypothesis? And why do we distinguish between loaded and true dice? Nothing follows from the principle of insufficient reason; we have, to be sure, no reason to select any one side of a true die, but this fact still gives us no ground to declare all the sides equally probable. On the contrary, we are led to assert the equal probability upon a very positive basis i.e., the homogeneous construction of the die. If preliminary suspension experiments demonstrate that the die's centre of gravity is not in the middle, we would certainly not assert equal probability for all sides. In accordance with quite definite principles, we select certain cases as equi-probable and express the view that this assumption will withstand the test of experience. The laws of probability, then, have an objective character. Only an objective theory of probability can do justice to the physical facts, and it will be the task of this theory to lay bare the presuppositions made with the assertion of statistical laws of nature.

The objective theory of probability has by now attained broad acceptance; most notably, it has been embraced by E. Kaila[80] and R. von Mises[81] in their more recent writings. On the other hand, J.M. Keynes[82] has further developed the subjective theory of probability. In what follows, we briefly set forth the results of the objective theory.

The assumptions needed for the verification of probability distributions divide roughly into two types, those concerning *causal factors* and those concerning *probability factors.* We can illustrate this by using the example of a game of roulette. The causal factors influencing the equal frequency of red and black are the equal sizes of the sectors, the equality of their number, and, further, the overall arrangement of the mechanism; but these alone can never effect the equal distribution, for a presupposition about probability must also be made. In this instance, this condition can be adequately filled by the assumption that the frequency of all values Ω for the angle of rotation of the pointer (reckoned by multiples of 2π) is regulated by a continuous probability function $f(\Omega)$ in an otherwise arbitrary form[83]. If we imagine the curve $f(\Omega)$ sketched into a system of orthogonal coordinates, the red and black sectors effect a division into narrow bands of ordinates possessing the equal width $d\Omega$. The probability of landing on red will then be equal to the sum of the bands 1, 3, 5, etc., while the probability of landing on black will correspond to the sum of the bands 2, 4, 6, etc. Since, owing to to the continuous nature of $f(\Omega)$, any two adjoining bands are of almost equal size, these sums are almost equal to each other. Thus the equally probable cases are a result of the interaction of causal factors with a law of probability which does not itself contain any assumption about equality of probabilities. The sole role of the causal factors is, then, to impart a definite form to the distribution of probabilities. The equality of the sectors causes red and black to occur with equal frequency. If, on the other hand, the red sectors were twice as big as the black, this causal factor would have, in conjunction with the same probability function $f(\Omega)$, the effect of making red occur twice as often as black – as is immediately apparent from the given derivation. Under certain circumstances, the causal factors may also cause a probability function $\varphi(\omega)$ deduced from $f(\Omega)$ to take on a quite definite form, independent of the form of $f(\Omega)$. Thus the probability function $\varphi(\omega) = \text{constant}$ holds in roulette for the angle of rotation ω of the pointer, enumerated only as far as 2π.

The distinction made here is vital for the problem of probability. A *metrical* probability assertion has been reduced to a *topological* one; the only genuine probability axiom required is the assumption that there is a law of probability, and its special form can be reduced to causal factors. Thus we have rendered an account of equally probable cases. The assertion that certain symmetrical cases, such as red and black in roulette (or the sides of a die, for the proof for the die can be given in just the same way as the proof for roulette) are equally probable is stripped of any puzzling or mysterious characteristics, for we are now able to *explain* it. We can now perceive clearly the senselessness of the principle of insufficient reason. We assert the equality of probabilities, not because we know *nothing*, but precisely because we do know the intervals $d\Omega$ to be equal. Symmetry is a testable causal factor, which must produce a certain metric of the distribution of probability if the topological probability assumption is valid.

It is upon the basis of the bipartite division into probability factors and causal factors that we are able, conversely, to infer from a given statistic – that is, from an observed particular form of probability distribution – the existence of a definite causal factor that brought about this particular form, e.g., from the prevailing frequency with which one side of the die occurs to the fact that the centre of gravity is not in the middle. This is the justification for using statistics to detect the presence of causes with weighted effects, both in physics and in other sciences, such as social statistics. The observed distribution stems from the interaction of a probability law of random distribution and causal factors. Taken by themselves, the causal factors could never bring about the statistical regularity; it is only as a result of the connection with a law of probability that they are able to determine a particular distribution and, as a consequence, to be known from these in turn.

It is an important task to separate through axiomatic investigations those assumptions underlying the statistical laws that are strictly probabilistic from those that are causal. To date, this task has been carried out in only a few fields. For instance, the aforementioned assumption of the existence of a continuous probability function has proved adequate for games of chance and for the theory of error. The latter is especially noteworthy. The Gaussian exponential function is a special form of probability function, and it is demonstrable that it need not itself be presupposed as an axiom, but can be reduced to probability functions of one form or another in the case of elementary error, given that many elementary errors of the same order of magnitude interact[84]. This achieves for the Gaussian function something similar to what was achieved for the function $\varphi(\omega)$ described by the roulette pointer in the earlier example: the *metrical* probability assumption of the exponential function is reduced to a *topological* probability assumption, and we have rendered an account of equally probable cases.

As for the kinetic theory of gases, we are concerned with investigations being carried out in connection with the *ergodic hypothesis.* This hypothesis, too, is unsatisfactory, because it presupposes a definite form of the probability function. It asserts that the frequency of the states of the gas is given by means of a certain function $f(x_1 \ldots x_n)$, the so-called ergodic density, where $x_1 \ldots x_n$ designate the canonical parameters of a system on the energy surface of phase space. It would thus be most helpful if we could also divide this metrical probability assumption into a purely topological probability assumption and the adjunct causal factors, which are precisely what provides the special form for the case under discussion. Investigations of this sort have been presented by E. Zilsel[85] and by R. von Mises[86]. Zilsel set forth clearly the epistemological problem connected with the principle of entropy and replaced the ergodic hypothesis with an *allogodic hypothesis,* which formulates the tendency in nature toward change. Further study is required to determine the extent of the success of this effort, which involves a special exposition of the basic idea presented in Zilsel's earlier essay on the 'problem of application' [see note 77]. Von Mises' studies employ the idea of presenting a particular metrical probability distribution as derived from primary probabilities, the numerical values of which are immaterial. Insofar as this process is successful, it, too, achieves the reduction of a metrical probability assumption to a topological. Unfortunately, however, the execution of this principle for the ergodic hypothesis has not yet been completely successful[87].

We are able to recognize, quite independently of the carrying out of this special investigation, that the laws of probability contain a particular model of regularity in nature that does not fall under the model of causality. Certain natural phenomena would be completely incomprehensible if we did not describe them in terms of probability laws. Specifically, this applies not only to the mechanics of molecules and their component parts, but according to von Mises[88], phenomena of this nature occur also in the mechanics of macroscopic bodies, particularly in the flow of liquids. As a consequence, the probability principle appears as an independent principle alongside the principle of causality; only the two principles taken together constitute the general assumption concerning regularity in physics. While causality makes a claim regarding the extension of an event in the linear direction of time, the principle of probability entails an assertion about a temporal cross-section of events, about the frequency of the initial and the boundary conditions of causal chains. The probability principle, as the *principle of regular distribution,* takes its place alongside causality, as the *principle of regular connection.* The precise formulation of this principle of distribution still requires special study[89]; if it is demonstrable that the given principle of the probability function is a sufficient assumption in every case, the import of the principle of distribution is that *conditions differing only infinitesimally from one another occur equally often.*

The division of regularity in nature into causality and probability corresponds to a classification of events in nature the principal significance of which is often misunderstood. It is always the preponderate factors alone that are singled out as the causal factors in events, but alongside these we find an inexhaustible remainder of minute influences that come from all corners of the universe and cannot be excluded, for closed systems cannot be realized. Consequently, every event is determined by rationally comprehensible factors plus an irrational remainder which, while it can be further subdivided, can never be exhausted. It is often held that this latter influence can be excluded by regarding causal laws of nature as being in every case valid only within certain boundaries; but this view is incorrect. It is impossible to ignore the possibility that someday the irrational remainder may bring about errors of arbitrary magnitude. Causality would not conflict with this possibility, which therefore cannot be disregarded, but only declared *improbable.* Thus the influence of the irrational remainder – and this is, precisely, the basic principle of the theory of error – can only be formulated in accordance with a probability assumption: *The irrational remainder influences events in such a way that the numerical values of all rational factors are subject to the laws of probability.* The principle of probability, then, entails an assumption concerning the irrational remainder of all occurrences, while the principle of causality presents an assumption about the rational factors; only the two taken together determine events in nature.

This applies to every individual event, and it is correct to say that the principle of causality would be worthless without the principle of probability, for there is no case in which it alone is applicable and therefore no case in which, taken by itself, it has predictive powers. We here meet up with the chain of ideas that guided us in characterizing the physical concept of truth in Section 8, and which takes account of the merely approximate nature of all knowledge about nature. If, in addition to this interaction of causality and probability in each individual case, we also acknowledge special statistical laws of physics, such as the principle of entropy, it is because here the probability regularity, with its universal influence, finds especially obvious expression as a result of particular relations in certain cases. Let us again take the example of roulette; if we put the mechanics of the roulette pointer into action, the probability function $f(\Omega)$ will express itself in our ability to fix the theory of error for the calculated value of the angle of rotation, but if we play a game of chance with the roulette wheel, we have created, by means of particular causal factors, such as equal red and black sections, a situation which turns the same probability function $f(\Omega)$ into a statistical law that takes its place alongside causal regularity as being a special form of regularity in nature.

Despite this explication of statistical regularity, the concept of probability still contains a special puzzle. Using the principle of the probability function, we reduce the probability concept of the statistical laws to that found in every causal assertion in physics – yet the probability concept itself is still not completely clarified by this process. While it is demonstrable that lawfulness of probability is a necessary condition of knowledge of nature, this does not really explain why it applies. For we have no means of proving that natural-scientific knowledge, even if possible heretofore, will always be possible[90]. For a discussion of the formulation of the presuppositions attendant on the concept of probability, we must refer back to Section 7, where we previously encountered this problem.

While we have developed the parallel nature of causal regularity and probability regularity, it is possible to imagine a further extension of the analysis in which the parallelism is abandoned and the concept of probability is regarded as the more fundamental of the two, leading to cause-like processes only in macroscopic processes. Consider, in this connection, the evolution of the law of entropy. It originally appeared as a purely causal law in thermodynamics, yet in the course of further development it revealed itself as a statistical law presenting the macroscopic form of many interacting elementary processes. We cannot exclude the possibility that this will turn out to be the fate of all causal laws[91]; recent conditions in quantum theory have, indeed, made a reality of this conjecture expressed earlier by philosophers of science. No *a priori* pronouncement is possible, of course; we must await the judgment of physical experience (cf. Section 24).

In no case should such a development be regarded as a failure of physics, as an unsatisfying lack of precision in our knowledge of nature. Probability laws are not 'less precise' than causal laws; in its application to reality, every single causal law contains the very concept of probability found in the statistical laws. There is no such thing as certainty in knowledge of nature; there is only probability. What causality has over statistics is simply the ability to ascribe a *higher degree* of probability to an *individual event,* while statistics is able to make predictions with a *high degree* of probability only for an *extended series of events*[92]. We cannot predict whether it will always be possible to single out individual events of a high degree of probability from the total reality in the atomic sphere, for this depends upon the nature of reality. Relations in nature are not subject to our wishes; no matter how attractive our concepts of precision are, their applicability is not under our control. All we can do is modify our conceptual system so that it reflects nature as faithfully as possible. Only future developments will teach us to what extent this is possible and in what, in the end, this 'faithful reflection' consists.

23. THE SIGNIFICANCE OF INTUITIVE MODELS

In the preceding discussion, we analyzed the method of physical knowledge from the standpoint of its logical structure. We paid little attention to the intuitive models generally employed by the physicist as a valuable aid in both the process of research and the process of comprehension, and which now call for investigation[93].

Consider, as a beginning, the models of space and time. Work on space-time concepts is always accompanied by intuitive images of the concepts, which, according to our earlier discussion, can only mean that whenever we are operating with the conceptual systems of space and time we have before us images of the perceptual experiences we have when occupied with rigid rods, light rays, and causal processes. The intuitive nature of Euclidean geometry and absolute time is based, then, upon the fact that deviations from them fail to show up precisely in daily experience – that is, in lower-level facts. As a result, we are accustomed to these conceptual systems and find abstract mathematical thinking psychologically easier if we can recall pertinent illustrations in our perceptual experience.

However, intuitive models are also to be found elsewhere in physics. A thermodynamic diagram, or the characteristics of an electron tube, are intuitive models, and anyone working with thermodynamic machines or electron tubes has before him the intuitive image of this diagram, thus imbuing his technical manipulations with a content that is theoretical and nonetheless intuitive. In this connection we should also note the models constructed for atoms, molecules, and crystals. By using the modest term, 'model', we wish to indicate that the image by no means corresponds in every respect to reality and that we regard as very much open to question the extent to which it is possible at all to create intuitive images in these cases.

Aware as we are of their analogical character, such images seem to be unavoidable; indeed, there are physicists who hold that the acquisition of intuitive models are a necessary constituent of theoretical knowledge, attributing to it still greater significance than they do to mathematical formulation. Conversely, others lay the greatest emphasis upon mathematical formulation and regard intuitive formulation as a mere aid to learning that itself possessses no cognitive value. We must now consider to what extent the epistemological ideas we have worked out above place these views in a new light.

We know that the system of concepts and relations we attribute to nature cannot be interpreted as being, as it were, a photograph of nature, but simply refers to objective situations by means of its type of order. Thus there is, as such, no objection to employing models that reflect nature in only some respects, for complete agreement would be an impossible demand. But we must still investigate more thoroughly just what is involved in models such as those mentioned above.

Consider, for instance, a thermodynamic *p–v* diagram. Pressure and volume are given as the coordinates; i.e. each is *coordinated* with one of the dimensions of space. The thermodynamic relation between pressure and volume will then correspond to a geometrical relation between the two dimensions of space, and if we wish to have a clear image of this relation, we need no longer use the entities, pressure and volume, as a basis, but can satisfy the relation by means of the intuitive content of spatial elements. Now pressure and volume are, to be sure, themselves intuitive concepts, yet the purely geometrical concepts are not familiar to us, especially where the relation is of a more complicated character, and herein lies the *value* of a geometrical diagram. The *possibility* of constructing it, however, rests upon the fact that the same 'framework of relations' is suited both to the entities pressure and volume and to the dimensions of space. The discovery of axiomatic mathematics that the system of geometrical axioms does not necessarily prescribe the use of such intuitive elements as point and straight line, but can equally be satisfied by many other elements, has found extensive use in the diagrams of physics.

Obviously, the applicability of geometry to physical space depends upon certain characteristics of rigid bodies that satisfy the geometrical axioms. Adding this to the preceding results, we find that *p–v* diagrams assert the existence of an *analogy* between two realms of reality, namely, between the entities pressure and volume on the one hand and rigid bodies on the other. The analogy consists in *the validity of the same framework of relations in both realms of reality.*

That the process in question renders an intuitive image of the one realm can obviously stem only from the fact that the other realm is more familiar. This applies particularly in those cases in which, by analogy, higher-level facts are represented through lower-level facts, and this is, therefore, the commonest procedure in the construction of intuitive models. For in daily life we deal solely with lower-level facts, and we experience the attendant habituation as intuitive [clarity]. On occasion the relation may be reversed, as when an individual's activities happen to entail a greater familiarity with particular higher-level facts. An example of this dual possibility is to be seen in the analogy between the laws of an electric field and currents in a liquid. This analogy completely fits the above description; both spheres of reality are governed by the same conceptual relations – in this instance, equations. Thus it is common to 'intuit' the electrical field via water currents. Yet anyone who deals extensively with electrical apparatus but has little to do with hydrodynamics reverses the process and 'intuits' hydrodynamic processes by using electrical models.

While the merely analogical nature of diagrams is so well known that no one believes p and v to *be* spatial coordinates or thermodynamic work to *be* a plane section, the situation is different with the model of the atom. In this instance the connection appears to be closer, for the atom itself is an entity in space, just like the model, and some people doubtless believe that the model is a *similar* representation on a large scale. Nonetheless, what we have here is again a mere analogy between two realms of reality in the same sense as before. The model is constructed out of rigid bodies and wires; the atom, while it exists in space, does not consist of rigid bodies but, on the contrary, is itself the constituent of a rigid body. If we nonetheless speak about similarity in this connection, we can only mean that the same framework of relations applies to macroscopic and to atomic material. The model is not a 'photographic enlargement' of the atom; rather, the relation between the two is the same sort as that between water currents and the electric field.

It is because of this analogical character that diagrams and models can be correct only with certain restrictions. To begin with, complete agreement between the conceptual structure of two spheres of reality is simply not to be expected. This is why physicists have hedged their models about with so many proscriptions. For instance, we are forbidden to think of molecules as having color, like the wooden balls in the model, or of the chemical valence as being concentrated in particular ray-directions, although they are presented in this way. The analogy is valid only in certain respects, but it *is* valid in these respects, and there is nothing to stop our using them along with the necessary restrictions. It is not wrong to picture a gas as being like a cloud of dust, the particles of which swirl about and collide with one another.

This image certainly captures some essential features of gaseous matter, and we need only be careful not to make use of such features as cannot be transferred to a gas. Thus the idea that models are superfluous appendages is excessively pessimistic. If, by means of an analogy with the structure of relations in a familiar sphere, such as that of colliding billiard balls, we are able to discover the structure of relations in an unfamiliar sphere, such as that of gases, then the coordination of the structure of relations with the new sphere is an established result that is correct, at least in its general features. The only error is to introduce into the new sphere too many details that do not make any sense there (as, for instance, Boltzmann's representation of the collision of molecules).

On the other hand, we realize it is quite impossible to demand *a priori* that we restrict ourselves to familiar models. It can easily happen that a model correct in its general features nonetheless fails so badly in its particulars that we must employ, for the understanding of them, entirely different models having nothing to do with the original model. Once we are quite clear as to the fact that a model is nothing but an *analogy between two spheres of reality,* this phenomenon becomes totally comprehensible. This point applies particularly to the case of the atomic model. It has often been held that it must always remain possible to construct atomic models of the usual sort, that we must continue to modify these models until the relations between the wooden balls of the model correspond exactly to the relations between the atoms in the molecule or the electrons in the atom, as the case may be. But this view rests upon the assumption that the geometrical relations are independent of the rigid bodies and that they must therefore be equally valid in the most minute spheres. This assumption is unjustified. It is quite possible that the relations in the atomic spheres are so completely different that it is no longer possible to coordinate them with the geometrical relations of the rigid body, no matter how many restrictions are introduced.

Experience with quantum mechanics arouses the suspicion that this is indeed the case, and Bohr expressed the view that "... in the general problem of quantum mechanics ... we are faced with a profound failure of the space-time models by means of which we have always sought to describe natural phenomena heretofore"[94]. This assertion is to be understood in the sense described here; it can only mean that the logical framework of relations for macroscopic and for atomic matter are no longer comparable.

And yet it would be a mistake to speak here of a loss of spatial intuition. When physicists find that, despite all restrictions, a model is no longer adequate, they usually decide that they will have to do without any intuitive models from then on. But it turns out that, after a while, they again give the sphere in question an intuitive form, simply coordinating it with *new* models and using a *different* analogy. This process of making recently opened-up spheres of reality intuitive with the help of analogies has proved very fruitful and must be deeply anchored in the psychological nature of human thinking, which is only able to carry out an extension of its territory by relating the new elements to the old after the manner of an analogy.

On the other hand, it is clear that we must not attribute an excessively great epistemological value to models. We require them principally for the process of absorbing new knowledge. Once we have been dealing with the new area for some time, we gain enough familiarity with the new elements to be able to use them themselves as the contents of the conceptual relations and need no longer intuit them with the help of a model. The epistemological value of a model consists solely in the fact that the same conceptual relations pertain to two different spheres of reality. Which of these is to be regarded as primarily intuitive is a matter of habit.

Along with the aforementioned admissible models, we also find inadmissible models in natural science. For instance, we may make a physical force intuitive by likening it to the feeling of power we have in straining our muscles, or we give the concept of motion an intuitive form by imagining the feeling of motion that we have when we move our bodies in certain situations, or we think of a causal chain as a sort of mysterious coming-into-being. Such models are nothing but forms of *anthropomorphism,* resting upon the totally meaningless assumption that events in nature must be thought of in just the same way as the psychological experimental events that we find in ourselves. 'Intuitions' of this kind can only do injury to scientific comprehension; indeed, we can rightly declare that exact natural science only became possible after such primitive notions were overcome. As the preceding investigations have shown, a knowledge of nature means the coordination of a conceptual system with reality, and nothing but purely logical concepts are to be permitted within this system of relations. He who finds this manner of comprehending nature inadequate may seek satisfaction in poetry and in painting – may seek it, in a word, in those activities that do justice to the emotional experiences of man. But anyone who has pursued natural science in its strict form and has ever become deeply immersed in it knows that here, too, lie experiences, not inferior in strength or profundity to those arising from the artistic enjoyment of nature, that blossom fully only in the pure atmosphere of logical lucidity.

24. THE EPISTEMOLOGICAL SITUATION IN QUANTUM MECHANICS

The expansion of this new discipline in the last few years requires a special summary of the attendant epistemological problems, since epistemological viewpoints play a major role in this theory. We can distinguish three groups of problems: the problem of the minimum number of parameters, the problem of models, and the problem of regularity.

The first group of problems was broached by Heisenberg in his first work on quantum mechanics[95], in which he put forth the requirement that unobservable quantities be omitted from descriptions of nature. He mentioned, in particular, the phase of the electron in its revolution around the nucleus as being an unobservable magnitude, and admitted essentially only frequencies and intensities of the emitted radiation as observable quantities. The term 'not observable' is not very aptly chosen, for we know from our discussion of the ranking of facts by levels in Section 9 that no physical quantity may be called directly observable. Indeed the frequency of the emitted light is not actually 'observed', but is inferred in a rather complicated fashion from the light and dark lines on photographic plates. Heisenberg's idea must be correctly reformulated as the stipulation that *dispensible* quantities should be eliminated from the description of nature, i.e., that we should regard as substantiated by nature only those quantities that are *necessary* for the theoretical interpretation of perceived experimental findings. Given this form, it is a recognized principle of physical research that has found frequent application in the past. For instance, we can view the Machian and Einsteinian attacks on absolute space from this standpoint, and the opponents of the atomic theory appealed to it – largely erroneously, to be sure – in that they believed themselves able to interpret the observed phenomena without using the atomic hypothesis. This principle is to be called, more precisely, the principle of the minimum number of parameters: Of two physical theories about the same factual situation (that is, about the same lower-level facts), the one employing fewer parameters is preferable.

The justification of this principle is related to the nature of inductive inference, which may be regarded as the reverse of strict implication. If we know that event C follows necessarily from event A, we are able, conversely, to infer from the presence of C the probable presence of A. The same is true if, in another instance, the event C necessarily follows from the two joint events A and B, in this case, we may infer from C the probable presence of A and B. However, one restriction must be laid down: if we know that A

and not-*B* also necessarily give rise to event *C*, then we can no longer infer *B*, but only *A* from *C*.[96] *B* then becomes a dispensible parameter, the calculation 'running idle', as it were, at the place at which it appears.

We can see that this minimum principle, much as it resembles a 'principle of economy', has nonetheless nothing to do with economy. Only considerations from probability theory play a role here.

We are not going to investigate here the question of how much justification there is for applying this principle to the phase of the revolving electron. However, Heisenberg himself evidently regarded his initial claim as too sweeping, for he has recently described basic methods of observation which allow the position of the electron to be determined. It is only for the interpretation of the spectral series that the position of the electron is a dispensible parameter; in other experimental situations it is not.

Our treatment of the second group of problems, the problem of models, may be connected with the discussions in the preceding section. There we have shown that the possibility of an atomic model, i.e., a depiction of the interior of the atom through macroscropic relations, cannot be required *a priori.* Here we wish to present current findings concerning the pertinence of such intuitive models.

Views about models have varied considerably, for they involve the *interpretation* of mathematical theories, which, as we have seen, is not directly given along with the theory. The *matrix mechanics* founded by Heisenberg, Born, and Jordan[97] takes the view that the construction of a model is to be avoided in principle. Nonetheless, a form of kinematics was developed along with other facets of the theory, and it remained an open question whether the equations designated by that name were not, after all, directed at hidden images of space-time events. Schrödinger[98], on the other hand, hoped to be able to preserve by means of *wave mechanics* our elementary images of space-time events. He traced the difficulties connected with the Bohr model back to erroneous ideas regarding the arrangement of matter *in* space; that is, he traced the uncertainty of all science regarding the strict location of the electron in space back to the fact that the electron does not exist at all as a sort of corpuscle-like construct, but is instead to be resolved into a field filling the whole area around the nucleus. This interpretation would, indeed, preserve the space-time model in its entirety. However, new difficulties arose for Schrödinger in that the wave event is embedded, not in the space of three-dimensional coordinates, but in higher-dimensional parametric space. The strict interpretation of his fundamental idea that the motion of the mass-point is really an interference process of electric fields would have had to give rise to a reinterpretation of parameter space as 'real' space[99] – a consequence that no one was prepared to draw. And after Born[100] developed the idea that Schrödinger's wave fields are not fluctuations of electric density but fluctuations of the state of probability in the vicinity of the nucleus, the electron maintaining its corpuscle-like nature, all hope vanished of using this route to maintain the space-time framework for atomic events.

But then Heisenberg began to develop considerations that largely supported the spatio-temporal character of the atom. In connection with the results of collision experiments, he worked out the idea that the position of the electron at a particular time can be established with any desired degree of precision by observing the electron in, say, a 'γ-ray microscope'. According to this idea, atomic events would once again have taken on a 'model character', had it not been for a singular complication pointed out by Heisenberg which is essential precisely for the quantum character of natural events.

In order to determine the position with the greatest possible precision, we require light of shorter wavelength, such as γ-rays, but light of this sort has high energy quanta $h\nu$, and therefore illumination with such light throws the electron off its orbit. Thus it is impossible to observe the same electron in its path within the atomic complex more than once. At the same time there is a discontinuous change in the momentum that makes impossible a precise calculation of the momentum. On the other hand, if we select radiation of long wavelength light, the momentum will not undergo much disturbance, but the determination of the position becomes indefinite instead. In general, we can determine with precision *either* the position *or* the momentum[101]; this relation is known as the *Heisenberg indeterminacy relation.*

Some have attempted to characterize this curious state of affairs by saying that, in contrast to the situation with macroscopic events, the *influence of the instruments of observation* cannot be ignored; but this explanation is not viable. First of all, the influence of the means of observation is taken into account for macroscopic observations, e.g., measurements of temperature, as well and is expressed in the form of correction factors. The division into 'independent event' and 'instrument of observation' is an idealization that has grown out of certain macroscopic events and is not entirely applicable here; it has shown itself useful in the acquisition of a simple description of nature without being a necessary presupposition of knowledge. The general procedure of natural science consists in inferring the existence of objective things from perceptions by means of theoretical approaches, as was worked out in Sections 6 and 9 and expressed schematically in (1), Section 6. It is not necessary to set aside one portion of the objective event as the 'instrument of observation', minimizing by artificial means its influence upon the results of perception. It suffices, rather, to design a comprehensive theory which embraces equally all objective influences.

Applied to quantum mechanics, this approach has the following import. The disturbance of the electron by the light used for observation would be insignificant if, by means of theoretical calculations and taking due account of the impact of the light, we were able to establish the position of the electron at the beginning of the impact and the accompanying momentum. But because, according to Heisenberg and Bohr, this is impossible, we are faced with a difficulty. The difficulty is to be understood in this way: *It is impossible to make a univocal inference concerning the events at hand from the perceptual findings.* We can, to be sure, by choosing the experimental conditions, select perceptual findings such that the inference regarding *either* the position *or* the momentum coordinate is sufficiently unambiguous, but then there remains for the other coordinate, whichever it may be, an interval within which it cannot be unambiguously determined. This is, indeed, an entirely new kind of restriction to our knowledge of nature, the existence of which was never before suspected.

Yet we would find it very unsatisfying to rest content with the thought that this is a limit only of our *knowledge* of nature, while objective events continue to take place with strict determinacy. Our aim must be to find a formulation that transfers the indeterminacy to nature and acknowledges *objective determinacy* only to the extent that there is also *subjective determinability.* Refusal to extend the application of Heisenberg's idea in this direction would constitute a violation of the *principle of the identity of indiscernibles*, for a justification of which we may appeal not only to Leibniz but also to its constant application in modern physics, especially in Heisenberg's own quantum mechanics, where it appears as the principle of the elimination of 'non-observable' quantities (cf. the beginning of this section). We suspect that the solution will consist in maintaining a merely probable connection between position and the momentum coordinate, so that the objective 'thing' is no longer characterized by means of a strict combination of the two parameters, but solely through a probability function that ascribes a certain probability to each combination and which, in special cases such as 'observations made with light of a long (or a short) wavelength', can be split in such a way that the two-dimensional domain of probability (one for the position parameters, the other for the momentum) develops into a one-dimensional band.

It would appear, accordingly, that the macroscopic *concept of space* can be adopted for atomic events and that only the *regular determinacy* of events needs to be conceived of in a new way. Whether the first of these claims is accurate remains an open question; the extent to which the basic concepts of geometry, such as point and straight line, can be transferred by means of coordinative definitions to the atomic level still requires study (cf. Section 15). Discussion of this problem about models gives way to the third group of problems, to which we now come: *the problem of the regularity of elementary processes.* It was felt for some time that quantum mechanics would strike the final blow to the concept of strict causality long predicted by philosophers of science: the substitution of an elementary event subject to probability for one subject to strict determination. Even Bohr's theory of electron jumps in the atom gave rise to the suspicion that the individual jumps might no longer be subject to causal explanation, and this suspicion has recently attached itself to Schrödinger's wave mechanics, which has been given a statistical interpretation by Born[102]. According to this interpretation, Schrödinger's wave equations do not constitute a causal description of the temporal development of the electron system, but instead determine directly only the temporal development of a probability function which indicates the degree of probability with which the occurrence of a certain state can be predicted for the electron system. Because these matters have not yet received an adequate physical explanation, however, we are unable in this philosophical discussion to do any more than sketch out the general epistemological framework within which such investigations must be placed.

At this point we may return to Section 20, in which we discussed the claims of determinism, substituting for them a more cautious formulation. In place of the strict assertion that natural events are totally determined by causal laws, we set the less sweeping assertion that the probability of prior calculation of events can progress indefinitely toward 1. Only when related in this manner to the concept of probability is the assertion of causality meaningful, and only in this form is it accessible to criticism. At the same time, this formulation allows us to express that modification currently under discussion in quantum mechanics, namely, the proposition *that the probability in the calculation of events cannot be made to approach arbitrarily close to 1 but is instead restricted to a limit below 1.* This limit is to be conceived of as a monotonic function of the quantum numbers, so that it only deviates discernibly from 1 when applied to elementary events, while being barely distinguishable from certainty in the case of macroscopic events. From an epistemological standpoint, we can only remark that a modification of this kind in the formulation of the concept of natural regularity is entirely possible; whether or not this change is materially justifiable is not an epistemological but a purely physical problem. Think, for purposes of comparison, of Einstein's doctrine of the limiting character of the speed of light. From an epistemological point of view, a theory in which there is no upper limit of the diffusion of causal influences is just as admissible as a theory in which light constitutes the upper limit. The material judgment can be handed down only by physics, on the basis of its store of experience. The situation in quantum mechanics is just the same. All that can be done from the epistemological side is to present physics with the conceptual means for instituting such an innovation, and this has already been largely accomplished through the criticism of the concept of causality and the expansion of the concept of probability. Beyond this, we can only recommend to quantum mechanics that it ignore any sanctimonious philosophical exclamations denouncing abandonment of the principle of causality as a violation of *a priori* laws, as a renunciation of natural science, or as an introduction of general chaos. We believe, on the contrary, that only deliberate stress of the concept of probability can smooth the path to a physics capable of handling the demands for the greatest rigor in natural philosophy.

NOTES

[1] Cf. H. Thirring, 'Ziele und Methoden der theoretischen Physik', *Naturwiss.* 9, 1025ff. (1921).

[2] We could assume this to be the situation if we were to regard Driesch's theories about the division of sea urchins and mature creatures to be decisive (*Philosophie des Organischen,* 2nd ed., Leipzig, 1921) [*The Science and Philosophy of the Organism*, London, 1908]. The author cannot, however, accept this view; cf. also M. Schlick, 'Naturphilosophie', in *Lehrbuch der Philosophie,* ed. M. Dessoir (Berlin, 1925), p. 463ff. [A major portion of this article was published in English translation as the Appendix to M. Schlick, *Philosophy of Nature,* tr. by A. von. Zeppelin (Philosophical Library, N.Y., 1949) – Ed].

[3] (Berlin, 1922); cf. the article by the same author in *Symposion* **1**, 61 (1925).

[4] Becher, E., *Geisteswissenschaften und Naturwissenschaften* (Munich, 1921).

[5] Oppenheim, P., *Die natürliche Ordnung der Wissenschaften* (Jena, 1926). Oppenheim wants to order the sciences into a continuous two-dimensional schema, comparing the logical aspect of his system with the periodic system of the elements.

[6] Rickert, H., *Die Grenzen der naturwissenschaftlichen Begriffsbildung* (Freiburg, 1896, and later editions).

[7] Whitehead, A. N. and Russell, Bertrand, *Principia Mathematica,* 2nd ed. (Cambridge, 1924).

[8] Cf. Bertrand Russell, *Our Knowledge of the External World* (Open Court, Chicago, and Allen & Unwin, London, 1915) and *The Analysis of Mind* (Macmillan, New York, and Allen & Unwin, London, 1921); also K. Gerhard, 'Der mathematische Kern der Aussenwelthypothese', *Naturwiss.* **10**, 423 and 466 (1922).

[9] More precisely: I can call up the mere image *under any circumstances*, but the perception only *under certain particular circumstances*, for of course I can, in many cases at least, establish at will conditions in which the desired perception takes place; I can, for instance, go and look at a house. In the terminology of Section 6, we would say: Mere imagination can realize *any* implications, but perception only a certain selection of implications.

[10] Here we are dealing with probability implications, not strict implication – hence this symbol. Cf. Reichenbach, 'Die Kausalstrukter der Welt' [1925d; translated in this volume as Chapter 47].

[11] Mach, E., *Analyse der Empfindungen*, 9th ed. (Jena, 1922), p. 23 [Trans. by C. M. Williams and S. Waterlow as *The Analysis of Sensations*, 3rd ed. (Dover reprint, New York, 1959), p. 29]. Mach says here, as in other writings, "Bodies are made up of elementary complexes (complexes of sensations)." Related conceptions are found in Richard Avenarius. Yet these founders of epistemological positivism did not have modern logic at their disposal.

[12] Russell, B., *Our Knowledge of the External World, op. cit.*, sections 3–4.

[13] Carnap, R., *Der logische Aufbau der Welt* (Berlin, 1928). [Trans. by Rolf A. George as *The Logical Structure of the World* (University of California Press, Berkeley and Los Angeles, and Routledge and Kegan Paul, London 1967]. Here, for the first time, a positivist theory of the universe is actually carried out, making extensive use of the apparatus of modern logic; for the term 'logical complex', see Sections 4 and 36.

[14] Schlick, M., *Allgemeine Erkenntnislehre* (Berlin, 1918), p. 170 [Trans. from the 2nd German edition by A. E. Blumberg as *General Theory of Knowledge* (Springer-Verlag, New York and Wien, 1974), Ch. 26].

[15] Mach himself made the error of reading his positivistic concept of existence as telling against the existence of the atom. He speaks of the "artificial hypothetical atoms and molecules of physics and chemistry," which are "only symbols of those peculiar complexes of sensible elements that we encounter in the narrower spheres of physics and chemistry" (*Analyse der Empfindungen, op. cit.*, p. 254 [pp. 311–312 in Eng. trans.]; cf. also *Scientia* 7, no 14 (Milan, 1910). Boltzmann, in the introduction to his *Vorlesungen über Gastheorie* (Leipzig, 1892, vol. 1, sec. 1) [Trans. with vol. 2 by S. G. Brush as *Lectures on Gas Theory*, University of California Press, Berkeley and Los Angeles, 1964], found it necessary to defend the atomic theory against its opponents, who were still influential at the time. He compares it very aptly with the hypothesis of the fixed stars, which is also only culled by means of conceptual construction from the "scanty visual perceptions" of those tiny points of light in the nocturnal sky. One could, he writes, accuse the authors of this hypothesis, too, of "constructing a whole world of imaginary objects alongside the world of sense perception."

[16] Schlick, *ibid.*, no. 5.

[17] Hilbert, E., 'Neubegründung der Mathematik', *Abhandlgn. d. Ham. Math. Sem.*, p. 163, (1922).

[18] Concerning Gestalt theory, cf. Max Wertheimer, *Drei Abhandlungen zur Gestalttheorie* (Wilhelm Benary, Berlin, 1925) [Two of these essays are partially trans. in

W. D. Ellis, ed., *A Source Book of Gestalt Psychology* (Routledge & Kegan Paul, London, 1938; 3rd ed., 1950) as 'Numbers and Numerical Concepts in Primitive Peoples' and 'The Syllogism and Productive Thinking' – Ed.], and Wolfgang Köhler, *Die physischen Gestalten in Ruhe und im stationären Zustand* (Benary, Berlin, 1920).

[19] This identification is quite consciously expressed by Carnap, *ibid.*, p.6 [p.10 in the English trans.].

[20] Hume, David, *An Essay Concerning Human Understanding*, sec. IV.

[21] Cf. Reichenbach, 'Metaphysik und Naturwissenschaft' [1925a; trans. in volume 1 as Chap. 31].

[22] This conception of truth is developed as far back as Plato's *Theatetus*, where Socrates explains error to consist in coordinating the wrong symbols with the object or perceptions, as the case may be (Stephanus 193–196). He summarizes the theory by saying, "False opinion consists neither in the relation of perceptions to one another nor in the relation of ideas to one another, but in the coupling of the perception with the idea contained in the mind." Here, too, the perception is to be understood as the object of perception. Kant also uses this formulation, in the *Critique of Pure Reason* (A 58, B 82).

[23] Schlick, *ibid.*, no. 10.

[24] Helmholtz, H. von, 'Die Tatsachen in der Wahrnehmung' in *Vorträge und Reden*, vol. 2 (Braunschweig, 1884), p.226 [Trans. by M. Lowe as 'The Facts in Perception' in Helmholtz, *Epistemological Writings* (*Boston Studies in the Philosophy of Science*, vol. 37: D. Reidel, Dordrecht and Boston, 1977), p. 122 – Ed.].

[25] Hertz, H., *Die Prinzipien der Mechanik* (Leipzig, 1894), p.1 [Trans. by D. E. Jones and J. T. Walley as *Principles of Mechanics* (Macmillan, London, 1899; Dover reprint with an introduction and Hertz bibliography by R. S. Cohen, New York, 1956)].

[26] Helmholtz, *ibid.*, p. 226.

[27] Here we are using the word 'fact' in a general sense; i.e., we take it to include laws of nature as well as propositions that merely report.

[28] Kirchhoff, G., *Vorlesungen über Mechanik*, 4th ed. (Leipzig, 1897), p. 1.

[29] Cf., e.g., *Analyse der Empfindungen, (op. cit.)*, p.40 [p.49 in the Eng. trans.] and *Erkenntnis und Irrtum*, 3rd ed. (Leipzig, 1917), pp. 164ff. and 449ff. [Trans. from the 5th German edition by P. Foulkes and T. J. McCormack as *Knowledge and Error*, (*Vienna Circle Collection 3*: D. Reidel, Dordrecht and Boston, 1976)].

[30] Leipzig, 1908 [*Physikalische Zeitschrift* **10**, 62–75 (1909)]; cf. also *Physikalische Rundblicke* (Leipzig, 1922), p. 28 [Trans. by R. Jones and D. H. Williams as *A Survey of Physics* (Methuen, London, 1925); reprinted as *A Survey of Physical Theory* (Dover reprint, N.Y., 1960)].

[31] Schlick, *ibid.*, p. 61.

[32] A more thorough investigation of these two forms of transition was undertaken by F. London, *Jahrb. f. Philos. u. phänomen. Forsch.* **6**, 335 (1923). A corresponding study of the third case would be welcome.

[33] Cassirer, E., *Substanzbegriff und Funktionsbegriff* (Berlin, 1910) [Trans. by W. C. Swabey and M. C. Swabey as *Substance and Function* (Open Court, Chicago and London, 1923; Dover reprint, N.Y., n.d.)].

[34] Sometimes the original particularizing assumptions are in principle incapable of being fulfilled because they signify a self-contradictory limiting case (e.g., the ideal gas). As this state is approached, a semi-convergent series emerges that ceases to converge

beyond a certain degree of approximation and diverges instead. In such instances the superiority of the more general law with respect to precision becomes particularly important.

[35] Cf. also the pointed criticism directed by Planck against Mach's principle of economy in *Physikalisch Rundblicke, (op. cit.)*, pp. 31ff. and in 'Zur Machschen Theorie der physikalischen Erkenntnis', *Vierteljschr. f. wiss. Philos.* **34**, 499ff. (1910) [Apparently this same article appeared in *Physikalische Zeitschrift* **11**, 1186–1190 (1910) – Ed.]. Schlick, too, rejects the principle of economy, *ibid.*, p. 81.

[36] The author has given a detailed exposition of his arguments against the Kantian philosophy of the *a priori* in *Relativitätstheorie und Erkenntnis a priori* [1920f; for English trans. see 1965a].

[37] Thirring, *Naturwiss.* **9**, 1024 (1921).

[38] Cf. E. Freundlich, *Die Grundlagen der Einsteinschen Gravitationstheorie*, 2nd ed. (Springer, Berlin, 1917), pp. 20–21 [Trans. by H. L. Brose as *The Foundations of Einstein's Theory of Gravitation* (Cambridge University Press, Cambridge, 1920)].

[39] Cf. Schlick, *ibid.*, no. 7, and H. Reichenbach, *Philosophie der Raum-Zeit-Lehre*, sec. 14 [1928h; for English trans. see 1958a].

[40] Reichenbach, *Axiomatik der relativistischen Raum-Zeit-Lehre* [1924h; for English trans. see 1969a], and *Philosophie der Raum-Zeit-Lehre, (op. cit.)*. Both of these works should be consulted for the detailed foundations of sections 15–18.

[41] Put more precisely, the unit line segment must be given for all directions at every point. Thus we must imagine a cluster of unit segments at every point. Riemann's line element fulfills just such a definition.

[42] Thirring, *Phys. Zs.* **19**, 33 (1918) and **22**, 29 (1921).

[43] Cf. Schlick, 'Die philosophische Bedeutung des Relativitätsprinzips', *Zs. f. Philos. u. philos. Kritik* **159**, 129–75 (1915), [Trans. by P. Heath as 'The Philosophical Significance of the Principle of Relativity' in M. Schlick, *Philosophical Papers, 1910–1936* (*Vienna Circle Collection*: D. Reidel, Dordrecht and Boston, forthcoming – Ed.], and Reichenbach, *Philosophie der Raum-Zeit-Lehre, (op. cit.)*, p. 215.

[44] Reichenbach, 'Bericht über eine Axiomatik der Einsteinschen Raum-Zeit-Lehre', [1921d], and *Axiomatik* (*op. cit.*) [1924h], and Carathéodory, C., *Berl. Ber.* (1924), p. 12.

[45] Reichenbach, 'Über die physikalischen Konsequenzen der relativistischen Axiomatik', [1925c].

[46] Helmholtz, 'Über den Ursprung und die Bedeutung der geometrischen Axiome', 'Über die Tatsachen, die der Geometrie zugrunde liegen', reprinted in Helmholtz, *Schriften zur Erkenntnistheorie*, M. Schlick and P. Hertz, eds. (Berlin, 1921) [Trans. by M. Lowe as 'On the Origin and Significance of the Axioms of Geometry' and 'On the Facts Underlying Geometry' in Helmholtz, *Epistemological Writings* (*Boston Studies in the Philosophy of Science*, vol. 37: D. Reidel, Dordrecht and Boston, 1977)].

[47] The author has given an overview of this discussion, along with a bibliography, in 'Der gegenwärtige Stand der Relativitätsdiskussion', [1922f; trans. in this volume as Chap. 44].

[48] Einstein's theory made physical use only of the relativity of motion, while Weyl attempted to include the relativity of geometry; cf. H. Weyl, *Raum-Zeit-Materie*, 3rd ed. (Berlin, 1920), sect. 9–13 [Trans. by H. L. Brose as *Space-Time-Matter* (Methuen, London, 1922; Dover reprint, N.Y., n.d.)].

[49] The author has recently presented a thorough study of this point in *Philosophie der Raum-Zeit-Lehre, (op. cit.)*, sect. 9–13.

[50] Einstein, A., *Geometrie und Erfahrung* (Berlin, 1921), p. 8 [Trans. by G. B. Jeffery and W. Perrett as 'Geometry and Experience' in *Sidelights on Relativity* (Methuen, London, 1922), pp. 27–56], and M. Schlick, *Raum und Zeit in der gegenwärtigen Physik*, 4th ed. (Berlin, 1922) Chap. 5 [Trans. by H. L. Brose as *Space and Time in Contemporary Physics* (Oxford, 1920)].

[51] Cf. H. Dingler, *Phys. Zs.* **23,** 52 (1922), and Reichenbach, *Axiomatik der relativistischen Raum-Zeit-Lehre (op. cit.)*, pp. 41 and 68.

[52] Cf. Reichenbach, *ibid.*, sect. 12.

[53] The subjective conception is advocated by Poincaré in *Science and Hypothesis*, Chap. 4 [Trans. by G. B. Halsted in H. Poincaré, *The Foundations of Science* (The Science Press, N.Y., 1929; Dover reprint, N.Y., n.d.), pp. 27–197)]; for the objective conception, on the other hand, see P. Ehrenfest, 'Welche Rolle spielt die Dreidimensionalität des Raumes in den Grundgesetzen der Physik?', *Ann. d. Phys.* **61,** 440 (1920) and H. Weyl, *ibid.*, p. 245. The connection between the number of dimensions and the principle of causality and the grounding of this connection in the principle of contact action was first presented by the author (*Philosophy of Space and Time, op. cit.*, sec. 44), along with a detailed comparison with higher-dimensional parametric spaces.

[54] Laue, M. von, *Das Relativitätsprinzip*, 2nd ed. (Braunschweig, 1913), vol. 1, p. 49.

[55] Cf. Reichenbach, 'Über die physikalischen Konsequenzen der relativistischen Axiomatik', [1925c].

[56] Einstein, A., *Äther und Relativitätstheorie* (Springer, Berlin, 1920), p. 10 [Trans. by G. B. Jeffery and W. Perrett as 'Ether and the Theory of Relativity' in *Sidelights on Relativity*, p. 15 (see note 50 above)].

[57] A comprehensive presentation of this development was given by H. Weyl in 'Was ist Materie?', *Naturwiss.* **12,** 561, 585, 604 (1924) [Incorporated in H. Weyl, *Space-Time-Matter (op. cit.)*].

[58] Landolt, H., *Berl. Ber.*, 1892, p. 301; 1906, p. 266; 1908, p. 354.

[59] Cassirer, E., *Zur Einsteinschen Relativitätstheorie* (Berlin, 1921), p. 65 [Trans. by W. C. Swabey and M. C. Swabey as 'Einstein's Theory of Relativity' and published as an Appendix to *Substance and Function* (see note 33 above)], and K. Bollert, *Einsteins Relativitätstheorie* (Dresden, 1921), p. 63.

[60] Bohr, N., Kramers, H. A., and Slater, J. C., *Zs. f. Phys.* **24,** 69 (1924).

[61] Mach, in particular, was of this view; cf. e.g., *Analyse der Empfindungen, (op. cit.)*, p. 74 [p. 86 in English trans.].

[62] For these concepts, see B. Russell, *Introduction to Mathematical Philosophy* (Allen & Unwin, London, 1919), p. 31ff. Regarding the presentation of causality and the space-time order by the concepts of the theory of relations, cf. R. Carnap, *Kantstudien* **30,** 331 (1925).

[63] Reichenbach, 'Die Kausalstruktur der Welt und der Unterschied von Vergangenheit und Zukunft', [1925d; trans. in this volume as Chap. 47], and 'Stetige Wahrscheinlichkeitsfolgen' [1929l].

[64] M. Schlick, 'Naturphilosophie' in Dessoir's *Lehrbuch der Philosophie (op. cit.)*, pp. 451–462, and E. Zilsel, 'Über die Asymmetrie der Kausalität und die Einsinnigkeit der Zeit', *Naturwiss.* **15,** 280 (1927).

[65] A similar point can be based upon the recurrence objection; cf. P. Hertz,

'Statistische Mechanik', in Weber-Gans, *Reportorium der Physik* (Leipzig, 1916), vol. 1, pt. 2, p. 464.

[66] In order to understand the probability relations stipulated here, we must keep clearly in mind that A, B and C, do not represent individual states, but classes of states possessing equal probability; thus we ought, strictly speaking, to refer to "a state belonging to the class A" rather than to "state A."

[67] This point is also emphasized by P. Hertz, *ibid.*, p. 467, and in *Ergebnisse d. exakten Naturwissenschaften*, vol. 1 (Berlin, 1922), p. 76.

[68] Boltzmann himself tried this solution, in *Vorlesungen über Gastheorie*, vol. 2 (Leipzig, 1892), pp. 257–58 [English trans. by S. G. Brush; see note 15 above – Ed.].

[69] For a comment on this point, see M. Born, *Zs. f. Phys.* **40,** 177 (1926).

[70] Reichenbach, [1925d], p. 133.

[71] Reichenbach, *ibid.*, p. 150.

[72] That is to say, in the terminology of Section 17, that the case in which E_1 has the mark and E_2 does not cannot be declared impossible, but only very improbable; thus it is very improbable that the particles or chalk on the stone upon which there is a mark will all just happen to attach themselves to the chalk as it touches them.

[73] L. L. Whyte has brought the conception of the asymmetry of causality to the fore for general philosophical reasons, with the help of biological considerations; cf. *Archimedes, or the Future of Physics* (Kegan Paul, London, 1928).

[74] Planck, *Dynamische und statistische Gesetzmässigkeit* (Leipzig, 1914), and *Physikalische Rundblicke, (op. cit.)*, p. 82.

[75] Stumpf, C., 'Über den Begriff der mathematischen Wahrscheinlichkeit; Über die Anwendung des mathematischen Wahrscheinlichkeitsbegriffes auf Teile eines Kontinuums', *Ber. d. Bayer.-Akad., philos.-hist. Klasse* (Munich, 1892).

[76] Kries, J. von, *Die Prinzipien der Wahrscheinlichkeitsrechnung* (Freiburg, 1886; 2nd ed., 1927). In contrast to the other two authors, von Kries, by means of his 'principle of indifference', takes a mediating position between the subjective and the objective theory; cf. also the preface to the 2nd edition.

[77] Zilsel, E., *Das Anwendungsproblem* (Leipzig, 1916).

[78] Reichenbach, *Der Begriff der Wahrscheinlichkeit fur die mathematische Darstellung der Wirklichkeit,* [1915b] and [1916a]; cf. also the introductory paragraphs of 'Die physikalischen Voraussetzungen der Wahrscheinlichkeitsrechnung' [1920c] and 'Philosophische Kritik der Wahrscheinlichkeitsrechnung' [1920e] [trans. in this volume as Chaps. 52a and 53].

[79] This fact was long ago pointed out by A. A. Cournot in his *Théorie des chances et des probabilitiés* (Paris, 1843), p. 104.

[80] Kaila, E., 'Der Satz vom Ausgleich des Zufalls und das Kausalprinzip', *Ann. Univ. Fenn. Aboensis B*, vol. 2 (Turku, 1924), and 'Die Prinzipien der Wahrscheinlichkeitslogik', *ibid.*, vol. 4 (Turku, 1926). [Trans. by Peter Kirschenmann in the forthcoming volume of Kaila's writings in the *Vienna Circle Collection* with an introductory essay by G. H. von Wright (D. Reidel, Dordrecht and Boston) – Ed.].

[81] Mises, R. von, *Wahrscheinlichkeit, Statistik, und Wahrheit* (Springer, Vienna, 1928) [Trans. by Neymann *et al.* as *Probability, Statistics, and Truth* (Hodge, London, 1939; 2nd revised edition, Allen & Unwin, London and Macmillan, N.Y., 1957)], and 'Fundamentalsätze der Wahrscheinlichkeitsrechnung', *Math. Zs.* **4,** 1 (1918) and 'Grundlagen

der Wahrscheinlichkeitsrechnung', *Math. Zs.* **5**, 52 (1919). Von Mises stresses the problem of the logical construction of the mathematical probability calculus, treating the epistemological problem as secondary.

[82] Keynes, J. M., *A Treatise on Probability* (Macmillan, London, 1921).

[83] This idea stems from Poincaré, *Calcul des probabilités* (Paris, 1912), p. 149. Cf. also Reichenbach, 'Über die physikalischen Voraussetzungen der Wahrscheinlichkeitsrechnung', [1920b] and 'Nachtrag', [1921a].

[84] This is a familiar fact in the theory of error. Cf. Reichenbach, *Der Begriff der Wahrscheinlichkeit für die mathematische Darstellung der Wirklichkeit, (op. cit.)*, p. 105.

[85] Zilsel, E., 'Versuch einer neuen Grundlegung der statistischen Mechanik', *Monatsh. f. Math. u. Phys.* **31**, 118 (1921).

[86] Mises, R. von, 'Ausschaltung der Ergodenhypothese in der physikalischen Statistik', *Phys. Zs.* **21**, 225 and 256 (1920).

[87] P. Höflich also fails to derive the ergodic hypothesis without using a metrical assumption of probability, in *Zs. f. Phys.* **41**, 626 (1927); cf. also p. 650.

[88] Mises, R. von. 'Über die gegenwärtige Krise der Mechanik', *Naturwiss.* **10**, 25 (1922).

[89] Kaila has proposed a 'contingency principle' in its place; cf. note 80 above.

[90] I ought to insert this correction in my writings on probability from 1915 to 1920, in which I still regarded this problem from the Kantian viewpoint and held that a proposition is proven if it is established as a necessary condition of knowledge.

[91] This possibility is indicated by F. Exner, in *Vorlesungen über die physikalischen Grundlagen der Naturwissenschaften* (Vienna, 1919), pp. 688–91. In contrast, the statistical laws were considered merely provisional by Planck, *Physikalische Rundblicke, (op. cit.)*, p. 98.

[92] Setting out from this thesis, the author has given a presentation elsewhere in which the causal connection of individual events is regarded as no more than a probability relation and the concept of strict causality is avoided altogether (cf. note 63 above).

[93] Cf. Frank, Philipp, 'Über die "Anschaulichkeit" physikalischer Theorien', *Naturwiss.* **16**, 121 (1928).

[94] Bohr, N., 'Atomtheorie und Mechanik', *Naturwiss.* **14**, 4 (1926).

[95] Heisenberg, W., *Zs. f. Phys.* **33**, 879 (1925).

[96] In carrying out a more precise justification, we must proceed on the basis of probability implication rather than strict implication; as a result, the inference to *B* will generally still possess a certain degree of probability, after all. The above argument has been greatly schematized.

[97] Cf. the presentation by P. Jordan, *Naturwiss.* **13**, 618 (1927).

[98] Schrödinger, E., *Ann. d. Phys.* **79**, 508–09 (1926).

[99] From a logical standpoint it would be entirely possible for natural events to have more dimensions at the micro-level than at the macro-level. Think, for instance, of a flat layer of sand particles which, strictly speaking, has three dimensions in the small, yet can generally be regarded as having essentially only two. The assertion that Schrödinger's q-space is not 'real' space requires special justification, which can only be given after it has been shown that ordinary space has three dimensions. We indicated in Section 16 that the principle of action by contact is required to establish the number of dimensions; and, in fact, Schrödinger's q-space is characterized by the presence of actions at a distance. It is, then, not 'real' space; cf. also Reichenbach, *Philosophy of Space and Time, (op. cit.)*, sec. 44.

[100] Born, M., *Zs. f. Phys.* **38**, 803 (1926). On the other hand, the real nature of the Broglie-Schrödinger waves would appear to be *supported* by the well-known experiments concerning apparent diffraction phenomena involving electron reflections. Cf. C. Davisson and L. H. Germer, *Nature* **119**, 558 (1927), and *Phys. Rev.* (2) **30**, 705 (1927); also Rupp, E., *Naturwiss.* **16**, 977 (1928).

[101] This idea was first presented by Heisenberg, in *Zs. f. Phys.* **43**, 172 (1927), and later established more precisely by Bohr, *Naturwiss.* **16**, 248 (1928).

[102] Cf. note 100 above.

49. CURRENT EPISTEMOLOGICAL PROBLEMS AND THE USE OF A THREE-VALUED LOGIC IN QUANTUM MECHANICS

Dedicated to Erich Regener on his seventieth birthday

[1951g]

Today, physics plays a more important role in philosophical discussion than ever before, for physical research has led to a revision of certain very general principles of our knowledge of nature. The philosophical results first became apparent in Einstein's theory of relativity, which taught us to regard the spatial and temporal dimensions of the physical world in a new light. It turned out that the excessively simple concepts of space and time that have been developed in connection with our physical surroundings in everyday life can no longer be reconciled with the results of precise optical and electromagnetic measurements. Yet the transformation of basic concepts that was consequently demanded of us now appears trivial compared to the changes in our vision of the physical world brought by quantum mechanics. For the physics of quanta demands that we give up the idea of universal causality – the idea that was regarded, all the way up through the theory of relativity, as the most fundamental principle of classical physics, without which natural science appeared to be impossible.

To be sure, the development took place in stages. At first, giving up the principle of causality entailed merely replacing strict physical laws with the laws of probability. This prospect had already surfaced in connection with Boltzmann's interpretation of the second law of thermodynamics. But what at that time seemed a mere possibility became a certainty with the advent of Heisenberg's principle of indeterminacy. The limitations to precision in the prediction of quantum mechanical phenomena are related to Planck's quantum of energy and compel us to conceive of quantum phenomena as being merely statistically determined. This limit to precision has nothing to do with the imperfections of the human observer; rather, it is grounded in the structure of the physical world and can be formulated as a relation between physical events, without making any reference to an observer. Laplace's superman, who was supposed to be able to observe precisely even the smallest particle in nature and to possess boundless mathematical powers, would here find himself no better off than the rest of us. He, too, would have no alternative but to apply Schrödinger's ψ-equations for the description of natural events. And thus he would become committed to the principle of the inverse relation of the distribution of probabilities of canonically conjugated parameters, which found its expression in Born's statistical interpretation of the ψ-function and in Heisenberg's indeterminacy principle.

This modification of the concept of causality does not appear very troubling. Even if the behavior of the smallest particles in nature cannot be predicted, nature remains predictable in the large, for the statistical regularity of large numbers guarantees sufficiently precise laws for the behavior of macroscopic objects. Whether the most minute corpuscles emulate the movement of the planet or the course of a game of dice seems immaterial so long as the overall results are the same. In any case, such a change gives rise to no difficulties concerning our image of the world. The law of causality is a result of experience, and we must be prepared to replace it with a more general law whenever physical observations demand it.

However, it turns out that this extension of the law of causality does not suffice for an understanding of quantum mechanical phenomena. Not only must we sacrifice the idea of a strict causal connection between every particle in nature but the concept of an elementary physical corpuscle is called into question. Today, it seems impossible to speak of elementary particles of matter in the same manner in which scientists from Democritus to Boltzmann spoke of atoms.[1]

The difficulties to which I refer are well known. They arise from the duality between waves and corpuscles, which first became known through de Broglie's discoveries and later were formalized as Bohr's principle of complementarity. Once it proved possible to produce the same phenomena of interference with an electron beam that had been used by von Laue to prove that X-rays are waves, this complementarity became an experimental fact. Any complete sketch of physics today must take due account of this fact.

It has been said that this is all merely a question of the mental images we create of quantum mechanical phenomena. That images are necessarily incomplete need not concern us, so the argument runs; it suffices that such images are a help to the physicist in his work, although he should not take them overly seriously. But I do not believe that the philosophical question at issue can be circumvented in this manner. If the physicist thinks of the electric current running through a wire as a stream of water and of the voltage as the different degrees of water pressure at various points along the pipe through which it is flowing, he has an image or model. Yet this image does not give him an answer to the question of what the electric current really is. Likewise, when the physicist visualizes the atoms in a crystal as large colored beads strung on a wire, he has an image. But this image does not permit him to dispose of the question whether the real atoms are objects – smaller, of course, and certainly not colored – taking up and located in space. Such questions are quite reasonable. And it is my belief that they can be answered, provided we are prepared to analyze them logically and to develop methods of giving them a more precise form.

For the physicist simply to reject questions of this nature would be short-sighted. After all, it is his aim not merely to possess formulas but also to understand what they mean, not merely to make predictions about experiments but also to become acquainted with the universal characteristics of nature that are displayed in experimental data. This desire to draw conclusions on the basis of observation as to the universal characteristics of physical reality is entirely justified. To be sure, quantum mechanical analysis has taught us that such inferences have a more complex logical structure than earlier generations had supposed. The conceptual difficulties of quantum physics stem from the attempt to apply to the microscopic sphere the excessively simple relationship between observation and the physical world that applies to the macroscopic sphere. It follows that the difficulties can only be resolved if we are prepared to revise our methods of philosophical reasoning.

The difference between what we observe and what we infer is encountered even in daily life. The inference usually takes a very simple form; we say that things are the same whether or not we are observing them. There is, of course, no way to prove this latter conclusion. If we are not looking at a thing, we cannot see it, and hence we cannot claim to have experimental proof of the fact that it is the same when under observation as when not under observation. I am not suggesting that it would be better to assume that assertions regarding the existence of objects not under observation presuppose certain logical assumptions that cannot themselves be the result of observation.

Kant has long ago taken note of these facts, but I do not wish to fall into his mistake of attempting to deduce universal principles of reality from pure reason. We need no metaphysical theories in order to speak of physical objects; that is, we do not need to assume the truth of any principles in order to draw inferences reaching beyond the sphere of observation. The principles required are more of the nature of definitions. We assert merely that things continue their existence unaltered when we are not looking at them. We have introduced into language an extension rule that permits us to make the transition from observed objects to unobserved objects.

But is this a permissible procedure? Naturally, we must be careful in handling such definitions. If a pickpocket steals my wallet while it lies unobserved in my pocket, I cannot by means of a definition cause it to return to my pocket. In its more precise form, the definition, or extension rule, runs as follows: *I shall assume that the same physical laws apply to both observed and unobserved objects*. Then I will be able to infer that, if I see my house in the morning and again when I return in the evening, it has continued to exist in the same location during the intervening time. And I can likewise infer that my wallet has been stolen if I put it in my pocket and cannot find it there later. Sometimes we can even conclude that making observations alters the state of affairs, as, for instance, when we place a thermometer in water. Then we can figure out the temperature of the water before the observation was made.

The foregoing point appears to be very trivial, and yet it is extraordinarily important. These considerations demonstrate that we must employ a particular form of language if we wish to discuss a physical world existing in itself independent of our observation. This form of language is by no means prescribed by reality. If someone prefers to say that objects disappear every time he looks away, he has opted to use another form of language; i.e., he has rejected the above-mentioned extension rule and replaced it with a different rule. But he must not fall into thinking that the contents of his world are different from the contents of ours. There exists for the physical world a class of equivalent descriptions that are all equally true and are only differentiated according to the extension rule upon which they are based. The common, realistic mode of language is merely the simplest; it is simpler in the same sense in which the metric system is simpler than the system based upon feet and inches. We could call the latter the 'normal system'. This is not to say that there are no true results, but only that these results must be expressed in a more complicated manner. A system of language together with its extension rule is either true or false. But without a statement as to which extension rule has been used, the system is incomplete and it is impossible to determine whether it is true or false.

Here, too, we must add a word of caution. We cannot predict *a priori* whether our extension rule can be used successfully. In classical physics it was always taken for granted that physical laws applying to observed objects could be applied to unobserved objects, yet it is an empirical question whether or not this extension of language can be implemented without contradiction. This is the juncture at which quantum physics becomes differentiated from classical physics.

It is commonly said that all we ever observe in quantum physics are coincidences. This is true enough, but the question remains, what are entities that coincide. For this reason I prefer to use the neutral term *phenomena*. To say that corpuscles coincide is already to overstep the boundaries of the phenomena, for it is to claim that what we observe to be precisely located was also precisely located before we observed it, to say something about how the entity looks when it is not under observation. I will give the name *interphenomena* to things not under observation.

It would be a mistake to suppose that we cannot make any assertions about interphenomena, for we can. We simply need to add extension rules to our language. For instance, a very useful extension rule would be that the value of an entity was the same before it was measured. This rule can be implemented without contradiction, and it can be shown that, in accepting it, we admit the corpuscle theory.[2] To be sure, it can also be demonstrated that the measurement, while not altering the observed entity, did alter its canonically conjugate value. The disturbance caused by observation, which, according to Heisenberg, is inevitable, has simply been transferred to the conjugate value. But it is clear that the corpuscle interpretation has thereby been assigned a precisely delineated sense and yet is not for that reason to be described as an image. It is as well defined as every other description of reality in macroscopic physics, e.g., as the language of reality, in which we say that houses exist even when we do not see them.

The wave interpretation can be just as clearly defined. It is tantamount to the claim that the entity, when not under observation, possesses simultaneously every value which it will ever be capable of having.[3] There is nothing paradoxical in such a definition.

However, if we wish to go further and carry out completely the corpuscle theory which we have introduced, we meet with singular difficulties. We become compelled to assume the existence of action at a distance spreading without an intervening field and requiring no time to reach a distant point. This is a result of the well-known experiment in which a particle going through a slit behaves differently depending upon whether another slit is also open, and so forth. I call these relations 'causal anomalies'. They also arise if we pursue the wave interpretation, although in connection with different experiments. It is well-known that, with plates of glass properly set up so that one ray is reflected while the other goes through the glass, it can be demonstrated that a wave which has just arrived at the moon will be annihilated as soon as someone working in a laboratory on earth makes the appropriate measurement.

Our first inclination is to reject these reflections as nonsense, but that would be to throw the baby out with the bathwater. A profound physical insight lies hidden in these logical investigations, as follows. No matter how we construe the extension rules of the language, we cannot construct a description of reality in which the principle of causality can be implemented in its normal form; in other words, there is no normal system contained in the class of equivalent descriptions of quantum physics. I have designated this result the 'principle of anomaly' and have offered a general proof of it on the basis of quantum mechanical principles in another work.[4] This is the reason that every exhaustive description of reality in quantum mechanical terms leads to unwelcome consequences and that physicists often prefer to maintain silence on the subject of interphenomena.

No objection can be made to a physicist's refusal to speak about the subject, but we should be clear as to why he refuses. It is not because assertions about interphenomena are not verifiable; assertions about houses not currently under observation are also unverifiable. His reason is rather that if he construes quantum mechanical interphenomena on the same principles he uses for unobserved houses, he encounters causal anomalies, which are not encountered in the case of houses. If experiments performed with two slits and a single electron beam produced a pattern of interference which, when superimposed upon it, corresponded exactly with the pattern produced when each slit is used separately, no physicist would hesitate to regard electrons as perfectly ordinary physical particles that fly about the room like tennis balls – even though it might not be possible to observe an individual electron at the slit without disturbance. But because we know that the results of experiments with interference are, in fact, quite different, we avoid saying anything about these tennis balls so long as we are not observing them.

In this connection I would like to mention a further group of causal anomalies, the significance of which has been presented in several recent works. Investigations carried out by Stückelberg and Feynman[5] indicate that a positron may be conceived of as an electron that moves backwards through time. In this interpretation of interphenomena even the direction of time is reversed; i.e., in the class of equivalent descriptions the direction of time is no longer an invariant. Experimental data cannot teach us anything as to the admissibility of such an interpretation; it is equally compatible with either interpretation. Here, as always, each of the admissible interpretations has, together with its drawbacks, its own special advantages. Accepting the reversal of the direction of time has the advantage of eliminating pair production and pair annihilation. On this interpretation, there is only a single electron that runs a portion of its world-line in negative time, thereby simulating a positron. Thus the *genidentity* – that is, the identity of an individual entity along a world-line – is regarded as being a noninvariant characteristic and therefore a matter of definition, which is given different forms in different equivalent descriptions. No objections can be raised to such an interpretation on logical grounds. Quite the contrary, it represents an extraordinary enrichment of our knowledge of quantum mechanics and demonstrates that the principle of anomaly even includes certain reversals of the cause-effect relation.

These considerations lead us to the conclusion that quantum mechanics entails a peculiar dislocation of the reality content of physical theories. If we wish to give a complete description of reality, it is not enough to construct a single description. Only after the entire class of equivalent descriptions is given is the presentation of physical reality completed. As long as there is a normal system, as in classical physics, it is sufficient to give the description in the language of the normal system. But when there exists no such system, the very statement that there is none points to a major characteristic of physical reality, and our presentation would be incomplete if this state of affairs were suppressed. We are forced, then, to employ, in addition to physical language, a metalanguage in which we speak about the class of equivalent languages, and only in this way will we be able to express, indirectly, a certain peculiarity of the physical world.

It has now become clear why we run into difficulties when we wish to speak about unobserved objects in quantum mechanics. Every assertion regarding unobserved objects contains an implicit assertion concerning causal connections, and in the case of quantum mechanical objects these causal connections do not possess the simple form applicable to classical objects. When we say that houses exist even when not observed, we have thereby made an assertion about causality, and when we say that it is wrong to conceive of electrons as miniature tennis balls, we mean that such particles do not fit into the framework of a causal system that satisfies the principle that there can be no action at a distance. The answer to the aforementioned question as to whether matter is made up of spatially located particles is, then, that such an assertion is meaningless, taken by itself, and acquires meaning only in conjunction with assertions about causal structure. Since we know that causality as applied to quantum mechanical phenomena has different structural characteristics from those when it is applied to macroscopic objects, we cannot apply the notions of classical atomism to quantum physics.

It now also becomes evident why it is advisable, in quantum mechanics, to replace the usual two-valued logic with a three-valued logic. Three-valued logic systems, in which the principle of the excluded middle is abandoned, have been familiar for some thirty years.[6] But at first they were constructed as mere logical possibilities; like so many other mathematical constructions, they had to wait in the wings for the moment when an appropriate physical application was discovered. Quantum mechanics has apparently offered the first occasion upon which a three-valued logic can be used in physics.

Three-valued logic contains a category, indeterminate, that lies between truth and falsehood, and this category can be used with reference to interphenomena in quantum mechanics. The value of the entity before measurement is said to be indeterminate, which has the effect of also leaving undetermined whether there is a single definite value that could in principle be ascribed to the unmeasured entity, aside from whether or not we actually know it. By the same token, the question whether interphenomena are made up of waves or of corpuscles is presented as indeterminate. This conception of the situation makes it possible to exclude causal anomalies from the sphere of assertable propositions. Of course, it is only as applied to interphenomena that a three-valued logic can be considered suitable for use in quantum mechanics. It is not needed in discussions of observed entities, or phenomena, for it is possible to make true-false assertions about them, i.e., assertions that can be established as true or false. A number of physicists have misunderstood this point. For instance, Bohr writes "... the recourse to three-valued logic ... is not suited to give a clearer account of the situation, since all well-defined experimental evidence ... must be expressed in ordinary language making use of common logic".[7] And Born asserts that "... the mathematical theory, which is perfectly capable of accounting for the actual observations, makes use only of ordinary two-valued logic".[8] Pauli[9] makes similar assertions. These observations concerning the logical nature of observed facts are doubtless correct, but they are irrelevant. In the three-valued logic that I recommend, observation language is specifically left as two-valued; the three-valued logic is used only in the sphere of interphenomena,[10] where it cannot be rejected. Otherwise, causal anomalies arise just as soon as we go beyond what is immediately observed.

Some scientists are of the opinion that interphenomena need not be included in the theory. But the exponents of this view offer no explanation as to why unobserved values can be embraced by the theory of classical physics, but cause difficulties when included in quantum mechanical theory. In other words, a language confined to observed entities is not rich enough to express everything we know about the physical world. That is why I consider it quite wrong to say that interphenomena can be excluded from the theory. For instance, Bohr's principle of complementarity is an assertion about unobserved entities if it is stated as follows: observed phenomena may be regarded as the result either of the actions of waves or of corpuscles, and there can be no crucial experiment to show that one of these views is right and the other wrong. This assertion derives support from Heisenberg's indeterminacy principle, but is broader. For the principle of indeterminacy applies only to the results of measurement, while the principle of complementarity claims also that it makes no sense to ascribe a specific value to what has not been measured. But this is itself a metalinguistic statement which is covertly based upon the principle of anomaly.

This matter becomes more comprehensible if we consider Einstein's theory of relativity. The theory employs a certain definition of simultaneity, although there is no way to prove that this definition represents, in some sense or other, 'true' simultaneity. It is possible to employ this definition of simultaneity because it leads to reasonable results, i.e., to a temporal order that does not violate the principle of causality. That the same is true of a certain class of definitions of simultaneity can also be demonstrated; to choose a particular definition, then, is to establish an arbitrary convention. In quantum mechanics the introduction of any one particular definition of the value of an unobserved entity is avoided, for every such definition leads to causal anomalies. This observation is itself an important result in physics and ought therefore to assume its place in the object language of physics. It can be formulated in object language when three-valued logic is used. It then takes the form of a logical formula expressible in the symbols of mathematical logic and has roughly the following import: if a quantum mechanical entity has, or does not have, a particular value, the value of a canonically conjugate entity is indeterminate.[11] On the other hand, rejection of three-valued logic necessitates the formulation of this idea in the metalanguage, which gives the theory an extraordinarily complicated logical form.

Let us take another example. Imagine an arrangement in which a beam of electrons of a very low intensity is aimed at a diaphragm with two parallel slits and forms a pattern of interference on a screen placed behind it. Here we would naturally like to say that a single electron, which has reached the screen, has gone through one or the other of the two slits. When this assertion is phrased in two-valued logic, we become entangled in causal anomalies. These anomalies are excluded when the assertion is interpreted in three-valued logic.[12] Yet Bohr and Heisenberg would consider only the following assertion as admissible: if an observer had been posted at each of the slits, only one of the two would have observed something, and then nothing would have occurred on the screen. It is difficult to understand why we should not be permitted to make a similar assertion concerning the case in which no observers are placed at the slits and observation is made solely at the screen. The use of three-valued logic permits just such an assertion.

Although it is two-valued, the language of observation is logically connected to the more general, three-valued language that includes interphenomena. Every logic contains assertions belonging to a limited domain of true statements that can be constructed from elementary propositions belonging to a more general domain of true statements. In two-valued logic, for instance, it is possible to construct, with a proper combination of elementary propositions, single-valued assertions, such as tautologies, that can only be true. In three-valued logic it is likewise possible to construct assertions that can only be either true or false and are therefore two-valued. I have demonstrated elsewhere[13] that it is precisely those assertions known as physical laws that fall into this category. Thus the two-valued language of predictions concerning observations, and the mathematics of such predictions, appear here as a natural consequence of a more general logical system which also has a place for unobserved entities.

One last word about the alleged necessity of two-valued logic. In introducing the theory of three-valued logic, we employ a metalanguage that is itself two-valued. This process is not grounded in a vicious circle. Rather, it is a legitimate procedure that seems to be useful just because we are accustomed to two-valued logic. The procedure is also quite innocuous, for it can be demonstrated that each system of logic can be transformed into any other system of logic. And it is certainly an error to conclude from this fact that three-valued logic is a mere game of symbols. For the scientist who does extensive work with these symbols can derive from them as much clarity and comprehension as can be gained from non-Euclidean geometry, which is likewise incompatible with our everyday world.[14]

This thought introduces a broader problem. The question has been posed whether the logic of a language is a matter of arbitrary convention or whether it in some manner expresses the structure of the world to which the language refers. This question cannot be answered either affirmatively or negatively until it is formulated more precisely. Since all logical systems are mutually transformable, any world can be described in terms of any logic. Inasmuch as this is the case, logic cannot be said to give expression to any feature of the physical world. But the introduction of some further postulates alters the situation. For instance, the requirement that no causal anomalies be capable of assertion with respect to interphenomena necessitates the use of three-valued logic in quantum mechanics, whereas the same requirement with respect to macroscopic physics is compatible with the use of two-valued logic. This result demonstrates that, under certain circumstances, the logic of a language can reflect the structure of physical reality. We bring such a situation about whenever we combine a logic with certain other principles and demand that the language satisfy the resulting system of principles. In one sense, then, it is correct to say that three-valued logic bestows upon the language of quantum mechanics that form which makes it an adequate expression of physical reality.

* * *

I would like to close this brief presentation of the current situation in quantum mechanics with a personal comment. This is the first time in many years that I have expressed my ideas regarding the philosophy of physics in a German publication, and it pleases me most particularly to do so as part of a commemorative volume for my revered friend Erich Regener. Mr. Regener was one of the first people who showed real interest in my work in the philosophy of physics, and I owe him a debt of gratitude for promoting my work at a time when scientific philosophy was still fighting to establish its right to exist. In admiration of his achievements in physics and in gratitude for his support of a philosophy that progresses hand-in-hand with physics, I extend to Mr. Regener my very best wishes on his seventieth birthday.

NOTES

[1] For a more thorough proof of the following ideas, see [1944b]. A more general presentation of the philosophical ideas underlying this book can be found in [1951a].

[2] See [1944b], p. 118.

[3] *Op. cit.*, p. 130.

[4] *Op. cit.*, section 26 (pp. 122–9).

[5] Stückelberg, E. C. G., *Helv. physica Acta* **14**, 558 (1941) and **15**, 23 (1942), Feynman, R. P., *Phys. Rev.* **76**, 749 (1949).

[6] See [1944b], pp. 147–8, for the literature on this subject.

[7] Bohr, Niels, *Dialectica* **2**, 317 (1948).

[8] Born, Max, *Natural Philosophy of Cause and Chance*, Oxford University Press (London, 1949), p. 107.

[9] Pauli, Wolfgang, *Dialectica* **2**, 310 (1948).

[10] However, this is not true of all attempts to present quantum mechanics in terms of a multi-valued logic. For instance, G. Birkhoff and J. von Neumann propose to apply their multi-valued logic of quantum mechanics to observation statements (*Annals of Math.* **37**, 823 (1936). It is difficult to see what advantages such a logic holds for quantum mechanics.

[11] See [1944b], p. 160.

[12] *Op. cit.*, pp. 163–4.

[13] *Op. cit.*, pp. 159–60.

[14] This answers a question raised by Born in his *Natural Philosophy of Cause and Chance*, p. 108.

50. THE LOGICAL FOUNDATIONS OF QUANTUM MECHANICS* +

[1952d]

TABLE OF CONTENTS

1. INDETERMINISM

The indeterminism associated with the quantum theory has been considered, as well by philosophers as by physicists, as the strongest deviation from classical physics that can be imagined. I do not believe that this judgment is defensible. It rather seems to me that this deviation is not as great as it seems at first sight, because there exist developments in classical physics which allow one to suppose that determinism is not the last word of the man of science regarding that which concerns natural phenomena. Thus, some other results of quantum physics contradict the classical conception of the physical world to a greater degree than does the abandonment of strict causality. I wish to speak of the change in the concept of matter, of the substance which makes up physical bodies, of a profound change which has resulted from the discovery of M. de Broglie regarding a certain equivalence of waves and particles and which has completely overturned philosophical theories (which had) come forth from classical science, as well as from our everyday experiences. Permit me, by way of introduction, to tell you about these developments from a comparative point of view, that is to say, by comparing them with classical ideas from which they have parted, and which they have transformed into a new picture of the physical world. It is only after this study of the philosophical background that we will be able to understand the significance of the discoveries of physicists during the thirty years elapsed since the first publications of M. de Broglie, and that we will be in shape to examine some other consequences of the new physics, consequences comparable if not superior to those which I have just cited.

* Lectures given at the Institute Henri Poincaré, June 4, 6, 7, 1952.

+Translated from 'Les fondements logiques de la mécanique des quanta', *Annales de l'Institut Henri Poincaré* **13**, part 2: 109–158 (1952/53).

Classical determinism originated in astronomy. Astronomers have the advantage of occupying themselves with large objects, far apart from each other, objects which thus are neither disturbed by one another, nor by human observation. These celestial objects move according to very precise laws which allow prediction of future positions if present positions are known, or rather, which allow very precise forecasts if one possesses a knowledge, sufficiently exact, of their present state. This extreme regularity had been considered as the ideal of nature, so to speak, and if one could not always find astronomical precision among terrestrial phenomena, one concluded that it is only the limitations of human capacities that limits our predictions to the uncertainty of statistical forecasts. The supposed ideal of nature had thus become the ideal of science, and the search for strict causal laws had taken the form of a moral obligation whose validity was beyond doubt.

It is hardly necessary to cite here the famous passage in which Laplace assigns to a superior intelligence the capacity to gather together "in the same formula the movements of the largest bodies of the universe and those of the lightest atom: nothing would be uncertain for it, and the future like the past would be present to its eyes".[1] To a modern physicist, these words seem like a profession of faith of a long gone age, of an age when one attributed to the stars the power to reveal the laws of the lightest atoms, of an age of confidence in the harmony of nature as well as in the capacity of man to translate her into mathematical formulas. The physicist of today no longer shares this confidence; he can scarcely imagine a time when it was believed that the large-scale world offered the picture of the small-scale world.

However, the question of determinism is not a question of faith, a question of the psychology of the physicist or of his epoch. It is a physical or epistemological question; it is a question which demands a logical analysis and an answer based on reason and experience. The opinion that all natural phenomena are governed by strict laws is the expression of a physical theory which, as such, ought to be subject to the critical rules generally accepted for the discussion of theories. Let us begin this analysis by attempting to give a precise formulation to the theory of determinism, a formulation which allows (us) to judge it as true or false, or, at least, as probable or improbable.

The causality hypothesis can be put in two different forms: a conditional form and a categorical form. In the conditional form, the statement of causality begins with the word 'if': if the situation A is completely described at time t_1 by the values $u_1, \ldots, u_n$ of certain parameters, it will be followed at time t_2 by the situation B. In the categorical form, we consider it established that the values $u_1, \ldots, u_n$ exist; and, omitting the phrase which begins with 'if', we conclude that the situation B will be produced. It is the second form, the categorical form, which leads to determinism, for the determinist maintains that it is admissible to separate the consequent from the antecedent and to affirm the conclusion.

One would be unable to find fault with this procedure if the situation were as simple as I have just described. Unfortunately it is much more complicated.

Even if we are justified in considering the conditional form as true, we are well aware that the condition expressed in the premise is not fulfilled. The description of the situation A by the parameters $u_1, \ldots, u_n$ is not complete. Despite this, we use it; and we are obliged to use it because we do not possess a better description. Fortunately, the error committed in using an incorrect description is not too great; the prediction of the situation B will be approximately valid. This means that it will be true within some narrow numerical limits in most cases. The laws of probability come to our rescue; if we have to forego an exact prediction, these laws offer us as a replacement a statistical prediction.

The solution seems simple, but it shows that the problem of causality is inseparable from the problem of probabilities. It permits us to separate the conclusion from the conditional form at the cost of sacrificing the pretension of arriving at a true statement: the conclusion is only probable. The solution thus substitutes the concept of probability for the concept of truth; and it requires that we reformulate the principle of causality in such a way that it takes into account its link with the principles of probabilities. Here is the answer that a logical analysis gives to the question of determinism: if a determinism can be sustained, it must first be formulated as a theorem concerning probabilities.

If each prediction is restricted to a degree of probability, then the causal analysis can only increase this degree of probability. The principle of causality is based on the notion that this increase of probability is always possible; and the idea of determinism refers to a limiting process regarding the predictive probabilities. If the description D^1 predicts the future state of B with a probability p^1, one could increase this probability by employing a more detailed description D^2. This new description of the initial situation differs from the first in the following respects:

1. The new description includes an account of the physical conditions in the spatial surroundings of the situation A.
2. The new description uses more precise measurements of parameters, and adds some new parameters to describe A, which were hitherto neglected.
3. The new description uses causal laws which have been perfected.

Item 1 serves to diminish some unexpected interventions coming from the exterior of the volume v in which the situation A at time t_1 and the situation B at time t_2 fit together. Items 2 and 3 serve to render more precise the account of the physical relations in the interior of v.

By repeating this process, one arrives at a sequence of descriptions D^i and of probabilities p^i, which allow us to formulate the hypothesis of determinism: this hypothesis claims that the sequence of the p^i converges toward the value 1 and that at the same time the sequence D^i converges toward an ultimate description D. The idea of *determinism* can then be symbolized by the following schema:

$$\begin{cases} D^1, D^2, D^3, \ldots D^i, \ldots \rightarrow D, \\ p^1, p^2, p^3, \ldots p^i, \ldots \rightarrow 1. \end{cases} \tag{1}$$

The question now arises as to whether this schema is corroborated by the testimony of experience.

For the classical physicist, the convergence of the probabilities p^i was accepted without doubt. Let us then postpone the discussion of this point and examine the question of the existence of an ultimate description. As regards points 2 and 3 (above), classical physics gave the answer that the definitive natural laws had been established in Newton's mechanics, and that the definitive parameters had been given by the mechanical model of the atom. Yet there remained the difficulties in that which concerns point 1. If space is infinite, the ultimate description D should include an infinite number of parameters; a logician would have difficulty in accepting such an idea, doubting that it is really meaningful.

The theory of relativity seemed to offer a remedy: the speed of causal transmission being limited to that of light, all the parameters that can influence the situation B at time t_2 are included, at time t_1, in a sphere of finite volume V. Yet another difficulty arises: by the same token, one will conclude that it is impossible to know the values of all these parameters before the instant t_2. Hence one can not give at time t_1 an ultimate description which allows prediction of the situation B.

The difficulty would be eliminated if the universe were spatially finite. In this case, at least, the logician would not have objections against the existence of the ultimate description D. Let us then accept this supposition in order to arrive at a form of determinism which is not susceptible to logical objections.

Yet, even in this case, determinism implies a doubtful enough hypothesis. In fact it is quite possible that the sequence of probabilities p^i converges to a limit, while the sequence of descriptions does not. This conception, which may be called *classical indeterminism*, can be symbolized by the following schema:

$$\begin{array}{l} D^1, D^2, D^3, \ldots, D^i, \ldots \\ p^1, p^2, p^3, \ldots, p^i, \ldots \to 1. \end{array} \tag{2}$$

Is there a possibility of distinguishing between the two conceptions (1) and (2) by means of an empirical criterion?

I suggest employment of the following criterion: that we examine the change in the probabilities when the prediction refers to a time instant $t_3 > t_2$. Having been given the same description D^i of the situation A at time t_1, we are well aware that the probability will diminish if t_2 is replaced by t_3. By repeating this consideration we arrive at a probability lattice of the following form:

$$\left\{\begin{array}{l} D_1^1\, D_1^2\, D_1^3 \ldots D_1^i \ldots \\ p_2^1\, p_2^2\, p_2^3 \ldots p_2^i \ldots \to 1 \\ p_3^1\, p_3^2\, p_3^3 \ldots p_3^i \ldots \to 1 \\ \ldots\ldots\ldots\ldots\ldots\ldots\ldots\ldots \\ p_k^1\, p_k^2\, p_k^3 \ldots p_k^i \ldots \to 1 \\ \ldots\ldots\ldots\ldots\ldots\ldots\ldots\ldots \\ \downarrow \;\; \downarrow \;\; \downarrow \qquad \downarrow \\ p \;\; p \;\; p \qquad p \end{array}\right. \tag{3}$$

The subscripts refer to time; the superscripts refer to the degree of detail of the description. In each line, the probabilities converge toward the value 1. In each column, the probabilities diminish and converge toward a mean probability of the appearance of the situation B, independently of the situation A.

Although this lattice is convergent, it does not present a uniform convergence. Given an interval of magnitude ϵ, one can find, for the time t_k, a description D_1^i such that $p_k^i \geqslant 1 - \epsilon$; but for the same description D_1^i, there is a time t_m, $m > k$, such that $p_m^i < 1 - \epsilon$. Hence there is no description D_1^i which predicts the future state for all times with a probability $p^i \geqslant 1 - \epsilon$. Consequently, a definite value cannot be assigned to the lower right-hand corner: in following the last column, one would assign it the value 1; in following the last line, one would assign it the value p.

It seems to me that this probability lattice indicates that the hypothesis of determinism is not confirmed by experience, even in classical physics. It is true that the notion of an ultimate description D does not logically contradict the structure of the lattice. But if this ultimate description D existed, it would be essentially different from all the descriptions D^i, in that it would establish a column of non-decreasing probabilities, while the values of the probabilities in each column starting with a D^i decrease. The non-uniformly convergent lattice can thus be considered as an *inductive proof* against the hypothesis of determinism, or at least, as the expression of the absence of any inductive proof in favor of this hypothesis.

To be sure, a *deductive proof* against this hypothesis cannot be given; in other words, one does not have a logical contradiction if one adds the supposition of the existence of an ultimate description to the observational facts included in this lattice. But this hypothesis will be introduced at the price of sacrificing continuity, because there is no continuous transition from the observable columns D^i to the non-observable ultimate column D. And continuity is the essence of inductive inference. The situation would be different if the lattice exhibited a uniform convergence: if so, one could consider this as an inductive proof of the existence of an ultimate description.

Employing the language of the verifiability theory of meaning, one can translate this result as follows. Determinism, if it exists, is not manifested in relations among observable quantities. These relations, to be sure, do not exclude determinism; but neither do they confirm it. Determinism, in classical physics, represents a vacuous addition to the system of observable relations; if it is omitted, if one renounces speaking of an ultimate description, nothing changes in the ensemble of verifiable statements.

Finally, we come to consider the situation in quantum mechanics. It is well known that, because of Heisenberg's uncertainty relation, it is impossible to increase the probability of a prediction beyond a certain value $p < 1$. We thus arrive at the following schema:

$$\begin{array}{l} D^1, D^2, D^3, \ldots, D^i, \ldots, \rightarrow D \\ p^1, p^2, p^3, \ldots, p^i, \ldots, \rightarrow p \end{array} \tag{4}$$

If the descriptions D^i approach a certain limiting description, given by the function ψ characterizing the state A, the probabilities p^i increase and converge to a value $p < 1$. It is this schema which formulates quantum indeterminism.

Comparing the three schemata (1), (2), (4), one sees clearly that one is concerned here with a continuous extension of concepts. The determinist believes in the existence of an ultimate description with the help of which he would arrive at perfect knowledge, such that "nothing would be uncertain for him, and the future would be seen just like the past". The classical indeterminist abandons the notion of this ultimate description; he rests content to improve his predictions bit by bit, without ever pretending to arrive by this process at a perfect prediction. The indeterminist of quantum theory restricts the power of predictions to a limit below the probability 1; and he can admit the existence of an ultimate description because this limiting description offers no predictive certainty.

This logical picture of the situation is presented here in order to show that one is concerned with a question of physics, and not a question of philosophical views or of world-conception. I do not believe that there are supposedly philosophical questions exempt from scientific treatment. If philosophy deals with problems of the structure of the world, it should abandon the idea of deriving this structure from an intuition allegedly of eternal truths. The philosophical truths of yesterday have become the errors of today. The philosopher who wants to contribute toward making the universe intelligible can aid the physicist only in searching for the correct form of a question, but not in searching for the answer. That is to say, his contribution will consist in a logical analysis of the problems, which separates the physical content of a theory from the additions in the guise of definitions, and which clarifies meanings of terms instead of prescribing the ways of thought. The philosopher can define determinism, and also indeterminism; but he cannot choose one of these structural forms as the existing one. It is the physicist who ascertains the structure which corresponds to the facts of observations.

According to quantum physics, it is schema (4) which describes the causal relations governing the physical world; that is to say, quantum physics has decided in favor of indeterminism. This result derives from the principle that the ψ-function includes all that can be furnished by observations. If this principle, which may be called the '*synoptic principle of quantum theory*', is accepted, then all that remains is mathematics; this means that one can derive the uncertainty relations from it mathematically. These relations uniquely express the fact that the Fourier analysis of a wave packet furnishes a greater number of frequencies, or harmonic oscillations, the more the extension of the packet is reduced. Once the synoptic principle is accepted, there is no more need to discuss the existence of hidden variables, precisely because such variables would constitute physical quantities not subordinated to this principle. Hence the study of the problem of determinism leads to the question: by what right does quantum physics maintain the synoptic principle?

The answer can be given that this principle has been confirmed by many observations. A case has never been found where it was possible to take measurements outside the limits fixed by Heisenberg's relation; and it has always been possible to express the sum total of all observations by a ψ-function. Experiences of this kind surely offer reasons to accept the principle; they serve to indicate that it is advantageous to employ the principle, but they cannot establish the principle as reasonable beyond doubt. In other words, they can not confer upon this principle the status of a natural law. One could always look forward to encountering some day an experimental situation which admits of observations that cannot be included in a ψ-function. This is, for example, Einstein's belief; he considers quantum mechanics as a statistical theory comparable to the statistical interpretation of classical thermodynamics, a theory which allows us to predict mean values of certain quantities which can be verified by measurements, but which does not exclude the construction of a detailed theory which determines the individual values of all physical quantities with the help of strict causal laws.

That is, today the synoptic principle has two sides, a positive side and a negative side; it allows advance calculation of results of certain experimental arrangements, but, on the other hand, it excludes the existence of observations which violate Heisenberg's law. It is this second aspect which is attacked by the determinists who are not in doubt about the legitimacy of the first aspect.

To find a solution, let us proceed by a method which may be called the method of the reversed situation. Suppose that the positive part of the principle is correct, while the negative part has to be abandoned. What would be the derivable results for physical objects, for particles which would have a well defined existence like the molecules of classical physics?

One must now recall the logical state of the classical statistics of Boltzmann and Gibbs. The existence of probabilistic laws does not contradict the notion of a determinism governing the trajectory of the individual molecule. Boltzmann established his famous H-theorem while supposing that Newtonian mechanics determines the motion of each molecule and their collisions. In fact, Boltzmann was able to derive the probability metric existing in phase space from the canonical equations of motion with the help of Liouville's theorem, which furnishes equal probabilities for equal volumes in this space. And, more recently, von Neumann and Birkhoff have succeeded in deducing the ergodic theorem from these same equations, a theorem which Boltzmann had added to his calculations as an indispensible axiom, but which he could not prove. Now the classical statistics of gases and determinism are quite compatible; they are the children of the same father, so to speak, the father being Newton, and the children live in pre-established harmony thanks to the canonical equations. Classical physics, even if it were incapable of adducing many reasons in favor of indeterminism, could certainly not invoke decisive proofs against its existence.

Let us now examine the logical state of quantum statistics. The statistics of Boltzmann has been replaced by those of Bose and of Fermi, according to which elementary particles are indistinguishable, that is to say they do not possess any individuality. As a consequence, this theory yields different numbers for possible combinations of particles, a result which in turn furnishes new values for thermodynamical quantities. Now since these values have been confirmed by observations, one concludes inversely that the experiments inform us that the particles are indistinguishable, or at least they behave as if they were. One must study the significance of this result.

No one would have said, at the time of classical physics, that he was able to really distinguish one molecule from another, or that he would ever be able to do it. An individual molecule cannot be observed, much less be marked as the zoologists mark fish or birds. The claim of the classical physicist according to which each molecule possesses individuality should be understood indirectly: molecules have individuality because their statistical behavior indicates that two arrangements of molecules should be counted as different, if one of the arrangements differs from the other only by a change of position of several molecules. Global inference replaces direct verification. Now, applying the same inference, we shall say that in quantum physics, particles do not have individuality.

The determinists, by contrast, believe in the individuality of particles; they refuse to close the door to the possibility that some day we might succeed in localizing and observing in continuous fashion an individual particle. Let us examine what the consequence of this would be. We would have to interpret the Bose or the Fermi statistics, which would remain valid, as produced by an ensemble of individualized particles. This is possible, but it leads to very strange consequences.

Let us suppose that we are playing heads or tails. In throwing two coins simultaneously, we distinguish four possible combinations, which can be symbolized. Let A stand for heads, and B for tails, in the following way, letting the two coins be distinguished by subscripts:

$$A_1A_2, \quad A_1B_2, \quad B_1A_2, \quad B_1B_2 \tag{5}$$

The heads-tails combination, whatever be the order, is produced in half of the cases, while each of the two other combinations is produced only in one case out of four.

Let us now suppose that the game of heads or tails is governed by Bose statistics. We would then have the case heads-tails once every three times; and the same fraction would be observed for the case heads-heads and for the case tails-tails. Would we have to conclude that the coins are indistinguishable?

Not at all. They can be easily distinguished by direct observation. We would arrive at a different conclusion: we would conclude that the throws are not independent of each other. We would say that if the first coin shows heads, there is a tendency for the other coin to show the same side, such that the probability of the combination A_1B_2 is reduced to 1/6. The same reasoning applies to the combination B_1A_2, and the probability of the disjunction A_1B_2 or B_1A_2 takes the observed value 1/3. This interpretation can be extended, in a consistent manner, to a disjunction of r possible cases.[2]

The interpretation will be more complicated if one considers three outcomes which furnish Bose statistics. In this case, there exists not only a dependence linking one outcome to each of the others; there exists also a dependence between one outcome and the other two taken in combination. This means that we must consider relative probabilities which have a two-term reference class, such as the expression $P(A_1 . B_2, C_3)$, which denotes the probability of getting outcome C for the third item if the first shows outcome A and the second shows outcome B. We have here

$$\begin{aligned} P(A_1 . B_2, C_3) &\neq P(A_1, C_3) \\ P(A_1 . B_2, C_3) &\neq P(B_2, C_3) \end{aligned} \tag{6}$$

Let us examine the consequences of this probability consideration for the problem of the quantum statistics of gases. The fact that the statistics of gases furnishes the distribution calculated by Bose or by Fermi is well confirmed by experiment. If we are to put this in accord with the idea of distinguishable particles, we will have to introduce inter-particle forces which attract or repel them in such a way that their motions are no longer mutually independent. Each particle would depend not only on each other particle, but also on the totality of positions of the other particles. These forces would have a rather strange nature because they would be transmitted instantaneously across space and could be observed only by their effects on the statistics.

We clearly see that the problem of determinism is posed, in quantum theory, in a form essentially different from that of classical physics. The classical statistics of gases is compatible with determinism; and although determinism can be considered, in classical physics, as a hardly necessary addition, this physics offers no contrary indication. Quantum statistics, however, can scarcely be reconciled with determinism. This does not mean that determinism is absolutely excluded. If there were methods permitting one to observe the individual motions of the particles without disturbing them, and if these observations substantiated the existence of strict laws with the help of which one could exactly predict the particle trajectories, then determinism would be established. But what kind of physics would then have been bestowed on us! It would be a physics of mystical forces, far from the accepted idea of force in classical physics; and the resulting physics would hardly resemble that of Newton or of Laplace. In fact, this physics would be further removed from the principles of common sense than the indeterministic physics of quanta accepted today.

Here is the reason why the synoptic principle, according to which the content of observations can always be included in a ψ-function, is solidly established by quantum physics, in its negative part as well as in its positive part. It is not only the absence of other forms of observation which has persuaded the physicist to accept this principle. The system of quantum physics, in its entirety, constitutes an intrinsic proof in favor of the synoptic principle and hence against the validity of determinism. Any merger of the positive part of quantum physics and determinism would lead us to consequences so absurd that they cannot be accepted as plausible; conversely, the deduction of these consequences presents an inductive argument against determinism.

This situation can be compared to the one which exists in the theory of relativity regarding the principle of limiting speed of signals, which is equal to the speed of light. This principle of Einstein is not based solely on the fact that signals having a speed greater than that of light have not been found. Rather, it is founded on the testimony of a complete theory, crowned with success, whose consequences would be absurd if Einstein's principle were false. The negative part of this principle, the exclusion of signals speedier than light, is thus based on positive reasons; this is why the principle has been accepted.

Quantum statistics and determinism are natural enemies; they do not come from the same father. In the above, I have presented several ideas which can serve to clarify this divergence. Nevertheless, there is much to add. The difficulties that I have just described for the Bose and Fermi statistics form only one part of a more general problem, a problem which has taken an unexpected turn in quantum physics: the problem of unobserved objects. The following section is devoted to several ideas concerning this general problem.

2. UNOBSERVED OBJECTS AND THREE-VALUED LOGIC

The period when Newton and Huyghens were discussing the nature of light marks the beginning of a historical development which is clearly divided into three phases. During about a hundred years following Newton's work, the corpuscular theory was generally accepted. The second phase began when, with the discoveries of Young and Fresnel, the wave theory received unexpected support; and since Maxwell and Hertz have demonstrated the existence of electric waves, the wave theory became "a certainty, humanly speaking", if I may be permitted to use the words of Heinrich Hertz himself. The development that followed has shown that the profound intuition of this great physicist was never so lucidly manifested than in these words "humanly speaking". These words anticipated the third phase, the phase of the duality of waves-corpuscles, which is linked to the discoveries of M. de Broglie and which represents what we can call today 'the definitive form, humanly speaking'. Let us try to make precise the meaning of this solution, which is not restricted to the interpretation of light, but applies as well to that of matter.

On several occasions it has been the case that physical questions have invited the physicist to become a philosopher. To be sure, his philosophy is not of the kind of systems which have in stock ready-made answers to all questions that can be posed. The physicist is satisfied if he can find an answer to the one question which occupies him at the moment; but he insists that the answer be given in a precise language, a language as precise as the equations of physics, and which does not disappear in a fog of words and pictures. Fortunately, such a language exists today, and it has been furnished with tools similar to mathematical equations, namely the formulas of mathematical logic.

The problem of waves and particles is a problem of the logic of knowledge. It refers to the relation between observation and inferred object, to the method of extending the knowledge acquired by observation of macroscopic objects to non-observable objects, a method of crucial importance for physics, by the employment of which it has made its great discoveries and which finally has drawn it into some unprecedented difficulties. Let us study this method, at first in a completely general form, and then in its application to the problem of waves and particles.

The exigencies of practical life require us to add to our observations a theory of unobserved objects. This theory is quite simple: we suppose that unobserved objects are similar to those that we do observe. No one doubts that a house remains the same whether or not we observe it visually. The act of observation does not change things – this principle seems to be a truism. But a moment's thought shows that it is wholly impossible to verify it. A verification would require us to compare the unobserved object with the observed object, and hence to observe the unobserved object, which amounts to a contradiction.

It follows that the principle does not have the logical status of a true proposition. Rather, it is here a matter of convention; one introduces the definition that the unobserved object is governed by the same laws as those which have been verified for observed objects. This definition constitutes a rule which allows us to extend the observation language to unobserved objects; let us then call it an *extension rule of language*. Once this rule is established, one can find out whether the unobserved object is the same as the observed object or not; for example, we can conclude that the house stays at its place when we do not look at it, while the young girl does not remain in the box when the magician saws the box in two parts. Similar statements would have no meaning if a rule of extension of the observation language had not been added.

The last example exhibits a case where the rule allows us to infer that a change occurred in the unobserved object. Using a more scientific example, I could speak of the change in temperature which occurs when a thermometer is placed in a reservoir of water. We have here a case where the act of observation changes the object; despite this change, we have no difficulty in calculating the value of this change with the help of the laws of thermodynamics. Thus it seems that the extension rule of language allows us always to speak of unobserved objects in the form of meaningful propositions.[3]

Nevertheless, we must examine this thesis more closely. A rule of language is arbitrary, to be sure! But one can ask whether it is always possible to apply it. In other words, one must study the question of knowing whether the physical system constructed with the help of this rule is coherent. In classical physics, the answer is evidently affirmative; but this fact should be regarded as a result of experiment. It cannot be shown by purely logical considerations that it is always possible to construct a physical language satisfying the rule of extension. Here is the point at which quantum physics differs from the physics of macroscopic objects, a possibility which was overlooked in logical investigations prior to wave mechanics.

In this new mechanics, one must distinguish two different questions. The first concerns the disturbance of the object by the act of observation. The second concerns the state of the unobserved object. We would be mistaken if we wanted to conclude that the disturbance by the observation necessarily entails the indeterminacy of the unobserved object. The example of the thermometer placed in the water-reservoir shows that the disturbance by the observation does not exclude the determination of the state of the object before the observation: this state can be calculated with the help of physical theory. If the same method does not apply to quantum physics, one must adduce special reasons for this impossibility; it is here that the problem of the extension rule of language comes in.

Let us study the first question first. There is a very simple way to demonstrate the fact of the disturbance by the observation. This fact is deducible in a well-known manner from the theorem of the addition of probability amplitudes. If we have three non-commutative quantities u, v, w, this theorem gives the value

$$P(u_i, w_m) = \left| \sum_k \alpha_{ik}\beta_{km} \right|^2 \tag{7}$$

for the probability of measuring the value w_m after the observation of the value u_i. The terms α_{ik} and β_{km} represent the transformation matrices. On the other hand, the probability calculus yields a theorem of addition of probabilities in the following form:

$$P(u_i, w_m) = \sum_k P(u_i, v_k)P(v_k, w_m) = \sum_k |\alpha_{ik}|^2 |\beta_{km}|^2 \tag{8}$$

Since the two expressions (7) and (8) are not identical, it is necessary that we interpret the expression $P(u_i, w_m)$ in two different ways: in the case of relation (8), this expression denotes the probability of observing the value w_m after observation of the quantity u_i if a measurement of the quantity v has been made in between, a measurement the result of which has not been registered, while (7) yields this probability calculated for the case where no measurement has been made between the two observations. Now any measurement changes the physical conditions; this fact is completely expressed in the relations linking observable quantities.

Allow me to add a logical remark. Although it may be possible to show that observation disturbs objects, it can never be shown that observation does not disturb them. This means that in a different physics, which would furnish us with identical results for the two cases (7) and (8), we would not be able to conclude that the unobserved quantity is equal to the observed quantity; we could conclude only that it is permissible for us to define these two quantities as equal. There is then a certain asymmetry between the positive case and the negative case. In other words, we can demonstrate only, either the admissibility, or the inadmissibility of an extension of language such that the unobserved object is equal to the observed object. We shall encounter the same asymmetry again soon in a more general question.

Let us tackle the second question. Given that the act of observation disturbs the object, is it possible to calculate the value of the quantity in question before the observation? The answer depends on the admissibility of the rule of extension of language, on the convention that natural laws are the same for both observed objects and for those which are not observed. For, if this rule is admissible, we are permitted to employ physical theory for a relevant inference of the state of the unobserved object, as in the example of the thermometer inserted in the water-reservoir.

It is sometimes said that the disturbance by the observation excludes the possibility of meaningful statements concerning the unobserved object. This opinion is not correct. One is concerned here with a problem of the extension of language which exists as well for classical physics, because in this physics there are also observations which change the object. A similar extension offers no difficulty if the rule of extension is admissible. What distinguishes quantum physics from that of the classical statistics of molecules is the fact that this rule is inadmissible, that it can be shown that the laws of unobserved objects deviate from those that govern observed objects. Here we meet again the logical situation that we have just studied; on the basis of only observable facts, one can demonstrate a certain difference between observed objects and unobserved objects.

For simplicity of notation, let us call observable phenomena simply *phenomena*; those that cannot be observed we shall call *interphenomena*, because they are interposed between the phenomena with the help of an inference procedure. We shall use the term 'observable' in a broad sense, in such a way that it includes as phenomena the coincidence of two particles, indicated by a measuring instrument. Such quantum phenomena are inferred from macroscopic observations by the use of classical theory only; this is why they can be treated on an equal footing with macroscopic phenomena.

While phenomena always have the character of narrowly localized objects, interphenomena offer two divergent possibilities of interpretation: it is here that the alternative appears, waves or particles. These concepts have no significance if we speak only of phenomena; they belong to a language which comprises interphenomena. In speaking of particles, we attribute to interphenomena a narrow localization similar to that of the phenomena; in speaking of waves, we consider the interphenomena as extended over a large space.

The two interpretations represent two different rules concerning the extension of language, and the question arises: which of the two satisfies the principle that physical laws are the same for both the phenomena and the interphenomena?

The answer is quite definitive: neither the one nor the other satisfies this principle. The attempt to assign to interphenomena a definite existence, that is to say to attribute to them precise values of position and speeds which exist simultaneously, necessarily brings us to strange consequences regarding the principle of causality. I am not speaking here of the substitution of probability in place of certitude of prediction. This change seems minor compared to certain deviations of a different kind. Classical causality included the idea of gradual propagation, of a continuous transmission of the causal effect from one point in space to another, a transmission which occurs as time goes on. By contrast, the interphenomena of quantum mechanics exhibit causal relations which occur abruptly, which take no time to propagate, and which thus represent a true action at a distance. In this regard, the interphenomena differ essentially from the quantum phenomena, because observable quantities are always governed by causal laws which, except for their probabilistic character, resemble those of classical physics and satisfy the requirement of continuous propagation.

I do not need to give all the details, because these results have often been discussed. I am thinking of diffraction experiments, of apparatus in which a beam of electrons passes through two slits and is projected onto a screen. A pattern of interference fringes is observed, and it is well known that the pattern produced by two slits is not made up of the superposition of the patterns produced individually by each slit if the other is closed. One has a macroscopic observation which can be translated into the language of interphenomena only under pain of violation of the principle of causality: a particle passing through one of the slits knows if the other slit is open or closed. Or rather, omitting the anthropomorphism of the word 'knows', we would say that the motion of a particle passing through one slit depends on the physical condition of the other slit, a causal dependence which contradicts the principle of gradual action. I have suggested that, in such a case, one speak of a *causal anomaly*, a term which indicates the deviation from the causal behavior of observable phenomena.

The wave interpretation, it is true, solves this difficulty in that it exhibits the pattern on the screen as a train of waves passing simultaneously through both slits and subject to interference. However, this interpretation brings us to causal anomalies of another kind. The wave arrives at the screen all extended; but its effect consists of a little flash localized at a single point of the screen. As soon as the flash occurs, the extended wave disappears. This means that the process at a point of the screen exercises an influence on the wave at each point of the wave front, an instantaneous influence which contradicts the principle of gradual action and which represents a form of causal anomaly. The transition from the particle interpretation to the wave interpretation does not eliminate causal anomalies; the anomalies merely change places, and the interphenomena called 'waves' are not more reasonable than those that we call 'particles'.

The word 'reasonable', it is true, also represents an anthropomorphism. Let us then say that neither of the two interpretations can satisfy the extension rule of language, according to which unobserved objects are governed by the same physical laws as observable objects. These latter do not present any causal anomalies; it is only in the world of interphenomena that such anomalies subsist.

This result is equally applicable to a third interpretation which has been studied in a critical way by M. Louis de Broglie, more than 20 years ago, and which has been advanced recently with many promises and little logical analysis in an article of D. Bohm:[4] the interpretation of the pilot wave. This interpretation represents a combination of the two others; it speaks of a field of waves which guide the particles. Insofar as it employs the idea of waves, this interpretation avoids the anomalies arising for particles which pass through a diffraction grating; interference is a property only of waves and the particle is guided along the trajectories of the field resulting from the superposition of waves.

Nevertheless, the anomalies of the wave interpretation are repeated for this interpretation; moreover, it entails another sort of anomaly which has been very clearly pointed out by M. Louis de Broglie ever since his first studies of this subject. In this theory, there is a dependence of the statistical distribution characterizing the assembly of molecules – on the path of the particle, a dependence resulting from the fact that the ψ-function, which represents the wave, expresses at the same time the probability of observing a particle at a certain place. The consequences are quite strange, in particular, if one considers radiation of weak intensity such that the particles are sent one at a time. The path traversed by the first particle would here indicate the statistical distribution of the particles which are to follow it. The dependence can be interpreted in two ways: either the first particle causally determines the behavior of the particles to follow, or the first particle is behaviorally determined by future events. This difficulty is not avoided by attributing to the wave a physical reality, as M. Bohm believes. This real wave can account for the individual path of the particle; but the fact that the wave reflects the distribution of the particles to follow expresses a pre-established harmony which would have enchanted Leibniz, but which, in today's physics, can only be called a causal anomaly.

Allow me to add here several words concerning the question of the interpretation of a physical theory. I do not believe that there are any forbidden interpretations. In favor of those who would want to interpret the quantum equations in a certain way, I would like to say that I do not allow myself the locution 'forbidden meaning' in physics. I would propose that this be replaced by another maxim which says: "If you like to choose an interpretation, go ahead and do it; but draw the consequences". Here is the point which distinguishes logical analysis from the language of pictures. Let those who prefer to conceive the microcosm as made up of waves, or of particles, or of the two together, do so! But let them not forget to describe in a precise language the properties of this physical reality which they have created. They will be surprised by the strange aspect that the world of their construction presents in a close-up view.

The problem of physical reality is posed, for today's logician, in the form of an analysis of language. The world admits of a plurality of descriptions; each is true, but each requires, in order to be verifiable, a statement of the conditions on which it is based. Hence there is a class of equivalent descriptions, among which one can make distinctions only according to descriptive simplicity, that is, a simplicity that concerns only the form of the description and which has no significance whatever regarding the question of truth. Yet one can consider the question of whether the class includes a *normal system*, that is to say, a description which satisfies the rule of extension according to which unobserved objects are governed by the same laws as observable objects. For classical physics, such a system exists; it is everyday language. By contrast, quantum physics does not admit of a normal system of description; each admissible description includes some causal anomalies. I have proposed the name *principle of anomaly* for this result.

Here is the decisive difference between the two physics, and here is the way in which modern logic treats some surprising properties of the microcosm revealed in quantum physics. A property of the world of small dimensions is expressed in the form of a property of the class of admissible descriptions. To illustrate this method, one could cite those geometric methods which characterize a space in terms of the invariants of a class of transformations; and one could recall the methods of the theory of general relativity which describe a gravitational field by some invariant relations regarding the transformations of systems of reference.

A description which attributes a well-defined existence to interphenomena can be called *exhaustive*; it gives an answer to each question that can be asked regarding their state at each moment. However, since these answers often seem disagreeable to us, dissolving into causal anomalies, it has been proposed that statements concerning the interphenomena be entirely omitted. One thus arrives at a *restrictive* interpretation, an interpretation which does not speak at all of interphenomena, and which is reduced to observable phenomena.

The interpretation advanced by Bohr and Heisenberg is of this kind. It seems to me, nevertheless, that the prohibition against speaking of the interphenomena cannot be justified by logical reasons; rather it is based on physical reasons. It is because of the absence of a normal system that it would be prudent not to speak of interphenomena; and that is all that can be said in favor of this prohibition. And the absence of a normal system expresses a physical fact. Therefore the restrictive interpretation arises from a physical fact.

In the language of the logician, the theory of Bohr and Heisenberg takes the form of a theory of meaning. A certain group of propositions is considered as meaningless and is thus omitted from the domain of admissible statements. The question arises as to whether this radical remedy does not go beyond the limits of a reasonable restriction. In fact, it seems doubtful that physics can completely renounce a description of the interphenomena. For example, if a beam of electrons passes through two slits and produces a pattern of interference fringes on a screen, one would like to say that something or other traverses the slits; no one doubts that this "something or other" cannot penetrate the solid matter of the diaphragm. Is it necessary to sacrifice this commonplace statement? If we take the conception of Bohr and Heisenberg seriously, then we would have to.

Mathematical logic offers a method of avoiding this unhappy consequence. It has presented us with systems of multi-valued logic, that is to say, systems in which the duality of truth-values, 'true' and 'false', is replaced by a multiplicity of values. In particular, it is the three-valued logic which presents itself for the interpretation of quantum physics. This logic includes an intermediate category between 'true' and 'false' which can be considered as denoting 'indeterminate', a category which can serve to include the statements concerning the interphenomena.

A very interesting interpretation of this kind has been constructed by Mme. P. Fevrier, who was the first to apply the results of multi-valued logic to quantum physics. The interpretation which I have proposed differs on certain points from that of Mme. Fevrier; and I permit myself to present here the general ideas on which this interpretation is based.[5]

The use of truth-tables, a method developed originally for two-valued logic, is very convenient for the construction of a logic. These tables indicate the relations between the truth-values of elementary propositions and those of propositions compounded from the latter with the help of logical operations, such as those expressed by the terms 'not', 'or', 'and', 'implies', etc. Their form is given in tables I and II. The letter "T" denotes 'true'; the letter "F" denotes 'false'; the letter "I" denotes 'indeterminate'. The theorems, or *tautologies*, of a logic are those formulas which possess a T in each place of their column. A comparative list of some tautologies of the two logics is given in table III.

TABLE IA
Two valued logic

	negation
a	$-a$
T	F
F	T

TABLE IB
Two-valued logic

		disjunction	conjunction	implication	equivalence
a	b	$a \vee b$	$a \,.\, b$	$a \supset b$	$a \equiv b$
T	T	T	T	T	T
T	F	T	F	F	F
F	T	T	F	T	F
F	F	F	F	T	T

I do not wish to enter into a detailed discussion of these tables. Allow me, nevertheless, to show you the method by which the three-valued logic treats the problem of radiation passing through two slits.

If no observation is made at the slits, one would say, using two-valued logic, that an individual particle observed on the screen has passed either through one slit, or through the other. This statement, which belongs to an exhaustive description, leads us into causal anomalies, as we have just discussed. The restricted description, in the form of a three valued logic, replaces this disjunction by another, constructed with the help of three values, and called *diametrical disjunction*. It has the following properties: if the particle has been observed in the vicinity of one of the slits, the disjunction is true;

TABLE IIA
Three-valued logic

	negation		
	cyclical	diametrical	complete
A	$\sim A$	$-A$	$\bar{A}$
T	I	F	I
I	F	I	T
F	T	T	T

TABLE IIB
Three-valued logic

				implication			equivalence	
		disjunction	con-junction	normal	alter-native	quasi-implication	normal	alternative
A	B	$A \vee B$	$A . B$	$A \supset B$	$A \rightarrow B$	$A \Rrightarrow B$	$A \equiv B$	$A \equiv B$
T	T	T	T	T	T	T	T	T
T	I	T	I	I	F	I	I	F
T	F	T	F	F	F	F	F	F
I	T	T	I	T	T	I	I	F
I	I	I	I	T	T	I	T	T
I	F	I	F	I	T	I	I	F
F	T	T	F	T	T	I	F	F
F	I	I	F	T	T	I	I	F
F	F	F	F	T	T	I	T	T

and if no observation has been made at the slits and the statement of passage is indeterminate, the disjunction is also true. By employing this meaning for the word 'or' we can thus say that the particle passes through one slit or the other, without arriving at causal anomalies. The diametrical disjunction, which can also be called the equivalence of contraries, is written in the form:

$$B_1 \equiv -B_2 \tag{9}$$

This statement, which is true, characterizes the passage of radiation across the two slits. But we cannot derive from it the ordinary disjunction

$$B_1 \vee B_2 \tag{10}$$

because this last disjunction may be indeterminate. It would be different in two-valued logic: here one could deduce relation (10) from relation (9).

The language of three-valued logic thus permits us to formulate, as admissible statements, all that we know as regards phenomena and interphenomena, without bringing forth causal anomalies. It is only the relation between the

TABLE III
Tautologies

	Two-valued logic	Three-valued logic
T1. Law of identity	$a \equiv a$	$A \equiv A$
T2. Law of double negation	$\bar{\bar{a}} \equiv a$	$A \equiv - - A$ $\bar{A} \equiv \overset{\equiv}{\bar{A}}$
T3. Law of triple negation	– – –	$A \equiv \sim \sim \sim A$
T4. Relation between negations	– – –	$\bar{A} \equiv \sim A \lor \sim \sim A$
T5. *Tertium non datur*	$a \lor \bar{a}$	– – –
T6. *Quartum non datur*	– – –	$A \lor \sim A \lor \sim \sim \underset{\bullet}{A}$
T7. Law of contradiction	$\overline{a \,.\, \bar{a}}$	$\overline{A \,.\, \bar{A}}$ $\overline{A \,.\sim A}$ $\overline{A \,.\, -A}$
T8. Laws of de Morgan	$\overline{a \,.\, b} \equiv \bar{a} \lor \bar{b}$ $\overline{a \lor b} \equiv \bar{a} \,.\, \bar{b}$	$-(A \,.\, B) \equiv -A \lor -B$ $-(A \lor B) \equiv -A \,.\, -B$
T9. First distributive law	$a \,.\, (b \lor c) \equiv a \,.\, b \lor a \,.\, c$	$A \,.\, (B \lor C) \equiv A \,.\, B \lor A \,.\, C$
T10. Second distributive law	$a \lor b \,.\, c \equiv (a \lor b) \,.\, (a \lor c)$	$A \lor B \,.\, C \equiv (A \lor B) \,.\, (A \lor C)$
T11. Contraposition Law	$\bar{a} \supset b \equiv \bar{b} \supset a$ $a \supset b \equiv \bar{b} \supset \bar{a}$	$-A \supset B \equiv -B \supset A$ $\bar{A} \rightarrow B \equiv \bar{B} \rightarrow A$ $A \supset B \equiv -B \supset -A$
T12. Dissolution of Equivalence	$(a \equiv b) \equiv (a \supset b) \,.\, (b \supset a)$	$(A \equiv B) \equiv (A \supset B) \,.\, (B \supset A)$ $(A \mathrel{\overline{\equiv}} B) \mathrel{\overline{\equiv}} (A \not\rightleftharpoons B) \,.\, (-A \not\rightleftharpoons -B)$
T13. *Reductio ad absurdum*	$(a \supset \bar{a}) \supset \bar{a}$	$(A \supset \bar{A}) \supset \bar{A}$ $(A \rightarrow \bar{A}) \rightarrow \bar{A}$

two kinds of phenomena which is expressed in the form of a true proposition, while a statement concerning the interphenomena separately receives the value 'indeterminate'.

Similar solutions for other problems involving causal anomalies can be developed, such as the problem of the energy barrier traversed by particles, and one can also formulate the relation of complementarity for two non-commutative quantities. These results indicate that the three-valued logic offers us a form of language which allows us to speak of the quantum world without undesirable consequences.

Let us try to summarize this analysis of the problem of unobserved objects. The alternative between corpuscular theories and wave theories, which marked the two first phases of conceptions of the nature of light, has been replaced, as a result of the discoveries of M. de Broglie, by a conjunction: the word 'or' has been replaced by the word 'and'. At the same time, this conjunction, which indicates the third phase of the historical development of investigations of the nature of light, has been extended in such a way as to encompass the nature of matter. Yet, a logical analysis shows that this word 'and', indicating the conjunction, does not belong to the language of physical objects; it refers to a duality of descriptions and hence belongs to the metalanguage, a language which deals with properties of linguistic systems with the help of which we describe the physical world. One can even pass from the duality of descriptions to a plurality: a class of equivalent descriptions can be constructed. It is in the form of this class that modern logic treats the problem of unobserved objects; and the choice of descriptions is presented as a choice among extension rules of language.

Although this plurality of descriptions is already applicable to classical physics, it is not of much importance there; among the equivalent descriptions there exists a normal system, which is convenient to employ, forgetting all the others. By contrast, in quantum physics, the class of equivalent descriptions does not include a normal system. Each description entails causal anomalies if it is exhaustive, that is to say, if it attributes a well defined state to the interphenomena. To avoid the anomalies, restrictive descriptions can be constructed which exclude statements about interphenomena from the domain of true assertions. The three-valued logic offers us an adequate form of this kind of description and allows us to speak of the interphenomena in an indirect way, such that the relations between phenomena and interphenomena are expressed by true assertions, while it is not possible separately to derive statements concerning the interphenomena.

Does this result show that the true logic of quanta is three-valued? I do not believe that one can speak of the truth of a logic. A system of logic is empty, that is to say, it has no empirical content. Logic expresses the form of a language, but does not formulate any physical laws. Nevertheless, one can consider the consequences for the language of the choice of logic; and, to the extent that these consequences depend at the same time on physical laws, they reflect properties of the physical world. It is thus the combination of logic and physics which indicates the structure of the reality which concerns the physicist. A quantum physics under the form of a two-valued logic exists; but it includes causal anomalies, while a quantum physics which employs a three-valued logic does not possess any such anomalies. This conclusion, which is the result of all the experiments encompassed by the wave mechanics, expresses the strange structure of the microcosm, a structure which has so worried the physicist, but which he has learned to master by means of an ingenious system of equations and of experimental apparatus. Here is the general result established by the work of the physicists; it remains only for us to accept it, although we are well aware that it is here again a question of certitude 'humanly speaking'.

3. THE DIRECTION OF TIME IN CLASSICAL PHYSICS

In order to understand the contribution of quantum mechanics to the problem of time, one must begin by studying this problem in classical physics. We shall see that, on the one hand, wave mechanics includes a development destructive of the concept of time, while, on the other hand, wave mechanics has given this concept a new foundation, a foundation that classical mechanics could not furnish.

The concept of time comprises metrical properties and topological properties. The metrical properties have been treated, in our day, by the theory of relativity; I do not need to speak here of these well-known discoveries which have profoundly changed our ideas of the simultaneity of spatially separated events. I wish to speak about the topological properties, and I would like to show that classical physics furnishes us with powerful instruments appropriate for analyzing the topological aspect of time, an aspect which determines so completely our conception of the physical world as well as the form of our psychological experiences.

The study of the topological problem of time requires that we carefully distinguish between *the order* and *the direction* of time. The order of time corresponds to the order of points on a straight line; this one-dimensional extension is ordered without possessing a direction. In other words, the points of a line are ordered with respect to the relation 'between', but one cannot say whether the line is extended from right to left or from left to right. In order to assign a direction to the line, one must use other means; for example, one can choose one point and specify that it is situated to the left of a certain other point. By contrast, if one is given three points, the line itself will determine that which is situated between the other two. In the same way, the order of events in time concerns relations expressed with the help of the word 'between', while the direction of time assigns a sense to temporal lines, a unique sense which manifests itself in the flux of time, in the conception that time goes from past to future.

It is well known that classical mechanics cannot furnish us with the *direction* of time. The differential equations of classical mechanics are of the second order; thus, given a solution, the replacement of the variable t by the variable $-t$ leads to another solution. A ball thrown into the air, a planet moving around the Sun, these are *reversible* phenomena: the motion can occur in one or the other direction. Thus, the observation of motion considered as a mechanical phenomenon gives us no information about the direction of the motion.

This result being well known, it has often been forgotten that classical mechanics can very well give us precision as regards the *order* of time. If a ball goes from point A, via point B, to point C, mechanics lets us consider this motion as a transition from C via B to A, but we are required to keep B between C and A. It is for this reason that classical mechanics presents us with a temporally ordered physical world. In the framework of the theory of relativity, this order has become the source of the causal theory of time, according to which the concept of time originates from the causal order and is based on the following definition: an event A precedes an event C if a signal sent from A arrives at C. We know that this definition, because of the limiting character of the speed of light, leaves undetermined the order of certain events situated on space-like world-lines, for which $ds^2 < 0$. But apart from this restriction, this definition determines a temporal order of the physical world which corresponds to the order of our psychological experience.

The causal time of the theory of relativity is ordered, but it does not have a direction. The Lorentz transformation is invariant with respect to time reversal: If t is replaced by $-t$, and t' by $-t'$, one obtains a new Lorentz transformation, describing a motion in the opposite direction. This is why relativity, despite its reduction of time to causality, has contributed nothing to the problem of the direction of time.

A definition of the direction of time requires us to distinguish between cause and effect, that is to say, to add to the concept of causal connection a criterion which gives meaning to direction. Such a criterion is furnished by the second law of thermodynamics, by the concept of entropy. Let us examine the significance of this concept for the direction of time.

At the time of classical thermodynamics, there was no doubt that one had found, in the concept of entropy, an instrument which permits the introduction of the direction of time in physical equations. This optimistic opinion ran into serious difficulty when L. Boltzmann gave the statistical formulation of the second law of thermodynamics. The difficulties came to be well known in the objection concerning the reversal of the motions of molecules: if it were possible to reverse the individual motions of all the molecules, the gas would pass from a state of higher entropy to a state of lower entropy. Such a situation cannot be excluded if entropy is interpreted as a statistical property, because the individual motions of the molecules are reversible. Moreover, such a situation ought to occur, in the history of an isolated system, as often as the corresponding situation of the passage from lower entropy to higher entropy. This result follows from certain theorems established by J. Loschmidt and H. Poincaré; it means that in the graph of entropy of an isolated system, there are, from time to time fluctuations, such that the number of peaks and of valleys is infinite and their quotient converges to unity.

It is very interesting to study the publications dealing with the objection of time reversal toward the end of the last century. Boltzmann assures us that the inverse transitions occur very rarely in the history of an isolated system; it follows that one would be justified in saying that if a system is observed to be in a low entropy state, then one could conclude that it will soon pass into a state of higher entropy. Unfortunately, the same conclusion follows as regards the preceding state; it can be shown that it is very probable that the system arises from a state of higher entropy. The inference is applicable, symmetrically, to the future and to the past; it is thus impossible for us to define a direction of time using the entropy of an isolated system.

Let us put this result in more precise terms. One must distinguish between an *inference from time to entropy* and an *inference from entropy to time*. The first pertains to the following question: suppose that the system is observed in a state A of lower entropy, what will be the value of the entropy in a subsequent state B? The answer is that it is extremely probable that this value is high. This result allows us to predict the future state of the system. The second inference pertains to the question: suppose that we have observed two states A and B of the system of which A has a lower entropy and B a higher entropy; which of the two states preceded the other? The answer is that it is just as probable that A preceded B as it is that B preceded A. This result follows from the symmetry of the first inference, regarding which we concluded that a state of lower entropy would be preceded, in all probability, by a state of higher entropy. We arrive at the result that the inference from time to entropy is admissible, as it is commonly acknowledged; but that the inference from entropy to time should be rejected and that it is impossible for us to define a direction of time for an isolated system.

There exists an isolated system which is of particular interest to us: the universe. This system, without doubt, is completely isolated. Thus our universe does not have any direction of time; this conclusion cannot be rejected if it is admissible to speak of the entropy of the universe.

But, is it admissible to speak of it? One can doubt it. If the universe is spatially infinite, the concept of entropy is not applicable to it. We would, however, have no need to study this question. The direction of time is something which concerns the ensemble of our immediate environs; if we have to have recourse to distant nebulas, hidden in the profound depths of the universe, in order to resolve the question of the direction of time, we would not be able to obtain an interpretation of the experience of habitual everyday time. We have to examine anew the inferences which the physicist makes when he passes from entropy to time.

In fact, when he observes a given entropy state, the physicist refuses to conclude that this state has been preceded by another state having a higher entropy. Suppose that you are shown two pictures cut from a film, one of them showing a house in good order, the other showing the same house after the effects of an explosion; you would know well in what direction the film was taken. You would say that it is very probable that the orderly state, of lower entropy, preceded the state of disorder, of higher entropy, thus rejecting the conclusion of Boltzmann according to which the probability of this affirmation is but 50%, the same as for the inverse order of time. How does one explain this refusal to accept the results of physical theory?

One runs up against difficulties when one wants to render precise the meaning of the word 'probable', since the latter is used so often in daily language. A probability denotes a frequency, and a frequency refers to a class of events, or, more exactly, to a sequence of events, that is, to an ordered class. Differences in the meaning of the word 'probable' are often explained by the differences among the sequences to which the word refers. Let us examine the question from this point of view.

The probability calculated by Boltzmann refers to the history of an isolated system, and thus to a *time ensemble*. By contrast, the probability used in the example of pictures of houses is not of this kind. In the history of a house, there are not enough changes of state to warrant constructing a frequency; an explosion happens once, and the history of the house is finished. If a frequency is to be reckoned, it concerns a plurality of houses; that is, it concerns a *space ensemble*. The probability that we use to infer a direction of time does not refer to a time ensemble, but to a space ensemble.

Let us study this ensemble. There are many physical systems which originate from larger systems and which subsequently get separated from the latter; once separated, they remain pretty much isolated for a long enough period. These systems, which I shall call *branch systems*, begin by being in an ordered state of lower entropy, and they proceed toward a state of disorder, of higher entropy. The ordered state is not presented here as a result of chance in the history of the isolated system, but as a result of a mutual action between the system and its surroundings; the larger system proceeds toward a state of higher entropy, while a system of low entropy is created in the form of a part of the larger system.

The universe has plenty of branch systems of this sort. Milk is poured into coffee and thus a branch system of low entropy is created; the sun's rays heat a rock which, together with the snow around it, represents a system of low entropy tending toward temperature equalization, etc. It is the ensemble of such branch systems which offers the possibility of explaining the form of probability used in the inference from entropy to time: instead of saying that it is probable that the ordered system is the unlikely product of a process of separation in the history of the isolated system, we say that it is probable that the system had not been isolated in the past and is a product of outside intervention. For example, if we find a cigarette burning on an ashtray, we do not believe that this state was preceded by a state where the cigarette was all ashes, a state from which the present state had developed by an improbable separation of the chemical components of the ashes. We would prefer to suppose that somebody had lighted the cigarette several minutes ago, and thus that the present improbable state is the result of outside intervention. We would say that this explanation is more probable than the preceding one. This second occurrence of the word 'probable' refers not to a frequency in the time ensemble, but to a frequency computed in the space ensemble of branch systems.

The probability calculus provides methods for treating these two kinds of probability.

We construct a probability lattice:

$$\begin{array}{ccccccc}
x_{11} & x_{12} & x_{13} & \dots & x_{1i} & \dots & \rightarrow p \\
x_{21} & x_{22} & x_{23} & \dots & x_{2i} & \dots & \rightarrow p \\
\dots & \dots & \dots & \dots & \dots & \dots & \\
x_{k1} & x_{k2} & x_{k3} & \dots & x_{ki} & \dots & \rightarrow p \\
\dots & \dots & \dots & \dots & \dots & \dots & \\
\downarrow & \downarrow & \downarrow & & \downarrow & & \\
p_1 & p_2 & p_3 & & p_i & & \rightarrow p
\end{array} \tag{11}$$

Each horizontal line represents the history of a branch system, and thus a time ensemble. By contrast, the columns represent space ensembles. Let us consider the probability of a lower entropy state: for each line, this probability p has a very small value. It is quite different with the columns: in the first column, the probability p_1 of a state of lower entropy is quite high because in the column one is concerned with states of separation, of initial states of branch systems. Little by little, the probability values $p_2, p_3, \ldots$ diminish and converge eventually to the probability p.

Using a notation developed elsewhere,[6] I write

$$P(B^{ki})^i = p \tag{12}$$

for the probability of a state B in a horizontal line, and

$$P(B^{ki})^k = p_i \tag{13}$$

for the probability of B in a column. The repeated index thus indicates the direction in which the frequency is determined. The convergent lattice (11) is characterized by the relation

$$\lim_{i\to\infty} P(B^{ki})^k = P(B^{ki})^i \tag{14}$$

In the theory of probability lattices it is shown that such a lattice which describes a mixing process is characterized by certain special relations which cannot be deduced from the axioms of the probability calculus and which have to be considered as definitions of types of lattices. From the standpoint of physics, this means that the type of lattice constitutes a form of empirical hypothesis; that is, that only experience can justify the choice of lattice used as a description of physical phenomena. What is in question here, in particular, is a property which I have called *lattice invariance*[7] and which can be formulated, in a simplified form, as follows:

$$P(A^{ki}, B^{k,i+n})^i = P(A^{ki}, B^{k,i+n})^k \quad (n > 0) \tag{15}$$

These expressions denote relative probabilities. On the left-hand side, we have the horizontal probability, that for which a state A of lower entropy is followed by a state B of higher entropy. On the right-hand side, we have a vertical probability of the following form: in the column $i + n$ one chooses all the elements $x_{k,i+n}$ (here k is variable, $i + n$ is constant) for which the element x_{ki} of column i is found in a state A; and in the partial sequence thus defined, one counts the elements belonging to state B. If this frequency corresponds to the frequency determined in a line, indicated on the left-hand side, the condition of lattice invariance is fulfilled.

It can be shown that if a lattice composed of horizontal series with diminishing memory (like Markoff series), if it satisfies the condition of lattice invariance, it is always convergent and thus characterized by the relation (14). In other words, whatever the distribution of a state B in the first column may be, the later columns have a tendency to reproduce the B-distribution existing in the horizontal lines.

Physically, this means that the space ensemble has a tendency to reproduce the time ensemble. Since this statistical tendency does not follow from the axioms of the probability calculus, it expresses a natural law which is verified by an immense number of observations and which is applicable to all the ensembles of physical systems, be they assemblies of molecules, or macroscopic systems composed of a large number of molecules, or stars. Let me speak here of a *statistical isotropy*, applying to the statistical conditions a term which, in optics, indicates an equivalence of directions. It is the statistical isotropy which yields an explanation of the phenomena which express the direction of time.

For it is the statistical isotropy which determines that the branch systems all proceed in the same direction. Let us suppose that half of these systems proceed in the inverse direction, starting from a state of high entropy and proceeding toward a state of lower entropy. It follows that the lattice composed of all the branch systems could not be isotropic: the last columns, like the first ones, would include a large number of states of lower entropy, a result which contradicts the distribution of similar states in the horizontal lines. The relation (14) would not thus be fulfilled. We conclude that this supposition is false. Statistical isotropy expresses that property of the universe to which we owe *the parallelism of the increase of entropy*: the totality of branch systems defines a single direction of time.

Once this result is established, it is easy to account for the inference from entropy to time. If the frequency is determined vertically, that is, in a column, one finds only a small probability that a state of lower entropy be preceded, for the same system and thus on a horizontal line, by a state of higher entropy; at least, this result applies to the initial columns of the lattice. We have thus succeeded in finding the answer to the objection of the reversal of molecular speeds. The symmetry of the probabilities in regard to the future and to the past exists only for the time ensemble; it no longer exists for the space ensemble. And the direction of time can be defined in terms of entropy because the aggregate of branch systems has an asymmetry with respect to time, an asymmetry which comes from the parallelism of the increase of entropy.

These ideas, it seems to me, furnish the solution to the problem of the direction of time in classical physics. Whereas classical mechanics gives us only the order of time, statistical thermodynamics distinguishes between the two directions of this order and shows that they are qualitatively statistically different. The irreversibility of the course of macroscopic events is compatible with the reversibility of the course of elementary events.

You might ask me: just how does this solution differ from that given by Boltzmann? Let me add a few words about this. Boltzmann was well aware that the entropy curve of an isolated system cannot define a direction. He thus restricted the direction of time to an ascendant part of the entropy curve of the universe, and he remarked that the human inhabitants of the universe considered as positive time the direction for which their part of the curve is ascending. He was well aware that this conception entailed the possibility of different directions of time for different cosmic epochs.

I have nothing to say against this last conclusion; indeed, I believe that a direction of time can be defined only for one cosmic epoch, and I reject completely all attempts to treat cosmic time as a whole. Thus, it seems to me that although the existence of a rise in the entropy curve of the universe is a necessary condition for a direction of time, it is not a sufficient condition, for it. One must add the hypothesis of branch systems governed by the laws of statistical isotropy. As soon as we add this hypothesis to the ideas of Boltzmann, we come to understand the direction of time; we are able to account for numerous observed phenomena in our immediate neighborhood which manifest the direction of time. Without this hypothesis, which is not a logical consequence of the fact that the entropy curve rises, we would not be able to explain the existence of such a direction.

I would like very much to present several other results which are deducible from the ideas just developed, and which concern something new: the statistics of macroscopic phenomena. In statistics of this sort, one counts as elements not the states of molecules but the states of macroscopic systems; and one can define a macroscopic entropy for them which resembles in many ways the thermodynamical entropy. To give an example, one can take the process of shuffling in a game of cards; there are many natural processes which exhibit similar qualities. A statistics of traffic accidents, for example, can be treated from this point of view, if you agree to call a traffic accident a natural event.

There exists a consequence which follows from the hypothesis of branch systems and which accounts for a certain difference between the positive direction and the negative direction of time. If we observe an improbable coincidence, we conclude that the two events are the products of a common cause. For example, if two electric lamps are simultaneously extinguished, one concludes that a fuse has blown out, or that the current has been interrupted for the whole street; it would be too improbable to suppose that the two lamps burnt out at the same moment. There is also an effect common to the two events; it becomes dark in the room and one stops reading one's newspaper. But this common effect cannot serve as an explanation of the improbable coincidence; it is only the common cause which can offer an explanation; this is what I call *the principle of the common cause*.

This idea can be expressed in another way: the total cause of a partial effect can be inferred, but the total effect of a partial cause cannot be inferred. This is why it is easy to register the past, whereas it is very difficult to predict the future. For example, one can deduce yesterday's barometric pressure from the mark registered on the paper of a barograph, but if one wishes to predict tomorrow's pressure, one must know some meteorological data concerning a much more extended region. A definition of the direction of time can be based on these ideas.[8]

What is at stake is the familiar enough difference between the positive direction and the negative direction of time, a difference intimately tied to the distinction between causality and finality, between explanation with the help of causes and explanation with the help of goals. It is not easy to incorporate it into the general statistical principle formulated in the second law of thermodynamics. Nevertheless, by employing the hypothesis of branch systems, it can be shown that the principle of the common cause, which governs a major part of our inferences and reveals our belief in a direction of time, follows from the second law and represents an application of the principle of isotropy to macroscopic statistics. The proof makes use of the probability lattice already discussed. It then follows that the complete abandonment of the principle of final causality, an abandonment which is characteristic of modern science, does not represent an arbitrary decision to accept a certain way of thinking, but is demanded by the second law of thermodynamics, which requires us to look always for a *causal* explanation.

Finally, these ideas can be related to a new branch of applied mathematics, to information theory, which has been developed in the works of C. Shannon, W. Weaver, N. Wiener, and J. von Neumann. We can use the measure of information introduced in these works to define the direction of time in terms of a statistics of macroscopic events. By combining the concept of information with the principle of the common cause, one can explain the fact that the past can be registered, but not the future. One thus arrives at a theory of registering instruments and one can show that the increase of information indicates the flux of time, the same direction (as that) characterized by thermodynamics by means of the increase of entropy.

There is a mutual relation between information and entropy: information signifies negative entropy, and entropy can be interpreted as a measure of ignorance rather than of information. A high entropy value indicates an extensive class of possible molecular arrangements; and the inverse relation between information and entropy expresses but a well-known law of traditional logic, the law of the inverse relation between comprehension and extension.

This consideration requires that we add a remark concerning registering instruments. In these instruments, the increase of information indicates the positive direction of time; thus this direction is indicated here by a diminution of entropy. This is possible because registering instruments are not isolated systems. They are pieces of apparatus which accumulate order taken from their surroundings; in fact, their order reflects the order of the space ensemble of branch systems and informs us about acts of intervention in the past, these being causal processes registered on the dial of the instrument.

Why am I presenting you with these ideas in the course of an analysis of the logical foundations of quantum theory? Surely, quantum phenomena are not macroscopic. But it is not the quantum phenomena themselves which constitute the aspect of the physical world of our experience; it is the consequences of these phenomena for macroscopic objects which create the system of relations governing the world of our experience. And these consequences, as regards the structure of time, proceed through the intermediary of the statistics of measurements and thus of the statistics of registered information.

4. THE DIRECTION OF TIME IN QUANTUM PHYSICS

In classical physics, the problem of the direction of time presents itself in the form of a paradox: while the course of molecular events is reversible, that of macroscopic events is not reversible. The solution of the paradox is given by the statistical conception of macroscopic matter, a conception which offers us the possibility of constructing a specific distinction between the positive direction and the negative direction of time.

People have often hoped to find a different solution for quantum physics. People have thought of the possibility that elementary phenomena might be of an irreversible nature; the abandonment of determinism, it was believed, could be associated with a change of this kind because an indeterministic mechanics would be exempt from strict laws which link the future to the past in the manner of a unique correspondence. I would like to examine this problem. But before entering into the details, let us say immediately that the result will be negative, that the elementary quantum phenomena have been shown to be just as reversible as the molecular motions of classical mechanics. Moreover, certain recent developments show that the problem of time in small dimensions is much more complicated than that of the molecules of classical mechanics. These latter at least left us with the concept of the order of time; but it seems that the particles of quantum mechanics do not even admit of an order of time. For this reason, the statistical theory of time is indispensable to quantum physics as well as to classical physics. On the other hand we shall see that the combination of quantum phenomena with statistical theory brings us to some surprising consequences which present the concept of time in a different light.

Classical mechanics does not furnish a direction of time because its differential equations are of the second order. If we replace t by $-t$ in a solution of these equations, we arrive at another solution. But, things are different with Schrödinger's equation because this equation is of the first order with respect to time.

Let us write Schrödinger's equation in the form:

$$H_{op}\psi(q,t) = c\,\frac{\partial\psi(q,t)}{\partial t}, \quad c = \frac{ih}{2\pi} \tag{16}$$

In fact, if $\psi(q, t)$ is a solution of this equation, then $\psi(q, -t)$ is not a solution, because this function satisfies the equation

$$H_{op}\psi(q,-t) = -c\,\frac{\partial\psi(q,-t)}{\partial t} \tag{17}$$

By contrast, it is the complex conjugate $\psi^*(q, -t)$ which satisfies (16), while the complex conjugate $\psi^*(q, t)$ satisfies (17). This is easily seen if one puts

$$\psi(q,t) = \phi(q)\,e^{2\pi i\nu t} \tag{18}$$

where $\phi(q)$ is a complex function which does not depend on t. We thus have

$$\psi^*(q,-t) = \phi^*(q)\,e^{\pi i\nu t} \tag{19}$$

and this function satisfies (16). But the functions $\phi(q)$ and $\phi^*(q)$ differ from each other; one can thus distinguish between them physically for example by calculating the distributions of a quantity which does not commute with q. Can we conclude from this that Schrödinger's equation defines a direction of time?

This conclusion would not be valid. While observations can distinguish between ϕ and ϕ^*, they cannot distinguish between $\psi(q, t)$ and $\psi(q, -t)$, that is to say between (18) and the function

$$\psi(q, -t) = \phi(q)\, e^{-2\pi i \nu t} \tag{20}$$

which possesses the same factor $\phi(q)$ as does $\psi(q, t)$. As (20) satisfies equation (17), a distinction between $\psi(q, t)$ and $\psi(q, -t)$ would be possible only if we could discriminate between the forms (16) and (17). But such a possibility exists only on the condition that we knew the direction of time. The sign of the second member of Schrödinger's equation indicates the direction of time, it is true; but this sign presupposes the direction of time, and if we reverse this direction, we pass from the form (16) to the form (17) without encountering any contradiction with experience.

To summarize, any determination of the direction of time with the help of Schrödinger's equation would constitute circular reasoning: it furnishes nothing other than the direction that has already been introduced by other means.

To put it another way, in quantum physics we depend on the methods of classical physics if we wish to ascertain the direction of time. By applying these familiar methods, we arrive at a sign of the second member of Schrödinger's equation and there is no other method.

The course of elementary quantum events is reversible; the equations of quantum theory do not offer us the possibility of defining the positive direction of time. Here is the first result of our analysis. A second result can be deduced, which seems even more destructive for the concept of time.

The identity of a physical object in the course of time is a relation which must be clearly distinguished from that of logical identity. Everything is identical with itself; this tautological principle of logic cannot give us any information about a physical object, because a physical object is composed of a sequence of successive states. Each state is logically identical with itself, but if we consider the entire sequence of states as constituting a physical object, we are employing a concept of *physical identity* which can be called *genidentity*, making use of a term introduced by K. Lewin.

As regards macroscopic objects, the application of this concept offers no difficulty. One easily distinguishes one person from another and even one egg from another, because the eggs can be marked. In the first section, I have pointed out the difficulties which arise for molecules, and I have discussed indirect methods for defining the genidentity of molecules. The results of this discussion can be summarized as follows: in classical physics, there is a natural genidentity of molecules, although it is not possible to determine, with the help of observations, the individuality of each molecule among the others. In the Bose or Fermi statistics, one can speak of the genidentity of molecules; but this genidentity is arbitrary and can be introduced in different ways. This means that, if two molecules collide, one can identify them after the collision as one wishes; it is not possible to distinguish experimentally a physical continuity and a kind of 'crossing over'. Genidentity has thus become a matter of arbitrary definition. Nevertheless, there remains a genidentity in the following sense: the number of particles is constant. In this sense there remains a residue of individualism: although one cannot distinguish the individual molecules one from the other, one can at least count them.

A deviation from this principle has been observed for photons. But photons are not ordinary matter; they have no rest mass. The deviation would be more serious if one could no longer count electrons. This conclusion, which represents the end of individualism, has been drawn by E. Stückelberg and R. Feynman[9] in recent works. According to these authors, the world line of an electron can curve in such a way as to return toward the past; in certain periods of its existence, the electron would thus be displaced backward in time. This part of its motion, however, admits of a second interpretation: one can speak of a positron moving forward in time. Here then are two equivalent descriptions of the same phenomenon; one is just as true as the other, and there is no experiment that could discriminate between the two interpretations.

In the discussion of the problem of unobserved objects, we have spoken of the theory of equivalent descriptions, and we have shown that each exhaustive description of the interphenomena, in quantum physics, entails certain causal anomalies. The results of Stückelberg and Feynman offer a new illustration of this principle of anomaly. The path of an electron, in fact, belongs to the interphenomena; the same is the case for the question of genidentity. Observable phenomena give us no information concerning genidentity; if we speak of an individual (particle) which remains identical with itself during a certain time-period, we extend our language in such a way as to include statements concerning unobserved properties. According to Stückelberg and Feynman, this extension of language can be constructed in two different ways with respect to the direction of time: either we have one individual particle moving, in part, in negative time, or we have two individual particles moving in positive time. The first interpretation includes the causal anomaly of a particle going contrary to the flux of time; but the second interpretation has other anomalies of which we must now speak.

Feynman proposed his interpretation, in particular, with a view toward interpreting pair-production. In the presence of a gamma ray, an electron-positron pair is sometimes produced from nothing, the positron soon meets another electron, and both of them coalesce in such a way as to leave only another gamma ray. We thus have three individual particles, two electrons and a positron, and their history is quite strange from the point of view of causality, in that it includes the production of a pair from nothing, and then the annihilation of a pair. This anomaly is eliminated, according to Feynman, if one replaces the three individual particles by a single particle, by an electron which in one phase of its existence moves backwards in time. Obviously, one anomaly can be replaced by another, and one can choose, according to one's taste.

This duality of descriptions is extremely interesting from the point of view of the logical analysis of time. It indicates the fact that even the order of time is not an invariant property in the class of equivalent descriptions. In the first interpretation, the electron coming from event A and the positron coming from event B constitute two causal lines directed toward the event C, where the two particles coalesce; in the second interpretation, these lines constitute only parts of a single line ACB. For this second interpretation, C is thus 'causally between' A and B; for the first, it is not. In other words, the order of time is ACB, or if you wish, BCA for one of the interpretations; for the other, the order is either AC and BC, or CA and CB. It is thus a question, not only of change in direction, but also of a change in the order of time.

Classical mechanics was not able to give a direction of time, but it did give us an order. We would only have to assign a direction to a single causal line and a direction would be assigned to every line. Things are different with quantum mechanics; the direction of one causal line can be reversed without reversing the others. Thus one cannot construct a coherent order which permits the definition of a direction of time. One would arrive at contradictory results: in following the path ACB, one would find that A precedes B; in following another path, one would find that B precedes A.

It seems that in the quantum domain, the concept of time loses its direct significance. The variable t has always played a doubtful role in quantum mechanics. It acquires meaning only as one passes to macroscopic observations. If this is true, then not only the direction of time, but even its order will be a product of statistics. Time would emerge from the chaos of elementary events just as thread comes out of a skein with the help of a spinning wheel. One must await the development of the theory of the atomic nucleus before one can speak definitively of that which we can anticipate today only as a possibility.

As regards Feynman's ideas, another remark presents itself. The life of a positron is quite short; this is why electrons traveling backward in time do not play a great role in the statistics of electrons. If things were not this way, it would be doubtful that an ordered time would result from atomic chaos. Perhaps there would then be closed causal lines in the macrocosm. One could conclude that the existence of a linear time is related to the difference between negative and positive electricity, to the fact that the electricity called negative prevails over that called positive, as far as the number of free particles is concerned. According to the conceptions of Dirac, this superiority of negative electricity derives from the fact that all the negative energy states are occupied, while those of positive energy are generally free; the positron exists only in the form of a hole left open when an electron leaves its place in the sea of invisible matter. These are pictures; but they express ideas which possess equivalents in mathematical equations. A solution of these problems, which greatly occupies the thought of physicists, would be capable of clarifying the problem of time.

However, it is the relations of the time of the macrocosm which determine the appearence of the time of our immediate experience. In this respect, quantum theory presents us with a quite interesting conclusion, which resolves certain difficulties which arise in the analysis of the time of classical physics.

The measurement of a quantum entity is a process which projects quantum uncertainty into the macrocosm. This results from the fact that a measurement represents an act of detachment; the elementary phenomenon, the arrival of a particle, directs the macroscopic phenomena into a unique form, and these phenomena, dial readings, would be wholly different if the elementary phenomenon were different. This is why the measurement process can give us information about the elementary phenomenon. And this is why it is impossible for us to predict the result of the measurement if it was preceded by the measurement of a quantity which does not commute with it; that is excluded by Heisenberg's relation.

Hence there exist certain macroscopic phenomena which cannot be predicted, but which can be registered. Let us suppose that consecutive alternating measurements of two non-commutative quantities are made. We will have a sequence of macroscopic events which cannot be predicted, but which can be registered. This sequence offers us a very neat distinction between the past and the future: the past is determined, but the future is not.

This thesis needs to be explained. The quantum uncertainty, in so far as it concerns only the elementary phenomena, has a dual aspect: it turns toward the past in the same way as it turns toward the future. If we have nothing but measurements made at a certain moment, we can calculate neither the past values nor the future values of these quantities. If the situation is not the same in our example, it is because we have registered the results of past measurements. Hence we owe our knowledge of the past, not to quantum methods, but to macroscopic methods, to the methods involved in using registering apparatus.

But notice the surprising conclusion coming from this consideration. The analysis of classical physics has shown us that one can register the past, but not the future. The combination of this result with the uncertainty of Heisenberg brings us to the consequence that one can know the past, but one can not predict the future.

In order to understand the meaning of this result, let us return for a moment to the determinism of Laplace. For an intelligence superior to that of man, says Laplace, "nothing would be uncertain, and the future like the past would be present to his eyes." It is not everyday experiences which gives birth to this thesis; on the contrary, it is the opinion of the man in the street that only the past is determined, while the future is undetermined. Laplace drew his thesis from classical physics, and if science has spoken, common sense has to keep quiet. Nevertheless, modern science has switched sides; it has put itself on the side of common sense and furnishes us in a precise way the difference between the past and the future that the physics of Laplace could not recognize.

It is true that the physics of Boltzmann, if we add to it the hypothesis of branch systems, yields a certain structural difference between past and future, a difference which is expressed in the inferences directed, either toward past facts, or toward facts of the future. We have discussed the fact that the partial effect admits of an inference to the total cause, while the partial cause, in general, does not admit of an inference to the total effect. But this difference, although it permits us to distinguish between past and future, was not associated with a difference of determination: although one cannot register the future, one can predict it, by basing the prediction on the totality of causes. Hence the future can not be called undetermined; or, at least, if one is to consider it as undetermined because a prediction is never absolutely certain, one will have to apply the same conception to the past, which also can not be deduced with absolute certainty. Insofar as the question of certainty is concerned, there exists a symmetry between past and future, as long as one remains in classical physics.

It is no longer the same in quantum physics. It is true that there exist past facts which can no longer be known because they have not been registered;

and there are future facts which can be predicted quite well, such as the movements of the planets, which are exempt from quantum uncertainty. But notice the difference: there exist future facts which are impossible to predict, while there are not any facts of the past which are impossible to know. In principle, they can always be registered. And I would like to suppose that the number of future events which depend, by way of detachment relations, on unforeseeable quantum phenomena is greater than is commonly believed. I would not be surprised if it were possible to show that a great number of human actions are of this kind.

The distinction between the indeterminism of the future and the determination of the past has found, in the end, an expression in the laws of physics. Here is the important result which comes out of the union of classical statistics with the uncertainty relation of quantum physics. The consequences for the time of our experience, for the time of every day, are evident. The concept of 'becoming' acquires a meaning in physics: the present which separates the future from the past is the moment in which that which was undetermined becomes determined, and 'becoming' means the same thing as 'becoming determined'.

There remains one question for us to discuss. What is the relation between the time of physics and the time of our experience? Why is the flux of psychological time identical with the direction of increasing entropy?

The answer is simple: man is a part of nature and his memory is a registering instrument subject to the laws of information theory. The increase of information defines the direction of subjective time. The experiences of yesterday are registered in our memory, those of tomorrow are not, and they cannot be registered before 'tomorrow' has become 'today'. The time of our experience is the time which is manifested by a registering instrument. It is not the privilege of man to define a flux of time; each registering instrument does the same thing. What we call the direction of time, the direction of becoming, is a relation between a registering instrument and its surroundings: and the statistical isotropy of the universe guarantees that this relation is the same for all instruments of this kind, including the human memory.

Let us add a word concerning the term 'now'. Symbolic logic teaches us that it is necessary to distinguish between the individual sign and the class of signs, called 'symbol'. With regard to many words, the individual sign can be neglected; the word 'house', for example, has the same meaning in all its instances. It is different for some words, like 'now', 'here', 'me', whose meaning changes with the individual signs. Allow me to speak, in this case, of *reflexive signs*, and of a *token-reflexive sign*, in using the word 'sign' in the sense of 'individual sign'.[10]

An act of thought is an event and thus defines a position in time. If our experiences always occur in the framework of a 'now', this means that each act of thought defines a reference point. We cannot avoid the 'now' because the attempt to avoid it would involve an act of thought and thus would define a 'now'. A thought without a reference point does not exist, because thought itself defines it. Grammar expresses this fact by the rule that each proposition must contain a verb, that is to say a reflexive sign indicating the time of the event of which one speaks; for the tense of a verb has a reflexive meaning.

Token-reflexiveness applies also to the concepts 'determined' and 'undetermined'. The word 'determination' denotes a relation between two states A and B; the state A determines, or does not determine, the state B. It is meaningless to say that the state B, considered separately, is determined. If we say that the past is determined, or that the future is undetermined, it is implied that this is relative to the present situation; it is relative to the 'now' that the past is determined and that the future is not. These words, and many others, thus have a token-reflexiveness. It is true nevertheless that these words express an objective relation; for it is a physical fact that, if A is the state defined by the act of speaking, then a state preceding A is determined with respect to A, while a state which follows A is not.

There would be much to add, but it is time to stop; and since the subject of this article is limited to physics, I allow myself to forego going further into the discussion of the problems of subjective time. It was my intention to show that quantum physics has nothing to fear from logic; that a logical analysis of the this portion of physics can be given without sacrificing either precision or rigor; and that an excursion into the domain of speculative philosophy is not needed in order to understand quantum theory. By contrast, I believe that science requires us to construct a scientific philosophy, and that such a philosophy can exist only by an intimate cooperation with physics.

NOTES

1 P. S. Laplace, *Essai philosophique sur les probabilités* (Gauthier-Villars, Paris, 1921) p. 3.

2 See my book, *The Theory of Probability* [1949f], p. 156.

3 I use the word 'reasonable' to suggest specifically 'conforming to common sense'. A meaningful proposition is thus a proposition which has a meaning, yet it might not be at all reasonable.

[4] David Bohm, 'A Suggested Interpretation of Quantum Theory in Terms of "Hidden" Variables' *Phys. Rev.* **85**, 166 and 180 (1952). See also L. de Broglie, 'Remarques sur la théorie de l'onde pilote', *C.R. Acad.* **233**, 641 (1951).

[5] The bibliography on multi-valued logics has been cited in my book *Philosophic Foundations of Quantum Mechanics* [1944b], pp. 147–148. See this book for a detailed exposition.

[6] See *The Theory of Probability*, *op. cit*; and my article 'Les fondements logiques du calcul des probabilités', [1937b].

[7] *Ibid.*, Equation (60).

[8] I thus take up again several ideas developed in a previous publication, 'Die Kausalstruktur der Welt und der Unterschied von Vergangenheit und Zukunft', [1925d; this volume, Chap. 57 – Ed.]. The inferences of which I have spoken are formulated there with the help of forkings of world-lines. However, I have lately changed my opinions about the mathematical relations governing these forkings.

[9] E. C. G. Stückelberg, *Helv. Phys. Acta* **14**, 588 (1941); and **15**, 23 (1942); R. P. Feynman, *Phys. Rev.* **76**, 149 (1949).

[10] I suggest the word 'token-reflexive' in my book *Elements of Symbolic Logic* [1947c], p. 284.

51. THE PHILOSOPHICAL SIGNIFICANCE OF THE WAVE-PARTICLE DUALISM

[1953a]

The theoretical scientist makes his discoveries within the pursuit of questions concerning explanation. Given the results of observation and experiment, he looks for laws combining the observational data into a consistent causal net. The more general and comprehensive the laws, the greater their explanatory power. However, in order to be accepted the law must not merely combine known data by way of a single rule; it is required to predict new data accessible to observational test. The criterion of scientific truth is the verification of predictions; and explanatory power is thus reducible to predictive power.

Yet scientific discoveries sometimes include other than observational implications. They may lead to a revision of the foundation of science, of those fundamental principles in which we formulate the most general properties of the physical world. In short, they may include philosophical implications. Implications of this kind, however, are often not openly visible in the beginning; and it may take a long time until they are seen distinctly and formulated precisely. In fact, the scientist himself is usually not immediately aware of the philosophic content of his discovery. His attention is concentrated on observational predictions, that is, on the truth of the new law; and he has no time to ask for its philosophical significance. What his discovery is going to mean for philosophy, is a question whose answer he may rightfully leave to later investigation. Science is positivistic in the sense that it puts empirical content first and logical analysis second. Let us be happy that science is so inconsiderate in its attitude to philosophy. If it were not, if the scientist were always fully conscious of all the logical implications of his work, he might be afraid to express ideas which contradict accepted philosophical principles – yet for whose discovery, at some later time, the philosopher is grateful to him, crediting him with revolutionizing philosophical thought.

When Louis de Broglie first formulated those principles which introduced a new era into the physics of the microcosm, neither he, nor anyone else, could realize what these principles would mean for the philosophy of nature. De Broglie was concerned with finding better laws for the explanation of observational data, laws that connect known data and predict new ones. The dual nature of radiation, its combination of wave and particle features, suggested to him the idea that a similar duality might exist for the more solid matter from which elementary particles are built up. The mathematical theory supplied the framework within which this supposed new law of nature was plausible; and when his predicted waves of matter were confirmed by later experiments, it was obvious that his scientific intuition had guided his guesses in the right direction. But the philosophical nature of these new waves was still completely unknown.

More than a quarter century has passed since de Broglie's epoch making discovery. The combined work of a number of distinguished scientists has eventually transformed his creative idea into a new physics of the microcosm. The names of Schrödinger, Born, Heisenberg, Dirac, Bohr, and many others come into our minds when we look back to this quarter century of physics. And we welcome the opportunity to express our best wishes to Mr. de Broglie on his sixtieth birthday, in recognition for the discovery which marks the beginning of the dual conception of matter as being both a matter of particles and of waves.

The physical implications of de Broglie's discovery, which have been derived during this period of continual change and improvement of the original version, are well known. I should like to give at this place a short summary of the philosophical implications of his discovery, which have been developed within the course of these twenty-five years. For the philosophical result has turned out to be no less significant than the contributions of this period to physics. In fact, if the physics of the atom is said to have revolutionized philosophy, this judgment is meant to summarize the consequences to which the duality of particles and waves has led.

I will begin with the changes in our conception of causality which have arisen from quantum physics. The idea that all physical occurrences are strictly determined by causal laws had been regarded as the principal result of the physics of Newton, a result which had been taken over and even made more credible by the physics of Einstein. Philosophers had seen in the principle of causality an *a priori* law of the physical world, whether they regarded it, like Kant, as originating from human reason, or, like Spinoza, as an ontological property of the physical universe. This so-called *a priori* law has now been abandoned by physics; with Heisenberg's relation of indeterminacy, the physics of matter waves has established the principle that nature is governed by probability laws, but not by causal laws. All that is left of strict laws is a differential equation determining probabilities; but it does not mean reestablishing determinism if it is regarded as possible to predict strictly with what probability an event will occur. And it does not help to say that ψ-functions can be regarded as physical states. They can be so regarded. But considered in terms of observables, ψ-functions are statistical states; and this means, once more, that observables can only be statistically predicted. If causality is to mean a rule governing observable phenomena, it has been abandoned by quantum mechanics – there is no escape from this conclusion.

However, the scientific philosopher will not find this conclusion as revolutionary as it may appear to philosophers of metaphysical inclinations. Ever since Boltzmann's statistical interpretation of the second law of thermodynamics, a transition from causal to statistical laws has been a possibility that could not be disregarded. In fact, a more careful wording of the principle of causality had led to the result that causality is meaningful only as a statement about a limiting process concerning probabilities. Given a certain description of the present status of the physical world, we can predict its future status with a certain probability; and this probability is increased if the initial description is replaced by a more detailed one. It was assumed that in this way the probability of the prediction could be made to approach the limit 1 as close as we wish. The principle of indeterminacy restricts this process of approximations to a limit below 1; this is the logical significance of Heisenberg's principle. Historically speaking, the quantum mechanical indeterminacy appears as an evolution rather than a revolution of physical principles.

This interpretation is made more evident when the convergence of probabilities toward 1, assumed in classical physics, is analyzed more carefully. It then turns out to be always coupled with a statement of divergence of probabilities. Given a description A_1 which predicts the future at the time t_1 with a certain probability p_1, we can find a more detailed description A_2 which predicts the state at t_1 with a probability p_2 higher than p_1. But we can also find a later time t_2 such that A_2 predicts the state at this time merely with the probability p_1, or an even lower probability. There exists both a convergence and a divergence of probabilities, in the face of which it appears highly implausible that there exists an ultimate description which, though unattainable for us, predicts the physical states at all times with certainty. But this is what determinism maintains. Even in classical physics, the evidence for determinism can only be called very meager; and the quantum mechanical transition to probability laws appears as a natural extension of a classical causality which is conceived as a statement about convergence of probabilities.

If it is the result of such considerations that the indeterminism of quantum physics is not surprising to the philosopher of science, there exist philosophical consequences of another kind which neither the philosopher trained in scientific method nor the scientist himself could anticipate. I refer to the peculiar result concerning the dual nature of matter, expressed in the dualism of particles and waves.

What is matter? This question is as old as western philosophy. The Greeks already answered it by the invention of atoms; and modern science, since Dalton's day, has accumulated overwhelming evidence for the atomistic structure of matter. Its answer to the question seemed therefore definitive; and nobody thought that it could ever be revised.

It is true, the physicist does not say that matter does not consist of atoms. He even has discovered quite a few new atoms, smaller than those of the chemist and therefore not atoms in the narrower technical sense. But he also found that all these particles, in certain conditions, behave like waves; and when he is asked whether they really are material particles, he answers that this is an embarassing question and that he would prefer not to be asked such questions.

This means that the answer requires more philosophy than the physicist needs for his practical research. It means that the answer cannot be given within the framework of philosophical conceptions that have arisen within every-day life or within classical physics. It means that we have come to a point where the traditional concept of a physical object existing in an undisturbed reality of its own has turned out to be inadequate. Only a philosophy which is willing to revise its fundamental concepts about physical reality can give an answer to this question.

The logical problem hidden behind the question is that of unobserved objects. Are the things still there, and are they the same things, when we do not look at them? This question has raised such philosophical problems as that of solipsism. Are things more than perceptions? Berkeley answered that they are not; they exist, however, when we do not perceive them because God perceives them. Kant answered that they are, but only because there are things in themselves, so-called noumena, behind them which must forever remain unknown to us. Do we have to accept these answers, leaving us only the choice between theology and irrational belief?

Fortunately, the scientific philosopher has found better ways of interpreting physical reality. He has seen that questions of the kind considered are questions of language and involve certain rules which allow us to extend the language of observables to that of unobservables. Such rules he calls *extension rules*. They have the logical status of conventions, and thus are not true or false. They can be replaced by different conventions; then we arrive, not at a different reality, but merely at a different description of reality. Given the same physical world, there exists a class of equivalent descriptions. Each of them is true, and it makes no sense to ask which of them is really true.

This does not mean that there are no false descriptions. A description is true, or false, as soon as we add to it the extension rules which are assumed for it, i.e., as soon as it is complete. And the question arises whether a certain extension rule can be carried through at all. Given a certain extension rule, is there any true description satisfying this rule? This is an empirical question.

In application to quantum phenomena, this means that we have to ask whether we can construct a language for them by the same rule that we use in the physics of everyday life. This is the rule: the unobservables are governed by the same laws as the observables. We can speak of unobserved houses because we introduce the rule that they follow the same laws as observed houses. Can we carry through the same rule for the unobservables of quantum physics?

It is here that the problem of waves and particles arises. When we say that matter consists of particles, we mean to say that not only the observables, the phenomena, show the discrete pattern of localized occurrences, but that the same applies to unobservables, to interphenomena. When we say that radiation spreads in the form of waves, we mean that, between the localized source and the localized place of observation, radiation is not localized, but fills a large volume of space. That is, the terms 'particles' and 'waves', in order to be meaningful, presuppose certain extension rules; and without specifying these rules, no description of matter can be given.

We can regard it as the result of quantum physics that it is impossible to describe unobservables by the use of the rule that they follow the same laws as observables. Every complete description of interphenomena leads to causal anomalies, to relations that can be regarded as action at a distance. This is what makes quantum reality essentially different from ordinary reality. And this is the logical issue which Bohr has attempted to describe in his famous principle of complementarity. When he argues that each interpretation, the particle as well as the wave conception, is only partially adequate and requires the use of the other to make up for its deficiencies, this conclusion can be translated into the statement that both interpretations have their causal anomalies, but each at a different place. This is why each complements the other.

Using the terminology of logical analysis, we would say that our usual extension rule, tacitly assumed for the language of everyday life, breaks down in the microcosm. That is why the microcosm appears so strange to us, so incomprehensible. We have to make this extension rule explicit; we have to make it clear why we can speak of unobserved houses, when we want to understand why we cannot, in the same sense, speak of unobserved quantum occurrences. This is the logical reason for the duality of waves and particles. Both constitute equivalent descriptions; but neither one is capable of describing unobservables in terms of the same laws that govern observables, because no such description is possible at all.

Those who ask whether matter consists of particles *or* of waves, forget that the answer may be that it is *both*; and those who say that matter can be described both in terms of particles and of waves, must not forget to add that neither description holds in the sense tacitly assumed for objects of the macrocosm. The reality of quantum objects defies the concepts developed in our daily business with objects of our environment. Speaking of things that remain the same when nobody looks at them is a reasonable language for the world of large dimensions; but we must not believe that the same language can be applied without qualifications to small dimensions. The rules of language do not represent philosophical truths; they originate from habits acquired in interaction with our environment. And when such rules turn out to be inadequate for the language of the microcosm, we have to replace them by better rules. This is what the philosopher has learned from the puzzles of quantum physics.

A language of better rules, which appears to be adequate for quantum mechanics, is constructed in terms of a three-valued logic. In addition to the categories of truth and falsity, this language employs a third category, called *indeterminate*, which applies to unobservables. Using this three-valued language, we can describe quantum mechanical reality in such a way that we say all we know and yet do not commit ourselves to misrepresenting the status of unobservables. That is, we do not use a terminology misleading the listener – and ourselves – to regard microcosmic entities as substances in the sense of the physical bodies of the macrocosm. The terms 'particles' and 'waves', in the macrocosmic sense, cannot be applied to quantum objects because they presuppose extension rules that are not applicable to such objects. If we use a three-valued language, these terms change their meaning to the extent that they become indistinguishable, because they no longer include determinate statements about what happens between observations, about interphenomena.

There is another philosophical issue upon which quantum mechanics has shed light: that is the problem of the direction of time. The solution which quantum physics offers for this problem is rather surprising, because in some sense it destroys time, in another, it confers upon time a direction which leads to a pronounced distinction between past and future, a more radical distinction than could be made in classical physics.

When we want to understand this situation, we have first to distinguish carefully between *order* and *direction* of time. The points on a straight line are ordered; but the line has no direction. Order refers to between relations: given three points on the line, we know which one is between the two others. But we cannot tell which of the outer points comes first, and which one comes last; either one can be regarded as being the first. A distinction as to first and last can only be made when a direction is known.

The equations of classical mechanics confer upon time an order, but not a direction. When a ball is thrown from *A* by way of *B* to *C*, the opposite motion is described by the same differential equations as the original one. This familiar result follows because the differential equations of mechanics are of the second order; if, in a solution of these equations, we replace '*t*' by '$-t$,' we arrive once more at a solution. This means that these equations do not determine a direction of time. But they do determine an order: for both solutions, the arrival of the ball at *B* is temporally between the arrivals at *A* and at *C*.

Classical mechanics, therefore, knows a time order, but its processes are reversible, and therefore it does not supply a time direction. If atomic processes are governed by the rules of classical mechanics, they are reversible and define for the microcosm merely an order of time. The problem of how to derive a direction of time for the macrocosm from a merely time-ordered microcosm was solved by Boltzmann: the macrocosmic direction of time is of a statistical nature. This means, it is not impossible, but very improbable that mixing processes go backward, even when they are composed of reversible elementary processes.

However, Boltzmann's solution cannot assign a direction to time as a whole. When we regard the universe as a closed physical system, developments from improbable to highly probable states, or from low to high entropy, are not more frequent than those in the opposite direction; in fact, both are very infrequent, the system being in states of equilibrium most of the time. Only while the system is on a long upgrade does its time possess a direction. It appears that the world in which we live is on such an upgrade; therefore we experience a time direction. Should our universe, eons from now, start to develop along a down grade, it would once more possess a time direction, which however is opposite to the present one. Boltzmann makes it clear that living beings in such a world would experience as future the direction which we call the past, although subjectively speaking, their experiences are like ours. In other words, there is no absolute time direction; a direction of time can only be sectionally defined. That these sections of upgrades or down grades are connected so as to be parts of a continuous line, has a meaning because time *order* is defined even for reversible processes and thus order extends through the horizontal parts from upgrade to down grade.

The time direction existing for the upgrade manifests itself, for the human observer, in many individual systems which for a short period remain practically isolated from the main system. When we find such systems in an improbable state, i.e., a state of low entropy, we conclude, not that they have developed into this state in isolation, but that shortly before they were still connected with the main system. This is the reason that isolated systems define the same time direction as the universe: they were not isolated all the time, but derive from interaction processes. It is impossible to define a time direction by reference to completely isolated systems. We owe our time direction to the existence of systems that are partly isolated, partly in interaction with their environment.

Although Boltzmann's solution accounts for a direction of time in our world, it leaves to time a certain symmetry: both the past and the future are strictly determined, because the elementary processes are controlled by mechanical laws. When it appears to us as though only the past is determined, whereas the future is not, it is merely our ignorance which makes us believe in this distinction. In principle, the future can be predicted to the same degree of exactness as the past; no limits are drawn to this exactness.

This is the structure which classical physics supplies for time. Time as a whole possesses merely an order. For sections of the total time, and only for the macrocosm, there exists a time direction. But even on these conditions, the flow of time does not represent a transition from a determined past to an undetermined future. It is an illusion springing from ignorance when we believe that the future is undetermined; as far as determinism is concerned, there is no difference between the future and the past.

Let us now examine the changes which quantum physics has introduced into this picture. They result from the fact that the elementary processes are no longer regarded as following the laws of classical mechanics. Rather, they are assumed to be controlled by the laws of quantum mechanics. Although these laws differ greatly from the previous ones, as far as their mathematical structure is concerned, they correspond to them in one essential point: they do not make the elementary processes irreversible. It is true, they introduce probabilities into a realm where classical mechanics spoke of certainty; the paths of elementary particles are controlled merely by statistical laws. But the probabilities for these paths are symmetrical with respect to past and future. Suppose that u and v are non-commutative quantities, and that a measurement has shown the value u_i to exist; then we can compute the probability p_{ik} of finding the value v_k in a measurement of the quantity v. Let p_{ki} be the opposite probability, i.e., the probability of measuring u_i after the value v_k has been measured; then the theory shows that $p_{ik} = p_{ki}$. For this reason, elementary processes do not define a time direction for quantum physics.

In this respect, quantum physics does not differ from classical physics. In another respect, quantum physics seems to deny the very property which classical mechanics had still left to the time of the macrocosm: it appears that the elementary processes of quantum physics do not even possess an order. They do not tell us between-relations. This, at least, is the result of certain investigations made by Stückelberg[1] and Feynman.[2] According to these authors, a positron can be regarded as an electron going backward in time. As a consequence, pair production and pair annihilation can be conceived as events in the history of only one particle, an electron, in such a way that there are no pairs of simultaneous particles, but 'successive' states of one particle. The electron travels first from A to B, where it changes its time direction so as to travel backward in time toward C, from where it turns forward toward D. Taken along the world line of this electron, the event B is between A and C, and the event C is between B and D. However, when we interpret B as pair annihilation, B is not between A and C, the world lines AB and CB being counterdirected. Similarly, when C is regarded as an event of pair production, C is not between B and D.

It is different, of course, if observations of these four events are made, for instance, if the whole process is photographed in a cloud chamber. The observations are macroscopic events, and as such are time ordered and are even distinguishable as earlier and later. But the elementary process itself does not supply a time order.

This means that the unobserved quantities, the interphenomena, of quantum physics do not supply a time order. I would like to regard this result as but another causal anomaly holding for interphenomena. Some of their world lines can be reversed, leaving others unchanged; and the lines can be pieced together so as to become a world line of one particle instead of being lines of three particles. However, since causal anomalies do not show in the phenomena, this result does not bear upon observables, and in so far, then, the loss of time order in the microcosm has no implications for the macrocosm.

Perhaps we shall have to conclude that, in quantum physics, not only time direction, but even time order is generated in the macrocosm. The microcosm is a chaos without a linear order of events; but out of this chaos is born time. This would make even time order a statistical concept. For macrocosmic time, this would not make much difference, although it may involve changes for the conceptual structure of quantum mechanics. If time order is not defined for quantum processes, the meaning of the variable t requires a new interpretation.

Turning to macrocosmic time, which still retains a direction of the kind constructed by Boltzmann, we now must point out an important conclusion derivable from the quantum nature of elementary processes. According to Heisenberg's indeterminacy, it is not possible to predict the result of measurements with more than probabilities. Measurements are macroscopic processes; therefore, there are macrocosmic events of the future, which at the present time are not knowable. There is no way to increase the probability of the prediction; we can only wait and see how the measurement will turn out. In this sense, such future events may be called undetermined at the present time.

Now it is true that the symmetry of the quantum mechanical equations leads to similar conclusions for past events. When we measure now the quantity v_k, we can only say with some limited probability what value u_i existed before. But here we have other ways of knowing: if the quantity u has been measured before, we can write the result down. Thus we can know the past, namely, by recording it; but we cannot know the future.

Why can we not record the future? Because recording involves irreversible processes; it is a macrocosmic affair. We are led back here to the statistical time direction of Boltzmann's gas theory. Quantum mechanical systems owe their time direction to an interaction with their environment, performed through the measuring process; in this respect, their time direction is defined in the same way as that of other subsystems in this physical world. But it turns out that this direction, in combination with the quantum mechanical indeterminacy, leads to a peculiar distinction between past and future, a distinction unknown to classical physics: certain events of the future are unpredictable, even to a superman of Laplacian abilities, whereas all events of the past can be strictly known if only they are recorded. In this sense, the past is determined, whereas the future is not.

It may perhaps be objected that this distinction applies only to such macroscopic events which depend on the quantum mechanical uncertainty. This is true. But the fact that there are such events is sufficient to draw a sharp line of demarcation between past and future. We do not claim that all future events are unknowable. Everyday life provides us with many examples of pretty reliable predictions. But this consideration shows that quantum physics leads to a macrocosmic time direction for which the future is distinguished from the past as the realm of undetermined events from the realm of the determined ones. And perhaps there are more events depending on quantum mechanical processes than is usually believed. Such processes, in the form of trigger action, may for instance play an important part in genetics. Since human history depends largely on the influence of individual personalities, the indeterminacy of quantum processes may thus be reflected in the impossibility of predicting historical events. In contrast, that historical events can be recorded is a familiar fact.

The time direction springing from quantum mechanics, of course, is subject to the limitations holding for that of gas theory. It exists only for the macrocosm, and only for sections of the total time of the universe. But it appears that quantum processes confer upon this time a certain property which the time direction of classical physics did not possess, that they make time-flow an act of becoming in which potentiality is turned into actuality.

These are some sketchy remarks in which I have tried to summarize the philosophical results of quantum physics. They all have grown from the conception of the dualism of particles and waves, formulated for the first time in de Broglie's great discovery. Like a tree grown from its roots, the theory of quantum physics derives from this fundamental discovery. And being itself an irreversible process, this evolution of ideas records the original idea from which it sprang. This is the award which the history of quantum physics confers upon Mr. L. de Broglie.

NOTES

1 E. C. G. Stückelberg, *Helv. Physica Acta* **14**, 588 (1944); **15**, 23 (1942).
2 R. P. Feynman, *Phys. Rev.* **76**, 749 (1949).

PART VI

PROBABILITY AND INDUCTION

52a. THE PHYSICAL PRESUPPOSITIONS OF THE CALCULUS OF PROBABILITY

[1920c]

I. APPLICATION OF THE LAWS OF PROBABILITY TO THE OBJECTS OF REALITY

At present, two different groups of scholars are working with probability laws. On the one hand, there are the mathematicians who, starting from the simple basic laws of the theory, develop complex forms of calculation, establish relations facilitating the solution of intricate problems, and also introduce new concepts, such as dispersion, mean-square error, compound probability, and so forth. On the other hand, there are the statisticians in every branch of science – physics, psychology, sociology, and so forth – who gladly take over the complex apparatus of the mathematicians, not, however, in order to carry it to completion, but in order to apply it to actual objects, in order to derive methods from it suited to the presentation of particular empirical states of affairs. This distinction between working methods corresponds to a profound and substantial difference, the same difference that separates pure mathematical research from all its applications. There can be no doubt but that the laws of probability represent a complete mathematical system, just as do the principles of the infinitesimal calculus or the principles of geometry; the strict certainty attendant upon these spheres must likewise be attributed to the principles of probability insofar as they represent closed relations, conceptual chains made up of combinations of elementary concepts. Let the reader recall Bernoulli's theorem for calculating the frequencies and the dispersion of simple repeated series, which, from the mathematical point of view, is nothing more than an enumeration of combinations. No one has ever called into question the correctness of this theory of combinations. All the greater, then, have been the doubts about the application of the laws of probability to the objects of reality. The statisticians of the various individual sciences have never been able to appeal to the mathematical rigor of theoretical probability, for the question whether real objects are subject to the relations calculated has not been resolved. The situation here is similar to that in geometry: no one doubts the correctness of the geometrical principles taken in themselves, but no judgment is possible on mathematical grounds as to whether they describe *real objects*, whether the space in which we measure physical objects is three-dimensional and Euclidean. This question can be decided only through the methods of physics and philosophy.

Geometry has the advantage of possessing a well-developed system of axioms, and it is possible today to replace the question as to the validity of its theorems with the question as to the validity of its axioms. If real space conforms to these axioms, it must also conform to all geometrical theorems; however, for purposes of investigation it is vastly simpler to pose the problem of the validity of the axioms alone. To be sure, the probability calculus also possesses a system of axioms, but these are complete only as the foundation of the purely arithmetical relations of this calculus. In their application, however, probability laws are intended to describe real events, to make definite assertions concerning temporal sequences, and therefore another group of axioms must exist, distinguished by its inclusion of the concept of time, that deals with the application of the probability laws to real events and which we will for that reason designate *axioms of applicability*. In contrast to the mathematical axioms, they may also be described as physical axioms, provided we construe physics in the most general sense, as the science of spatio-temporal events.

The present author carried out the construction of this group of axioms in another work, to which the reader is referred for a more thorough justification of the ideas discussed here.[1] In the present article, the results of past investigation will be given and the essential features of the method of derivation demonstrated. As laws of probability are invariably laws of approximation, such that with an increasing number of repetitions the required distribution is more and more closely approached, the axioms of applicability must also be formulated as laws concerning a relation of approximation. We will discover that these axioms are reducible to a single axiom. At present, to be sure, it is not possible to deny that certain physical hypotheses, such as Boltzmann's ergodic hypothesis for the basis of molecular statistics, may also contain other presuppositions, and to this extent the investigation may not yet be regarded as closed. Furthermore, the present investigation has been confined to physical problems, and we are therefore not in a position to make any judgments about the application of the probability calculus in psychology and sociology. Nevertheless, it will prove to be demonstrable that the axiom, once established, has a philosophical significance going beyond the boundaries of narrow probability calculus and that its validity is very closely bound up with the concept of knowledge in physics. This question will be discussed in a forthcoming issue of this journal [1920e; this volume, Chap. 53] the present article, however, will be confined to establishing the axiom.

II. THE AXIOM OF APPLICABILITY OF PROBABILITY PRINCIPLES; HYPOTHESIS OF A PROBABILITY FUNCTION

A game of dice offers a simple example of the physical realization of probability laws. The six possible positions of the die are designated as possible cases, and each position is considered equiprobable. This classification of the possible cases into a number of equiprobable cases is characteristic of all probability calculations. Cases are said to be equiprobable if, in a series of repetitions of the event, those cases are realized an equal number of times, that is, e.g., if each side of the die comes up an equal number of times. The problem is this: from whence do we get the right to claim that particular cases, e.g., the showing of the sides of the die, are equiprobable?

Attempts have been made to define equiprobability in such a way that it does not include any assertion as to the repetition of the event, in the belief that the problem of large numbers could thereby be separated from the problem of probability. In these theories, equiprobability is defined instead as consisting of a certain physical structure, for instance, the spatial symmetry of the die, and the proposition that upon repetition the number of occurrences of each side becomes more and more nearly equal is rejected. But, of course, the problem cannot be circumvented by such definitions. True, equiprobability, so defined, no longer poses a problem; its presence can be physically established, for we can tell by making measurements whether the sides are of equal size and whether the center of gravity is in the middle. But the striking fact remains that equal frequency over a series of repeated cases corresponds to precisely these geometrical-physical relations. That this is a fact is assumed in every application of statistics, and this is the fundamental problem with which we are faced. It is, too, only through the existence of such a frequency relation that we can justify that feeling of expectation which we usually associate with the concept of probability and which permits us to regard as improbable, e.g., the occurrence of the same throw twice in a row. Once it is established that such a combination occurs less often than others, the feeling of expectancy with which we anticipate the probable while dismissing the improbable is psychologically justified. The feeling cannot, however, be based upon geometrical-physical relations. Still less can this feeling be used in defining probability, say, by defining as equiprobable those relations that 'generate in us the same unhesitating sense of expectation'. Definitions of this sort, which use as their starting point the effect upon the observer, mislead us into viewing the establishment of probability as a purely subjective anticipation, the contents of which depend upon the state of our subjective knowledge at a particular time. The conception overlooks the fact that there actually exist objective states of affairs that can be exhaustively described by means of probability laws, for instance, the regularities shown in throwing dice. The job of explaining how knowledge of such objective states of affairs gives rise to certain feelings of expectation must be left to psychology; it has nothing to do with the regularity of probability.

We will therefore define as equiprobable those cases that, with repetition, converge more and more closely toward an equal number of realizations, and we must establish which physical presuppositions must be made before such a phenomenon can occur.

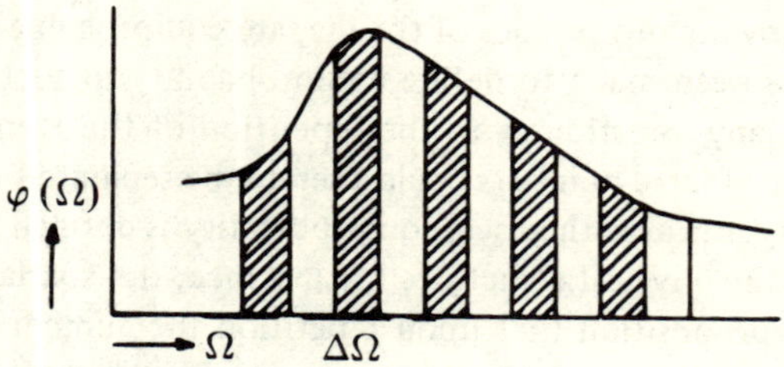

Fig. 1. Reduction of equiprobability to the continuity of a curve.

Let us take roulette as an example of a probability mechanism. Poincaré[2] studied this game, tracing it back to a very simple assumption. The rotating pointer will run through its cycle a large number of times before it stops at one position. We can measure its path by means of an angle if, after the first complete cycle, we start counting from 360° and continue until we reach degree Ω, at which the pointer has stopped. If Ω were slightly greater, by the amount $\Delta\Omega$, which corresponds precisely to the width of a colored segment, the pointer would stop on the other color. If we imagine this path to be increased once again by a factor of $\Delta\Omega$, the pointer will again stop on the first color, and so forth. If we spin the pointer a number of times, the path Ω will be different every time, for the force with which the pointer is propelled is never exactly the same twice. But whether red or black will be shown depends solely upon the interval $\Delta\Omega$ in which the pointer comes to rest. Let us now imagine that, during a large series of experiments, we count for every individual interval $\Delta\Omega$ exactly how often the pointer stops in that particular interval, and let us suppose this quantity h to be divided by the total number N of tries, so that we obtain the relative frequency h/N of the interval. This is graphically presented in Figure 1. Here the intervals $\Delta\Omega$ are shown as equal segments along the abscissa Ω, and the relative number of hits h/N is presented for every interval by the narrow rectangle-like stripes rising from it. The irregular form of the curve indicates that the pointer by no means shows each value Ω equally often; on the contrary, it shows a preference for a certain area in the middle, seldom indicating very small or very large values of Ω. The number of cases in which the pointer stops on a black sector is given by the sum of the shaded stripes, the number in which it stops on red, by the sum of the other stripes. Here each sector of the roulette wheel itself corresponds to a sum of stripes in the figure, and the colors, red or black, correspond to a yet larger sum. Nevertheless, any two contiguous stripes are of almost equal size. Every small shaded stripe corresponds to a smaller unshaded stripe, and every large shaded stripe corresponds to a still larger unshaded stripe; thus, when the two kinds of stripes are separately totalled up, the totality of shaded plane surface turns out to be almost equal to the totality of unshaded plane surface. The smaller the division $\Delta\Omega$ (and, therefore, the greater the number of intervals), the more nearly equal they will be, and it can easily be demonstrated that the two plane surfaces become precisely equal in the limit for infinitesimal $\Delta\omega$. The only necessary assumption is that the curve in Figure 1 is *continuous* throughout; that is, it may not be abnormally discontinuous. But its form is otherwise irrelevant; it may rise and fall and may have any kind of curvature. To be sure, the sum of the rectangles must always remain finite, and the curve at both ends must therefore run asymptotically relative to the abscissa axis; i.e., *very large* and *very small* values of Ω may only occur very rarely.[3] If these preconditions are fulfilled, it follows that the pointer will land on red as often as on black; thus the equiprobability of the two colors has been reduced to the existence of a curve such as is assumed above.

What is the significance of this curve? We must keep clearly in mind that its existence was in no way established. We simply said that we would count up the number of spins and record them in accordance with the procedure outlined above. Strictly speaking, we do not obtain any curve at all in this way. To begin with, we can only draw over each $\Delta\Omega$ a rectangle, the surface area of which is equal to the relative number of strikes h/N for this interval, and the top edges of these rectangles will then form a series of steps. If we increase the number of tries N, the steps will begin to even out; if we then imagine the intervals $\Delta\Omega$ to be smaller, the series of steps will more nearly approach a curve. We may not initially select excessively small $\Delta\Omega$ for this graphic procedure. If, for instance, the number of $\Delta\Omega$ is greater than the number N of spins, it will be impossible for the pointer to strike each $\Delta\Omega$, so that there will be many places in which a rectangle $= 0$ would need to be sketched in, making the steps highly irregular. Only when the number of tries is made greater will a regular series of steps reemerge, approaching more closely a continuous curve than the previous series. But we will never obtain a genuine continuous curve with a finite number N of tries – and finite quantities are all that we have at our disposal. If we nevertheless assume the existence of a continuous frequency curve for the spins of the pointer, it signifies our acceptance of the hypothesis that, in the outlined graphic procedure, an increase in N will be accompanied by convergence towards a continuous curve with asymptotic ends. Curves (Ω) of this nature are known as probability functions, since the probability P that Ω will lie in the interval between Ω_1 and Ω_2 – that is, the relative frequency h/N for any interval – is given by the following equation.

$$\left(\frac{h}{N}\right)_{\lim N=\infty} = P = \int_{\Omega_1}^{\Omega_2} \varphi(\Omega)\, d\Omega$$

(The integral for $\lim N = \infty$ emerges from the summation of the rectangles.) The equiprobability of the two colors in a game of roulette may be said to be reducible to the *hypothesis* that a *probability function exists* for the angle of revolution of the pointer.

This discovery constitutes substantial progress with respect to the problem of probability. We were previously faced with the problem of explaining the equiprobability of the red and black sectors. At that time, this equiprobability seemed to be some mysterious characteristic of the colored segments, and there was no apparent reason why such a sweeping assertion should be made about these segments. Indeed, some thinkers have attempted to make a philosophical principle out of the *absence* of any reason, asserting that the segments must be equiprobable because there is no reason for any segment to be preferred. This 'principle of insufficient reason' neglects to notice that there is also no reason to declare the segments equiprobable and that the opposite could therefore also be inferred; drawing inferences on the basis of the absence of reason is always very hazardous. The hypothesis of the probability function gets us out of this difficulty with one fell swoop. For this hypothesis removes the required characteristic from the segments and transfers it to the rotating pointer; it asserts something about the nature of the rotation, whereby the segments simply take on the role of making *visible* the peculiar nature of the phenomenon of the rotation, of depicting it in particularly sharp and intuitive form. We now have sufficient reason indeed to call the segments equally *probable*, namely, because they are equally broad. If they were of different sizes, the intervals $\Delta\Omega$ would also not be equal, and the frequency of the colors would be given not by 1 : 1, but by some other relation. The segments simply produce a particular division and coordination of the various angles of rotation of the pointer. The numerical value of the group relations that occur during the process is determined by the size of the segments. With this shift in the problem, the principle of insufficient reason vanishes.

The problem is not, however, thereby resolved, for the new hypothesis must first be justified. It is altogether different from other physical hypotheses; furthermore, it cannot be tested by employing the usual physical methods, as it cannot be established by means of measurements. It asserts the existence of a regularity in nature that is expressed in the enumeration of quantities, and in this, too, it goes far beyond all experience, for it assumes a limit for an infinite number of observations. But we may already mention, as the advantage of this hypothesis, that it possesses the form of a continuity presupposition and does not prescribe any quantitative relations. We need no longer assume that equiprobability is to be attributed to finite intervals. Our hypothesis does *not* claim that all values for the angle of rotation Ω are equiprobable, but only that infinitely contiguous values are equiprobable. The *superiority of this hypothesis* lies in its requiring an assumption not about the specific value of the probability function $\varphi(\Omega)$, but only about its continuity, in order to explain the *equi*probable cases that form the basis of the probability calculus – a fact that will prove significant for the philosophical aspect of the problem. However, our immediate task is to demonstrate that the same presupposition is also sufficent for the resolution of other problems.

III. CONTINUITY OF THE PROBABILITY FUNCTION AND THE PHYSICAL FOUNDATIONS OF SOME GAMES OF CHANCE

Games of chance have always been regarded as the ideal case of probability calculus. It has become common to use them as examples in explicating the laws of probability, for nowhere are the presuppositions of that remarkable theory of combinations presented in the probability calculus – the individual equiprobable cases and those exhausting all possible values, their unlimited capacity for combination and the regularity of their occurrence – so clearly and distinctly displayed as in this renowned arena of chance. Such games appear, indeed, to be the very symbolization of those rules for calculation which form the structure of probability calculus, while their physical nature remains of very little interest and importance. This situation changes only when we seek out, as we did for roulette, the physical presuppositions which promote these phenomena – which are, after all, only empirical – to models of mathematical operations. In the pursuit of this goal, we have succeeded in discovering that axiom of applicability of the probability principles which it was our aim to find from the very first and which we are now able to formulate as the hypothesis of a probability function.

Given the great similarity between all games of chance, it is easy to demonstrate that this hypothesis also applies to all other games of chance, in addition to roulette. Let us take, for instance, the game of tossing a coin. Here there are two possible and equiprobable cases: heads and tails. But once again, the essential hypothesis lies not in the coin, but in the nature of the phenomenon of its motion. The span of time from the tossing of the coin to its landing, which is different for each toss, is now represented by the factor Ω, which would be entered in the figure as the abscissa. The division into intervals results from the rotation of the coin; the coin lands either in the interval characterized as heads or in that characterized as tails, depending upon whether the time of the fall is relatively longer or shorter. To be sure, the rotation of the coin does not occur with uniform speed, and the intervals are therefore not of equal size. But we may assume that the speed changes *continuously*, so that contiguous intervals are very nearly of equal size. A figure illustrating this game would therefore look somewhat different from Figure 1, as the division $\Delta\Omega$ constantly increases in the direction of the right-hand side, but because of the approximate equality of contiguous $\Delta\Omega$, the same inference can be made as to the equality of the shaded and unshaded stripes. Like the sectors of the roulette wheel, the two sides of the coin simply take on the classifying of the landing times, their separation into two groups, and the existence of a probability function for the landing time remains the characteristic hypothesis. (The fact that we must also presuppose the continuity of the speed of rotation is not a hypothesis characteristic of the probability calculus. Physics always makes assumptions of this nature about the factors it studies: they signify that the magnitude of a physical factor cannot change discontinuously but must run through all the intermediate values. It is therefore permissible for us to make use of this presupposition without thereby introducing a new element into the problem. On the contrary, it is to be expected that the general assumptions invariably employed by physics will be applied to the field of probability laws as well. Our task is not to establish these assumptions, but to discover those *special* presuppositions required for the validity of the laws of probability.)

The same observations apply to throwing dice. The factor represented by the probability function is once again the time of fall, and the rotation of the die divides into continuously increasing intervals, the contiguous intervals being almost equal. The only difference is that the intervals must be divided into six groups, corresponding to the six sides. In the figure, we would have to shade a series of six contiguous stripes in different ways, starting again with the same shadings for the next six. But it is possible to give a completely analogous demonstration showing that the six areas consisting of identically striped bars become equal to one another in the limit for infinitesimal intervals. That the precision increases with smaller intervals is a fact acknowledged in practice, for it is generally supposed that the distributions become more regular as the die rotates faster. For this game of chance, too, then, the hypothesis of a probability function proves to be a sufficient presupposition.

It is vital throughout that the hypothesis demand only continuity of the probability function, while presupposing nothing about its special form.

There are cases to which we 'instinctively' attribute a probability function of the form:

$$f(x) = \text{constant}$$

It is important that we be able to deduce this form, too, from the mere continuity of a probability function, but for another factor. For purposes of illustration let us once again make use of roulette. Earlier, we measured the angle of rotation in excess of 360° and assumed the existence of a continuous probability function $\varphi(\Omega)$. (See Figure 1.) Every red or black sector already corresponds to a sum of intervals $\Delta\Omega$, and if we divide the sectors not into two groups, but, as in tossing dice, into as many groups as there are sectors, the same sum of shaded stripes and, therefore, the same probability, results for *every individual sector*. We need no longer go as far as the greater *sum* of the red and black sectors. Were we to make the sectors of unequal size, a proportionally larger striped plane segment and, therefore, a correspondingly larger probability would attach to the larger sector. That is, the probability is proportional to the angle of the sector, and if we do *not* count the angle in excess of 360°, designating it by θ, this signifies that:

$$\int \varphi(\theta)\, d\theta = k \cdot \theta,$$

where K represents a constant. From this it follows that:

$$\varphi(\theta) = k = \text{constant}.$$

For the factor θ, then, we obtain a probability function of the special form

$\varphi(\theta) = \text{constant}$, if we assume for the factor Ω a probability function $\varphi(\Omega)$ of any arbitrary form. Under certain circumstances, then, the hypothesis of continuity is a sufficient presupposition for the emergence of a quite special probability form.

IV. EXTENSION OF THE HYPOTHESIS OF A PROBABILITY TO THE COMBINATION OF SEVERAL MUTUALLY INDEPENDENT EVENTS

The probability calculus does not rest content with establishing the equiprobability of the individual cases. Rather, it really begins to take shape only when it constructs combinations of these cases and calculates the probability of any given combination from the known probabilities of the individual cases. Here, the 'rule of compound probability' is employed: this rule asserts that the probability of a combination is equal to the product of the individual probabilities if the cases in question are mutually independent (theorem of multiplication). The question now arising is, what physical hypothesis must we make in order to render this method of calculation valid? Yet before we approach it, we must explain in greater detail what independent events are, for this concept is vital to the validity of the law.

We describe a physical event by setting in relation to one another its characteristic elements. For instance, we note that a falling stone has a certain velocity and compare it with the duration of the fall so far; the resulting relation $f = v \cdot t$ presents a description of the fall. We may say that the velocity is a function of the duration of the fall, or more simply, that it is dependent upon the time of the fall, and the process of learning in physics can actually be described as a search for dependent factors and the nature of their dependency upon one another. The number of relations of dependency is very great, so great that it is quite impossible ever to exhaust them, but among them are certain ones that stand out in that they constitute the predominant characteristics of events in the universe; substantial knowledge is to be gained even if we confine ourselves to these relations. For instance, in our example of the falling body, there is also a relation between the velocity and the density of the air, which influences the degree of friction, yet it can be demonstrated that great changes in the density of the air correspond to only minor changes in the velocity of the fall, and it is therefore permissible to ignore this relation in favor of the previous one. We would, to be sure, include these influences if we were constructing a more precise theory of free fall, but they signify a dependency of relatively low degree and remain in this respect distinct from the first dependency. However, the degree of dependency can sink still lower. For instance, the velocity of the fall is also dependent upon the position of the moon with respect to the earth, for this influences the gravitational field. But large changes in the position of the moon correspond to such very small changes in the velocity of the fall that we do not, in practice, consider dependency to be present. In principle, of course, we must bear in mind that there is no such thing as an independent physical factor. For there are no closed systems; every system is connected at the surface with other systems, which are in turn connected with others, and so forth, so that, ultimately, all systems are related to one another. But there are demonstrably very low degrees of dependency in some cases – indeed, there is no limit to the lowness possible – and it has become customary in physics to designate such slightly related factors as independent. We will continue to define independence, or, more properly, a low degree of dependence, as being present when negligibly small changes in one factor correspond to great changes in another.

We can define the independence of events in an analogous way. Let us imagine, for example, two bodies falling to the ground side by side. We may characterize each of the two falls by presenting coordinates of the center of gravity as a function of the time by reporting, for instance, that at a particular time the body is at such and such a position, and so forth. In these relational equations, there will appear, in addition to time, other factors of particular magnitudes that influence the function, for instance, the initial height and the air's resistance. Similarly, the coordinates themselves are physical factors and are presented by means of the equation as independent of the other factors. What is *not*, however, shown in these equations is any dependence between the coordinates of the one body and those of the other; no matter what the position of the one body, the coordinates of the other are not influenced by it. Strictly speaking, this is not quite correct. For instance, one of the bodies will give rise to an air current that at the same time exercises lateral suction, pulling the other body off course, so that the coordinates for this body display different times that vary in accordance with the proximity of the first when it rushes by. Yet the degree of dependency between these factors is so very low that they may be regarded as independent. Accordingly, we will designate two events independent if the characteristic determining elements of the first, e.g., the coordinates, are independent of the determining elements of the second, i.e., if only infinitesimal changes result even though the determining elements of the second take on quite different values.

Armed with these basic definitions, we can now tackle the problem of compound probability. Let us suppose that two coins have been dropped on the ground side by side and inquire into the probability that both will come up tails. According to our definition of probability, this means the same as asking how often, in a series of repetitions, this combination will occur relative to the other possible combinations. Insofar as our hypothesis of a probability function is valid, we know that heads and tails are shown with equal frequency for each coin; now, is it possible to infer from this that the combination tails-tails will occur in exactly one quarter of the cases, as it ought, in accordance with the familiar formula? The following consideration shows that we cannot make such an inference. Suppose the first coin is tossed repeatedly and the results prove to be such that they correspond to the law of a probability function for this coin. Now, if the second coin were also tossed and invariably showed the same side as the first, the results would likewise correspond to a probability function; but for the combination there would result the very peculiar law that heads-heads and tails-tails would be the only combinations that ever occur. We would then have a distribution in which each individual event is regulated by a probability function, yet which does not correspond to the rule of multiplication for probabilities. This proves that the hypothesis of probability function is not a sufficient condition for the rule of multiplication.

Let x and y be the values for the time of fall of the first and second coins. Let us now make the following hypothesis: There exists a probability function

$$\varphi(x, y)$$

in which a probability is coordinated to every combination of values x, y in a way corresponding exactly to that we defined for the simple probability function; i.e., the probability that x will lie between the boundaries a and b and, at the same time, y between the boundaries c and d will be given by the double integral

$$P = \int_a^b \int_c^d \varphi(x, y) \, \mathrm{d}x \, \mathrm{d}y.$$

Given this presupposition we are able to derive the rule of multiplication of probabilities. Let us suppose the values x to be entered on the X-axis and the values y on the Y-axis, each axis being divided into intervals[4] to which there corresponds a network of rectangles [Figure 2]. Imagine the appropriate probability – that is, the probability that x will occur precisely in this interval Δx at the same time that y occurs in the corresponding Δy – to be represented by a prismatic column over each rectangle. The tops of these

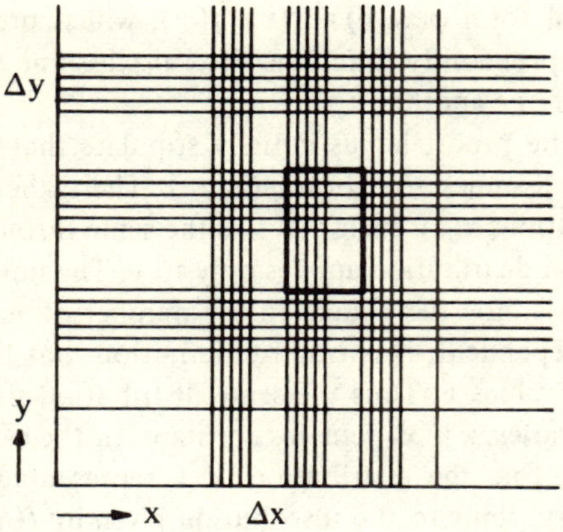

Fig. 2. Probability of the simultaneous occurrence of two mutually independent events.

columns will lie along a curved surface: also, we will presuppose that this surface is continuous, although the curvature may be of any sort, and that for remote distances it approaches the *x*-*y* plane asymptotically. The stripes for the intervals in the *x*-*y* plane are alternately shaded, as before. Let us now observe the portion of the figure singled out by the four heavy lines, where the white rectangle corresponds to the combination tails-tails, the criss-crossed rectangle to the combination heads-heads, and the two rectangles with one-way shading to the two mixed combinations. Since $\varphi(x, y)$ is represented by a continuous surface, the prismatic columns over these four rectangles become practically equal when the intervals are sufficiently small, and even if the columns at different points along the plane are of completely different heights, the four sums resulting from addition of the columns over rectangles of identical shading will be practically equal. At the limit for infinitely small intervals they will be exactly equal. These sums are finite because, owing to the asymptotic course of the surface, the space between it and the *x*-*y* plane is finite. The probabilities of the combinations correspond to these sums; for the combination tails-tails the result is $\frac{1}{4}$, for the mixed combination, $\frac{1}{4} + \frac{1}{4} = \frac{1}{2}$, and for heads-heads it is again $\frac{1}{4}$.

However, we can derive a further result from our presuppositions. To do so, we must stipulate that, along with the existence of $\varphi(x, y)$, our old hypothesis applies, i.e., that, *in addition*, for each factor, x as well as y, there exists a probability function $f(x)$ (or, as the case may be, $f(y)$) that *simultaneously* determines the distribution of these factors. We may then derive for

$\varphi(x, y)$ the special form $\varphi(x, y) = f(x) \cdot f(y)$, which prescribes the rule of multiplication for probability functions if we also assume that the two factors are *independent* of one another.[5]

To carry out the proof, let us initially stipulate that $x = x_0 = \text{const.}$, so that the function assumes the form $\varphi(x_0, y)$. There thereby arises a distribution of combinations x_0, y such that x is the same throughout and only y is a variable; thus the distribution applies only to y. The number of values y for any given interval is, however, equal to the number of values (x_0, y). Now if the events are independent, it is true by definition that the values y are not influenced by the values x. Thus the same distribution y must occur regardless of whether x varies, $x = x_0$ remains constant, or the event x does not take place at all. Therefore the distribution of y represented by $\varphi(x_0, y)$ must correspond in every point to the distribution given by $f(y)$, and thus be proportional to it. However, the factor of proportionality k may depend upon x_0 but not upon y. It is therefore characteristic of independent events that:

$$\varphi(x_0 \cdot y) = k(x_0) f(y).$$

As this is true of any $x = x_0$ (again the stipulation of independence), it must be an identity, so that we may write:

$$\varphi(x \cdot y) \equiv k(x) f(y).$$

The same holds if $y = y_0$ remains constant while x is a variable, so that it is also true that:

$$\varphi(x \cdot y) \equiv f(x) k(y).$$

From this it follows that:

$$k(x) = f(x); \quad k(y) = f(y);$$

That is to say:

$$\varphi(x \cdot y) = f(x) f(y).$$

The stipulation that the two events be independent is vital to this proof. Let us imagine, for example, a body rolled back and forth at random on a rough patch of ground, so that its position is determined by a probability function, and imagine that another body, which glides along the ground in jumps is joined to it by means of a flexible connection. A probability function will exist for the second body as well, and even a probability function $\varphi(x, y)$ for the combination of their positions. But this function will not possess the special form $f(x) . f(y)$. Instead, certain combinations in which the spatial distance from x to y corresponds to the average length of the connection will be favoured. For in this case the condition of independence is not fulfilled. It is only natural that this should be entered as a requirement for our proof, for it is also presupposed by the accepted probability calculus for the rule of multiplication.

Thus the rule of multiplication forces us to broaden the hypothesis of probability function and extend to combinations of several independent variables. It is evident that we may not stop at the number two, for combinations of several events must also conform to a rule. Here again, however, we need demand only that the function be continuous, not that it possess some particular value. The special form of the product for the particular instance of independence is derivable from the general form. Incorporating the extension into the original concept, we may now state the following principle: The existence of a probability function for one or more variables represents the sufficient condition for the applicability of the probability laws.

It is noteworthy that we can derive the rule of multiplication of probabilities that can be depicted after the pattern of the tossed coin merely on the basis of the continuity of $\varphi(x, y)$. That is to say, we do not, for this purpose, need to employ the special product form of $\varphi(x, y)$, viz., the rule of multiplication for probability *functions*. It appears, therefore, that the first rule of multiplication, unlike the second, does not require the presupposition that the events are independent. But this is erroneous. The very continuity of $\varphi(x, y)$ presupposes that there exists no relation of dependency between the intervals of the one quantity and those of the other, i.e., that a 'tails' interval in the one series does not entail a 'tails' interval in the other. If this were the case, a decrease in the intervals would never be accompanied by a convergence of the prismatic columns, this being prevented by the continuity of the function. This is the only form of independence required for the multiplication of simple probabilities. Otherwise, the values may indeed be dependent upon one another, and, given a suitable distribution of intervals, the schema for simple multiplication could likewise be derived from the mechanism with the flexibly joined bodies. It is only the rule of multiplication for the functions that requires the independence of all the events.

V. PROBABILITY FUNCTIONS IN THE THEORY OF ERROR IN MEASUREMENT (THE GAUSSIAN ERROR FUNCTION)

Probability functions play a prominent role in the theory of observational error. This theory has the task of calculating from numerous varied measurements the magnitude that agrees most closely with the value sought for. To this end, it must make an assumption regarding the distribution of the errors in measurement, and even though it can make only very general conjectures about these errors, it must derive quite particular forms of the law of distribution if it is to produce any results at all. We must bear in mind here that the error involved in a single measurement rests upon the interaction of a great many sources of error (elementary errors). The most significant solution to this problem is the Gaussian error function, which makes the probability of an error exponentially dependent upon its magnitude.[6]

Various presuppositions may be used for the derivation of this special form. For purposes of our discussion, the vital fact is that the Gaussian law can be constructed upon the basis of the following three conditions:

1. The frequency of every elementary error is determined by a probability function of some form or other.

2. These functions are calculated in accordance with the rule of multiplication.

3. A large number of mutually independent errors of the same order of magnitude must interact.

It is apparent that the first condition is identical with our first hypothesis and that the second can, after the stated derivation, be reduced to the expanded hypothesis for the probability function for combinations. In contrast, the third condition does not represent a presupposition resting on principle, but rather an assumption that pertains only under certain circumstances. Whether or not these conditions are present may be established empirically, and only where they do will the Gaussian law apply. For this law is a special form that in no way represents a universal principle and that ought not to be pushed to the foreground of our discussion. The law derives its practical significance from the fact that modern instruments for measurement embrace, by virtue of their complex structure, innumerable sources of error, thereby fulfilling the third condition – hence the frequency of its application.

The situation is much the same for the so-called hypothesis of the arithmetic mean, according to which the mean value of the measurements corresponds with a higher degree of probability to the required magnitude. This law is fulfilled wherever the Gaussian law of error is applicable, for it follows from this by means of a very simple mathematical operation. But this implies, too, that it is connected to the same conditions as the Gaussian exponential law; if the third condition does not apply, the arithmetic mean is also not applicable. There is therefore no point in referring to the hypothesis of the arithmetic mean. It is better to reserve the term 'hypothetical' for the more fundamental presuppositons.

It follows that the theory of error, too, does not include any fresh hypotheses for the applicability of probability laws. Rather, it leads us back to those presuppositions that we already developed from simple examples of probability distributions. With this we shall end our study of physical problems of probability. We have reached the conclusion that all these problems include a characteristic hypothesis, which we have been able to formulate as the principle of the probability function and which constitutes the law of probability. Our next task will be the testing of the justification of this hypothesis. The critique of the hypothesis will provide the answer to the philosophical problem of probability.

NOTES

[1] *Ztschrft. f. Philos. u. philos. Kritik* [1916a].

[2] Poincaré, H. *Calcul des Probabilités*, (Gauthier-Villars, Paris, 1912), p. 149.

[3] The proof may be carried out, provided only that the curve be integrable and that the integral from $-\infty$ to $+\infty$ possess a finite value. Thus the demand for continuity is somewhat excessive, yet it expresses the required characteristic most clearly and will therefore be employed throughout this article. We ought, more precisely, to speak of the continuity of the integral.

[4] For the sake of simplicity, we will show all the intervals as of equal magnitude; i.e., we imagine the velocity of rotation of the coins to be constant. The same result can, of course, be derived for the more general presupposition of continuity for this velocity.

[5] As I did not formulate this proof clearly enough in the work mentioned above, I am presenting it here in its complete form.

[6] This function runs as follows:

$$\varphi(x) = \frac{h}{\sqrt{\pi}} \cdot c^{-h^2(x-a)^2}$$

where x is the magnitude of error, a the systematic error, and h the degree of precision.

52b. APPENDIX

A LETTER TO THE EDITOR ON THE PHYSICAL PRESUPPOSITIONS OF PROBABILITY CALCULUS

[1920d]

A discussion with Mr. von Laue has prompted me to amend by the following note my article on the physical presuppositions of probability calculus, published this year in *Naturwissenschaften* [1920c].

In that article, I derived the rule of multiplication for probability factors from the existence of a probability function $\varphi(x, y)$ without presupposing the special product form

$$\varphi(x, y) = f(x) f(y).$$

I was able to demonstrate that this special form applies in the case where both events are *in*dependent, and I gave it the name, rule of multiplication for probability *functions*. However, I indicated that, if their intervals are classified in a certain way, *de*pendent events also lend themselves to the production of a mechanism conforming to the rule of multiplication for independent probability *factors*.

In carrying out the proof, I made use of a transition to a limit (cf. Figure 2 in the article), yet I did not stress clearly enough how this transition may be *physically realized*. If, for instance, the variable factors x and y are the times of fall for two coins, the decrease in the intervals Δx and Δy may not be carried out by means of geometrical operations, but only by increasing the velocity of the rotation of the coins; the case of falling coins is distinguished in this respect from a game of roulette, in which the intervals may be decreased by dividing the wheel into smaller sectors. As there are no infinitely great velocities, no physical mechanism will ever do more than approximate an equal distribution.

Further reservations may be aroused by the claim that the rule of multiplication of probabilities is to apply even to events that are not independent. Let us suppose, for example, the two falling coins to be joined by means of a rigid, guided connection in such a way that, say, the time of fall for the first is always smaller by the same amount than the time of fall for the second. The rule of multiplication will then not apply to the combinations of heads and tails. But there will also not exist a continuous function $\varphi(x, y)$; instead, this plane degenerates into a curve the projection of which in the $x - y$ plane is a straight line. Only when the connection between the two events varies itself in such a way that it assumes definite values through a probability function will it be possible to assign a frequency to every combination x, y, at least within a finite area of the plane; only then, that is, will there exist a continuous function $\varphi(x, y)$. On the other hand, it is evident that the rule of multiplication may be derived for heads and tails even for this type of variable independence, provided that, by increasing the velocities of rotation (which, by the way, need not be the same for the two coins), we reduce sufficiently the intervals Δx and Δy. For the degree of dependence at which the rule of multiplication for probability *factors* is still applicable is adequately characterized by the existence of $\varphi(x, y)$, and the total independence of the two events and, with it, the rule of multiplication for probability *functions* need therefore not be realized. Here, too, by the way, we will not find it possible, in deriving the rule of multiplication for heads and tails, to avoid the transition to the limit by increasing the velocity of rotation.

The same transition to the limit, carried out through an increase in the velocity of rotation, is also required for the derivation of the special form

$$\varphi(\theta) = \text{constant}.$$

For a precise presentation of this problem, I refer the reader to Section III in the afore-mentioned article [in Note 1].

I am all the more grateful to Mr. von Laue for occasioning this expansion in view of the fact that philosophical analysis of physics has fallen badly into disrepute through the blunders of some philosophers. The proofs required by such analysis cannot be too carefully formulated.

April 15, 1920

53. A PHILOSOPHICAL CRITIQUE OF THE PROBABILITY CALCULUS

[1920e]

I. REGULARITY AND CAUSALITY

The uncertainty of opinions regarding the validity of the laws of probability stems from the apparent contradiction between these laws and the accepted methods of acquiring physical knowledge. The basic method of physics consists in tracing observable events to relations of dependency, in depicting the present events as the effect of a past event and the cause of a future event. The resultant causal chains are considered to be unambiguously determined functional connections, and even where they have not succeeded in discovering causal chains, physicists steadfastly maintain that they exist in principle and will ultimately be discovered. In contrast, the probability calculus makes assertions about non-causal connections, demands, indeed, the causal independence of its objects as a condition of the validity of its propositions. In games of dice, for instance, it is presupposed that the individual throws are independent of one another; the causal chain leading to the result for each single throw is totally disregarded, and the resulting distribution of the throws is designated expressly as a chance distribution, as a 'game', in contrast to the causally connected course of other natural events. It looks, then, as though probability and causality are mutually exclusive, as though they introduce into physics the question, 'Chance or law?', forcing the physicist to declare himself for the one or for the other.

We must, therefore, state at the beginning of our investigations that it is a mistake to view these two alternatives as incompatible, that there is, in fact, no irreconcilable contradiction between these two concepts. Even if *causal dependence* is a methodological presupposition of physics, it is by no means the only possible form of a functional relation of dependency. The relation, 'If A, then B', does not, as such, assert that A is the cause of B. Rather, the concept of cause presupposes, in addition to the relation of temporal sequence, quantitative relations such that a quantitatively determined B invariably corresponds to a quantitatively determined A. If, for instance, the gravitational force of the sun is designated as the cause of planetary motion, this signifies that the magnitude of this force quantitatively determines the degrees of motion. But there are other conceivable functional relations, e.g., 'If A varies within an interval α, then B varies within the interval β.' The existence of this relation presupposes by no means that a particular value B within the interval β corresponds to each value A within the interval α. An assertion of this nature would also represent a law of nature, although it would certainly not be a causal law. If, along with causal laws, we were to formulate natural laws of this second type, no contradiction would be involved. For the second type of regularity certainly does not exclude the first; it is, for instance, entirely possible that, within the intervals, a certain B corresponds to each particular A. There are also other possibilities; for instance, there might be no causal connection between A and B, while an entirely different factor C might stand in causal correspondence to A, a factor D in correspondence to B, both factors being unaffected by this law. As long as the law of the second type does not specifically exclude the first (causal) dependency, as long as all contradictory presuppositions concerning dependence within the interval are avoided, the two kinds of regularity are compatible.

If, then, the laws of causality are *different in kind* from causal laws, it does not follow from this fact in itself that they stand in *contradiction* to causality. Regularity is a more general concept than causality. That there exists a law for games of dice that determines the distribution of the throws without at the same time assigning a cause to the individual throws does not contradict the principle of causality. For the law leaves quite open the question of which factor is to be coordinated with the individual throw as its cause, just as we left this question open in formulating the law for the varying values within the interval. On the contrary, no one doubts that a causal chain could be established for the individual throw which would lead precisely to this particular result. But, again, that does not alter the fact that the totality of throws is subject to a law of distribution. The laws of probability may be expressed in this form: If certain conditions vary within certain intervals (e.g., the time of fall, the velocity of rotation of the dice, etc.), then other values (the sides of the dice that turn up) will vary within certain intervals according to a certain law. The absence here of any assertion of causality does not signify denial of its existence.

To be sure, it is not enough to demonstrate that, in addition to regularity, other forms of regularity may exist in nature without contradicting it. It must also be shown that the *particular form* of probability relation, the 'special law' in the afore-mentioned definition, does not in any way contradict the principle of causality. We are in a good position to establish this point, for, in an earlier article[1], we set forth the physical presuppositions of the probability calculus and are quite familiar with that special law. We found it possible to segregate those of the presuppositions that are ordinary experimental physical results capable of being subjected in detail to the methods of scientific observation. These include the spatial equivalence of certain shapes, such as the sectors of the roulette wheel; the independence of the individual events (we can easily establish, for instance, whether two falling stones influence one another or not); and the presence of a greater number of disturbing influences, discussed in connection with the theory of error. We were able to demonstrate that, in addition to these, there is one remaining assumption that cannot be tested by physical methods, but which signifies a regularity in nature and may be regarded as a law of probability. We formulated it as the hypothesis of a probability function; because we find contained in it that which is actually problematic in the theory, we can reach a decision concerning the entire theory merely by subjecting this hypothesis to critical scrutiny. We will proceed immediately with the investigation of the compatibility between this hypothesis and the principle of causality in order to present the *possibility* of the laws of probability. Only afterwards will we turn to a demonstration of their *necessity* for physical knowledge.

II. THE COMPATIBILITY BETWEEN THE LAW OF PROBABILITY IN THE FORM OF THE HYPOTHESIS OF A PROBABILITY FUNCTION AND THE LAW OF CAUSALITY

What is the significance of the principle of the probability function? The import may be set forth as follows:

If, when a certain change is made in the initial conditions, a certain physical factor x is repeatedly observed, the frequency of its individual values may be coordinated with a continuous curve (x) through the expression

$$\left(\frac{h}{N}\right)_{\lim N=\infty} = \int_a^b \varphi(x)\,\mathrm{d}x\ ,$$

where $a-b$ are the limits of an interval of the factor x, h represents the number of values falling within the interval, and N represents the total number of repetitions. (Note that, for the sake of simplicity, we are discussing here only *one* variable; later on we shall extend our remarks to several variables, which will not alter the situation in any fundamental way.)

The first question is, what is meant by *repetitions of the same factor?* Only an *event* can be physically realized; thus only an event, a *temporal sequence of physical changes,* can be repeatedly realized. Nonetheless, we describe an event by establishing its determining factors, its parameters, and representing their fluctuations as a function of time or of their mutual values. For instance, the parameters determining the revolving roulette pointer are, besides the angle of rotation Ω (frequently reckoned in multiples of 360°), the length of the pointer, its mass, the friction, and so forth. Each of these parameters may be regarded as a factor repeatedly realized through the repetition of the event; an assertion corresponding to the above formula could be constructed for each one of them. That we single out the angle of rotation Ω in roulette stems solely from practical considerations; this factor is visible and can be intuitively grasped, its division into intervals is immediately given through a simple geometrical construction, the colored sectors, and facilitates an intuitive depiction of the law of probability[2]. Making a selection from among the observed factors does not, then, present any fundamental difficulties. But now another uncertainty arises. What right have we to designate the angle of rotation Ω of the roulette pointer as the 'same' factor undergoing repeated presentation? We know that the individual numerical values for this angle are different each time. Certainly we may say that this angle is always determined by the same parameters: the initial impulse of the revolution, the mass of the pointer, the friction between it and the axis. But that merely postpones the problem: the question then arises why these are 'the same' factors, since after all, their numerical values must fluctuate if different values Ω are to occur? In order to become independent of the fluctuations in the human strength with which the pointer is impelled, let us imagine a machine that undertakes the initial impulse; let us further imagine it to be constructed with such precision that the determining factors are held as constant as is possible. Let us even suppose that this machine is maintained by an engineer of superhuman skill, so that the parameters in question really do have precisely the same value every time. – Even then, no one will doubt that the angle Ω will vary. The reason is that these three parameters are not the sole determinants of the rotation. There are always other determining factors. For instance, the resistance of the air will have an influence; if we wished to maintain it at a constant level, we would have to hold the air pressure, temperature, and humidity constant. Even if we succeeded in doing this, there would still be other, yet more minor factors, such as the vibration of the machine and the attraction of adjacent masses. In a word, it is impossible to reproduce all the conditions without quantitative variations. We shall call the sum of this infinite number of factors the *irrational remainder of determinants* and their individual components the *remainders.* It will, accordingly, be impossible to give Ω the same value every time. Thus if it is to be meaningful to refer to the 'same' angle Ω, the identity may not be determined by quantitative *equality.* We shall declare, instead, that as long as the fluctuations of the machine remain within certain narrow boundaries, the resulting angle of rotation remains 'the same', and in this sense we may also refer to variations of 'the same' factor. This is not to say that these fluctuations must always remain below the level at which measurement is possible; the deviations in the angle of rotation Ω are, for instance, quite capable of measurement. This definition is, of course, of a temporary character and will eventually be replaced by a more precise definition.

We are led at the same time to explain what is meant by the *variation in the initial conditions* that is mentioned in the above definition. It consists of fluctuations arising from the totality of measured and unmeasured determinants which, therefore, always maintain their irrational remainder. If we demand, by hypothesis, that Ω assumes particular values through a continuous function, this implies (since Ω represents the product of all measured and unmeasured factors) a corresponding assumption regarding the interaction of all of the infinite number of determining factors, i.e., that this interaction produces, in a particular way, a *continuous* distribution of the values of Ω. Thus we find that the hypothesis of a probability function represents an assumption concerning the irrational remainder of the factors determining a value.

The above interpretation shows clearly that this hypothesis involves no contradiction to the principle of causality. For the principle of causality applies specifically to the *measured* individual determining factors, demanding their unambiguous dependency; as to the irrational remainder, it can only make the assertion that, as analysis progresses, the remainder will increasingly be resolved into individual causally dependent determinants, although it will never be exhausted. The hypothesis of probability, on the other hand, is an assertion about the *sum* of all remaining factors – and this is a matter beyond the scope of the causal principle. The assertions contained in these two principles concern, in fact, quite different objects; they can both represent laws of nature without contradicting one another.

Furthermore, no contradiction to the causal principle is to be found in the independence of individual events demanded by the probability calculus. We showed in the earlier investigation that we ought more accurately to speak of diminishing degrees of dependence rather than independence, and this concept is entirely compatible with causal dependence. Two events are independent in this sense if the determining factors (parameters) of the one event do *not* change substantially when the factors determining the other event assumes different values. Thus the position of the focal point of a lens is considered independent of the intensity of the light, even though a certain dependence does exist. The light absorbed by the lens will warm the glass in proportion to its intensity, thus altering its geometric dimensions and displacing its focal point; yet minor changes in the location of the focal point correspond to such great changes in the intensity of the light that we may, for all practical purposes, speak of independence. In probability calculus, too, such low degrees of dependence are to be considered tantamount to independence. For instance, two dice falling side by side are to be regarded as virtually independent, even though the air current brought about by one die influences the course taken by the other.

In this connection, a comment concerning the concept of chance is in order. A probability distribution such as occurs with the throwing of dice is called a product of chance because it has nothing to do with the causal determination of the individual throws. According to our analysis, the element of chance is to be sought essentially in the peculiar tendency of the physical remainders to display a continuous distribution. In order for this continuity to engender that distinctive regularity in the distribution of individual cases, it is further necessary to create a division into small intervals and a classification of these into several groups, as we have seen[3]; we must bear in mind that even a small increase $\Delta\Omega$ in the angle of rotation in roulette corresponds to a new color. Thus chance emerges when *minor* changes in the one value result in *major* changes in others. (Minor changes in Ω result in discontinuous changes in the color over which the pointer comes to rest, and a falling die turns over many more times if the time of fall increases ever so slightly). Note the contrast between this formulation and the definition of independence; there the reverse was the case, large changes in one factor resulting in only minor changes in another. (For instance, major changes in the intensity of the light produce only an infinitesimal dislocation of the focal point.) This contrast is striking, for chance, as opposed to causal determination, is frequently defined as independence. Clearly, there are two possible ways of passing to the limit: transition through the lowest degree of dependence leads to the concept of independence, transition through the highest degree of dependence leads to the concept of chance. Both of these, chance and independence alike, represent idealizations that cannot be fulfilled in reality because they essentially contradict the principle of causality. They are meaningful as limiting concepts because they may be approached without restriction. Yet the great difference between these two concepts remains undiscovered so long as the method leading to their limits is not pointed out.

III. THE PRINCIPLE OF REGULAR DISTRIBUTION AS A NECESSARY PRESUPPOSITION OF PHYSICAL KNOWLEDGE

Now that we have shown that the law of probability, in the form of the hypothesis of a probability function, does not contradict the law of causality – that is, that it is *possible* in conjunction with it – we shall demonstrate its *necessity* for the concepts of physical knowledge in an entirely different connection. For this purpose, we must analyse the basic meaning of a physical judgment.

There are two fundamentally different kinds of judgments from the epistemological standpoint. In the first kind, the object is posited as a mental fiction, and fictions are connected to one another by means of the judgment. Mathematical propositions belong to this category. Their objects are fictions, artificial constructs of the mind; they are not real objects like those existing outside the mind, which we can experience, but are established solely as a result of mental constructs, and all propositions about them form relations that are meaningful only within the mental sphere. It is therefore vital for these judgments that their object be precisely determined, that it be exhaustively defined by the mental construct. The relations expressed in the judgment can thus take into account all the characteristics of the object and attain unshakable validity. The compelling nature of the mathematical judgment is universally acknowledged; it is grounded in the fact that the object of the judgment is stipulated and that this mental stipulation permits of an exhaustive definition. We shall designate this class of judgments *stipulative judgments.*

In contrast to these are all those judgments having an object existing in reality. Physical judgments are to be accounted among these. What is real is fundamentally different from what is thought; it is something that cannot be defined any further – for that would again mean thinking – but can only be experienced in its distinctive nature. We receive information concerning the reality outside our ego through the perceptions of our senses, and it is therefore characteristic of natural objects that they come into our conceptual sphere upon the basis of some perception or other. The connection may also be indirect; no one has yet seen the side of the moon that is turned away from us, but its existence is considered a certainty because a mental connection is established between it and other perceptions, i.e., perceptions of the side that is visible. Here we note, to be sure, that logical relations enter into propositions about reality, and a more thorough investigation is needed into the nature of this connection. Yet the fact remains that the object itself does not represent a mental fiction, but something foreign to thought, something given. We will designate this second class of judgments *reality judgments.*

That physical judgments may also be clothed in the form of conditional propositions would appear to contradict this distinction: If the sun is out, it is warm. But it would be a mistake to believe that the object is posited along with the setting up of the presupposition. Only the assumption that it exists is posited; the object sun, however, remains that shining body which can be visually pointed out, which we can perceive. Unlike the mathematical object, this body is not defined by stipulation. Suppose we say, "If an angle at the circumference lies on the diameter, it is a right angle." The angle is defined by the antecedent, which adds to the concept of an angle at the circumference (which itself is likewise defined by a series of such propositions[4]) another distinctive mark, which, in conjunction with the others, exhaustively defines the object. In the above judgment concerning the sun, on the other hand, the antecedent does not express any determination of the object; rather, the hypothetical formulation simply expresses the fact that the realization of the consequent is connected with the realization of the antecedent. The hypothetical form, then, has a completely different significance in the two judgments, and we must not be misled by their grammatical similarity.

We might attempt to define the physical object similarly to the mathematical by listing its determining factors. For instance, we could define the terrestrial sphere as a sphere possessing a particular oblateness, yet we cannot thereby exhaust the actual terrestrial sphere, the form of which is far more complicated and is also subject to changes in time. A complete definition would require an account of the totality of the infinitely many laws of nature that play a role in determining it. But that is an impossible undertaking. If, in setting up the definition, we confined ourselves to the essential characteristics and set the resulting definition into the judgment – if, for instance, we transformed the judgment, "The earth imparts to the bodies at its surface the acceleration 981", into the judgment, "A sphere possessing the particular mass m and the particular radius r imparts to the bodies at its surface the acceleration 981" – the distinctive feature of the physical judgment would be lost, and the proposition would be reduced to a mathematical judgment. We are invariably confronted with this choice: either we define the objects exhaustively through their determining features – in which case we know nothing as to their occurrence in reality – or we take the objects to be those physical things that we can only indicate ostensively – in which case the judgment loses the compelling character of a mathematical judgment.

Nonetheless, physics employs a similar procedure. We are well aware that it makes use of the principle of the [ideal] force of a sphere in explaining the gravitational force of the earth. But the novelty of its presentation does not consist in the mathematical relation representing the force of a sphere, but in the *application* of this particular mathematical relation to the real, natural earth; this is the characteristically physical element in the principles of physics. It coordinates mathematical relations to objects presented to the senses; its method consists in this. To be sure, we must not imagine this coordination to be an exceedingly simple affair, in which, say, the mathematical form of a sphere is coordinated to the terrestrial sphere. The very term 'terrestrial sphere' assumes prior stipulation of the coordination. The earth is, in the end, nothing but a collection of sensible impressions, a visible 'something', and this 'that over there', which can only be ostensively indicated, is coordinated to the mathematical sphere. Physical knowledge consists in this: in the intellectual imposition of a certain mathematical structure upon the chaos of perceptions. What is ordinarily called a physical object is itself such a coordinated mathematical structure. We must bear clearly in mind that, even if all propositions about physical events consist of mathematical relations between various mathematical structures, that which is truly physical about them remains the coordination of the resulting mathematical complex to certain perceptions. The systems of physical equations represent coordinations of this kind. For instance, Maxwell's equations are coordinated to those real events that we describe as electrical, while the Einsteinian equations for gravitation correspond to the real events we call mechanical.

Why must we coordinate these particular equations to these particular sensible events? We know that we certainly cannot exhaustively represent the events by means of the equations. Why, then, do we not arbitrarily select some other equations? The answer is that there is one more peculiar fact, namely, that the real things behave *approximately* like the mathematical fictions with which they are coordinated. Even though the earth is not a sphere, the acceleration calculated for a sphere is approximately the same as that measured at the earth's surface. This characteristic fact forms the basis of all physical judgments.

Approximation rests upon a comparison of numbers. It is therefore essential to the physical judgment that it define numbers, the approximate correctness of which can be measured. This would appear to contradict the usual presentation, according to which physics searches for functional connections – according to which, that is, physical knowledge represents, not the magnitude of a factor, but the law of its variations in conjunction with another factor. We might argue, for instance, that the fact that the magnitude of the earth's acceleration, $g = 981$, is not an item of physical knowledge, which we achieve only with the law representing g as a function of the radius, for which 981 is no more than one particular value. It is true that this law represents *causal* knowledge, for we established earlier that the meaning of *causal* regularity consists in the quantitative dependence among factors, Yet in every function there appear certain *constants* which must be numerically determined if the function is to be defined, if the law is to have a definite meaning. In the function that represents g as dependent upon the earth's radius there occurs the gravitational constant k; it must be possible to assign to it a numerical value if the causal law concerning g is to have a certain defined significance. That is why the determination of numerical constants cannot be separate from physics proper; it belongs to the presentation of causal laws just as much as does the imparting of the functional form. And the factor g itself may play the role of such a constant; in the law of falling bodies $s = \frac{1}{2}gt^2$ it occupies just the same position as k in the Newtonian formula for gravitation. Its numerical value must be definite if the law of falling bodies is to represent the actual behavior of falling bodies.

This idea may be expressed in such a way that the physical judgment entails two kinds of coordinations for the objects of reality: first, a functional form, as, e.g., the coordination of the form $s = \frac{1}{2}gt^2$ to falling bodies, and, second, certain numerical values, as, e.g., $g = 981$. Causality guarantees that a functional form exists; but is it also able to guarantee that certain numerical values exist for a given group of objects?

It would appear that causality can achieve this as well. For it attempts to justify the particular numerical value by presenting it as, in turn, causally determined, as itself a function of other factors. Thus it represents g as being determined through the magnitude of the earth's radius and resolves the gravitational constant k into a function by pointing out in the more general Einsteinian equations for gravitation those special circumstances that bestow upon k the particular Newtonian value. This is, in fact, the process for expanding knowledge employed by the causal principle. But in the light of the ideas discussed in the preceding section, we must declare that the causal principle does not suffice for the establishment of the definite magnitude of special numerical values. There we showed that causality can never do more than indicate *individual* determining factors, while the definite outcomes invariably consist of the sum of infinitely many influences – and it is this very sum about which causality tells us nothing. This sum could be of any magnitude whatsoever, and every measurement could indicate a completely different value for the constants, without contradicting the law of causality, for the neglected remainders might at any time begin to exert an influence and, again in accordance with strict causal laws, alter the values. If, for instance, the sun were to suddenly spin off great masses in the direction of the earth, the magnitude of g would change. The mass of the sun is already contained as a value in the equation $g = 981$, but it appears there as a component in the sum of very minor 'disturbing' influences having no substantial influence upon g. To be sure, we could supply yet another causal explanation for the change in g. But this would fail to explain why there is not, in general, a sudden increase in the disturbing influences, why it is consequently permissible to ignore them and assign to actual bodies at the earth's surface the number 981. Yet, let us repeat once again, there is no point to giving the law a hypothetical formulation such as: If the influence of the disturbing factors remains small, the law of falling bodies is applicable and the numerical value of g is 981. For the physical judgment is thereby reduced to a mathematical judgment. The real earth is not a sphere possessing the radius r and the mass m, but 'that thing there', capable only of ostensive definition, which we perceive with our senses and which holds a position in the totality of events in nature.

Whence do we derive the right to coordinate certain numerical values to certain real things? It does not arise from the principle of causality, but requires the addition of another principle containing a hypothesis concerning the appearance of certain numerical values and, therefore, concerning the influence of the remainders. This principle is the hypothesis of a probability function.

In the previous section, we learned that this hypothesis presents, in fact, an assumption about the sum of the remainders. For it asserts that, while it is possible for this sum to be of any magnitude whatsoever – and hence is compatible with the causal principle – there exists a law for the frequency of the outcomes; certain individual values, the normal values, occur very frequently, while others, the extreme values, occur with infinitesimal rarity. This assertion is expressed in the asymptotic course of the curve at both ends. The hypothesis asserts further that infinitely contiguous values occur with equal frequency, that we may, that is, assign to every infinitesimal interval a definite relative frequency: this is the stipulation that the curve be continuous. This is a necessary addition, for without it the law would not be capable of confirmation. We cannot count the frequency of the occurrence of individual numerical points because we cannot show the values as numerical points, but can only include them within limits. This hypothesis produces an appropriate assumption concerning the numerical values of physical constants. If there existed a completely precise analysis of real things, we would be compelled, in accordance with the causal principle, to grant the proposition that a certain outcome, once determined, must be capable of repetition at any time and at any place. As we cannot attribute continuous equality to the outcome, the generalization that naturally comes to mind is that *some law,* even if not *this law,* exists for their distribution in space and time. This assumption is tantamount to the principle of a probability function. Let us note that it does not prescribe any definite form for the law of distribution, that particular forms, such as $f(x) = \text{constant}$, are rather to be inferred from the general presupposition in conjunction with particular circumstances. This principle does not follow from the causal principle, but neither does it contradict it. It must, rather, be placed alongside the principle of causality if physical knowledge in the form of the coordination of definite functional laws and constants is to be possible at all.

A corresponding law applies to the frequency of combinations of two outcomes, for if a law of nature contains two or more constants, a corresponding assumption must be made concerning the frequency of the occurrence of this combination. Otherwise, it will not be possible to coordinate these constants to the relevant group of real objects. This is the origin of the probability function of several variables.

IV. THE ANALOGY BETWEEN CAUSAL LAWS AND PROBABILITY LAWS AND THE IMPOSSIBILITY OF THEIR EMPIRICAL CONFIRMATION

We have now indicated the philosophical position of the laws of probability. It rests upon an epistemological principle completely parallel to the principle of causality. The principle of the regular connection of all events which is represented by causality is not sufficient as a foundation for physical knowledge. Another principle must come into play, binding events to one another laterally, as it were; this is the principle of regular distribution. It may also be formulated as the principle of a probability function and is identical with the hypothesis required for the standard probability mechanisms.

From this standpoint we are able to resolve a difficulty we encountered at the outset of our investigations. At the beginning of Section II, we inquired after the source of our right to designate a physical factor as the *same* factor in variously repeated events, given that its numercial value is different each time. We were satisfied for the time being with the answer that the deviations in numerical values had to remain within certain limits, although we also had to admit that very great deviations were occasionally possible. Now that we have established the place of the principle of distribution in relation to the concept of physical knowledge, we are at last in a position to give a precise answer to the question. By reversing the described connection, we may assert that a mathematical parameter represents one and the same physical factor if its observed numerical values conform to a continuous function of distribution. This signifies that the deviations are usually small, yet it does not exclude occasional major deviations so long as they do not fundamentally disrupt the schema of distribution. Thus the introduction of the laws of distribution as the basis of scientific classification is required for the solution of the problem of the identity of physical factors.

We might raise the question whether this principle is to be called empirical in the sense in which we regard ordinary physical observations as products of experience. The principle of energy, for instance, is a result culled from experience; observations have taught us that the physical factor known as energy maintains the same magnitude in all events. The reverse is also conceivable; energy might, for example, increase continuously, as is known to be the case with another physical factor, entropy. Likewise, attempts have been made to establish the causal principle on an empirical basis. It is claimed that all our observations show every event to have its cause and its effect; the universal law of causality was laid down because this was established in so many instances. In any case, the argument continues, causality is not logically necessary; it is equally conceivable for the same event to have different effects. This point must certainly be conceded. The principle of causality is not logically necessary; no more is the principle of distribution logically necessary, for it is certainly possible to conceive of nature running its course in a completely irregular fashion. But it is a mistake to suppose the philosophical possibilities to be completely exhausted by the two categories, 'logically necessary' and 'empirical'. The great service rendered by Kantian philosophy was the introduction of a new form of question into the problem of knowledge. Kant asks: Which principles are distinguished by being a necessary constituent of our knowledge of nature? He calls such principles *conditions of knowledge* because a knowledge of nature is possible only through them, and he replies as follows to the question regarding the position of the law of causality: Certainly it is conceivable that nature might run its course without any functional dependencies, but if there exists any knowledge of nature, the principle of causality is valid, for without it all such knowledge is impossible. These principles that are not logically necessary (analytic) but are nonetheless necessary for empirical knowledge are called by Kant *synthetic a priori judgments.* Their validity cannot be established empirically, by means of individual observations, but stands and falls with the possibility of knowledge as such and must likewise be called a transcendental fact. We could undoubtedly acquire physical knowledge if the principle of energy were not valid; we would simply end up with different equations. But knowledge would be impossible if the law of causality were not valid, for we would be unable to establish any quantitative functional relations whatsoever. This distinction produces a new classification of natural laws, singling out certain of them as *a priori* valid. Expanding the Kantian idea, we must now declare the law of distribution to be an *a priori* principle of knowledge in just this sense. For it is likewise a necessary presupposition of knowledge, and we may say: If physical knowledge exists, then the principle of distribution is valid.

The analysis undergone by the concept of probability through its formulation as the principle of distribution or as the principle of a probability function places the observed regularities in a new light. We now understand why physical laws are described as merely probable. It is because no assertion can be made regarding their special form in individual cases; a statement about their frequency can only be made after repeated realizations, and this statement leaves the numerical value of the individual case undetermined. The so-called philosophical probability of the validity of the laws of nature is brought together with physical probability by means of the principle of the probability function. On the other hand, we can understand why it is that the regularities of distribution, which we have observed in games of chance, the theory of error, and the like as laws of physical probability, must recur again and again, and why it is that in every investigation it has been necessary constantly to distinguish them from ordinary physical laws. Some thinkers have, to be sure, proposed various theories about these regularities, but have not been able to furnish a proper basis for them; they have failed to notice that what they have come up against is a principle of knowledge that can only be judged from an epistemological standpoint, that cannot be verified or refuted by individual experiences because its significance lies much deeper, in the very essence of knowledge. Seen from this viewpoint, all attempts at experimental investigation of the laws of probability calculus must appear absurd. Such attempts have, to be sure, been made. For instance, one scientist carried out 120,000 throws of the dice in order to establish whether an equal distribution would actually be produced. The astonishing result was that one side came up more often than any of the others – but the experimentalist concluded from this result that the die had an irregular center of gravity, rather than that the laws of probability were false. It is telling that this scientist was altogether incapable of resisting the *a priori* power of the principle. It is the same here as with the principle of causality: if we encounter an incompatible state of affairs, we alter the special form of the principle used for explaining this state of affairs, but not the principle itself. This possibility invariably exists for both the causal principle and the principle of distribution. In this connection, Marbe's attempts to refute statistical regularities through observations and to replace them with rhythmical laws[5] also seem fruitless. Even if he were to confirm the existence of such rhythms, Marbe would never be able to conclude any more than that certain conditions giving rise to the rhythms are present in special relations pertaining to the object under observation. There are indeed objects, e.g., in psychology, in which alertness is a function of success[6]; in these instances the stipulation that the individual events be independent, which we laid down as a special condition for *equi*probability, is not fulfilled, and consequently the particular form of regularity is altered. But Marbe's research is not even adequate from the standpoint of mathematical probability theory, as has been established after a most thorough fashion by R. von Mises[7]. As these investigations have not even been carried out in a methodologically satisfactory manner, philosophical criticism may bypass them.

Philosophical considerations led to our viewing the laws of probability as objective laws of natural events holding a position analogous to that held by laws of causality. We may therefore no longer see in them laws of embarrassment, escape routes sought out by the physicist when he lacks more precise knowledge of the connections involved. Laplace expressed the idea that a human being possessing perfect mental faculties would have no further need of probability laws, being able to comprehend the totality of events by means of causal laws. Let it be noted that this super-intelligent creature would be proceeding in a highly impractical manner if he took precise account of every single throw of the dice, refusing to make use of the regularity contained in the equal distribution of the sides. For not even the very wisest of intelligent beings could alter this state of affairs; his precisely calculated throws would also conform to the distribution schema. To limit ourselves to causal laws is to renounce one portion of the description of nature. In his well-known lecture on dynamic and statistical regularity, Planck carried out a thorough demonstration of this duality of methods; we are now able to understand it from a philosophical point of view, for we have established a parallel interpretation of the two principles of connection and distribution for the concept of knowledge. Our critique incorporates probability laws into physics as a branch on an equal plane.

We are indebted for the result to the combination of two methods of investigation: the axiomatic method, which guided us to a precise formulation of the axiom of the applicability of probability laws, and the critical method, whereby we scrutinized the position of this axiom with respect to the concept of knowledge. To be sure, we were only able to apply these methods to physics, and it is therefore not yet possible to form definitive judgments concerning the validity of probability laws in other fields, such as psychology and sociology. But we may safely predict that what has revealed itself to be a philosophical principle in the one case will not turn out to be an empirical law in the others. It would appear that here, as with other problems, physics has taken the lead solely because of its more advanced levels of mathematical formulation.

NOTES

[1] [1920c] (translated in this volume as 'The Physical Presuppositions of Probability'). This article develops the axiomatic foundations of the following philosophical investigation, and its results will therefore be presupposed here.

[2] *Ibid.*, section II, paragraph 4.

[3] *Ibid.*, Figure 1.

[4] Cf. the very clear discussion of implicit definitions in Schlick, *Allgemeine Erkenntnislehre*, (Springer, Berlin, 1918), p. 30.]English translation by A. E. Blumberg, *General Theory of Knowledge* (Springer, New York and Vienna, 1974) – Ed.]

[5] Marbe, K., *Die Gleichförmigkeit in der Welt*, (Munich, 1916).

[6] Cf. my review of Sterzinger's *Zur Psychologie und Naturphilosophie der Geschicklichkeitsspiele*, in *Naturwissen.* 7, 644 (1919).

[7] *Naturwissen.* 7, 168 (1919).

54. NOTES ON THE PROBLEM OF CAUSALITY

[A Letter from Erwin Schrödinger to Hans Reichenbach]*

[1978b–54]

January 25, 1924, Zürich

I can best delineate my attitude toward this question by characterizing in a few words my position respecting the problem of causality, without, of course, saying anything basically new.

The profound *problem* of causality seems to me to lie in the following question: Why do we always expect *completely* similar circumstances to produce results that are *completely* the same – not only after many repetitions, but even with the very first repetition? Why does a *different* result compel in us the conviction that the circumstances must have changed, if ever so little? We can reconcile ourselves to the conjunction of minimally altered circumstances with very strong influence upon the result, but can never admit to the slightest change in the result in *genuinely* unaltered circumstances.

I call this the puzzle of inductive inference. I do not believe that we can in any real sense resolve it. If we contemplate it for any length of time, we begin to get an extraordinarily uncomfortable feeling, arising not from half-witted musing but from a sort of mental dizziness; we keep thinking we have got hold of the problem, but then become aware that we are constantly moving in ever-narrowing circles.

We find ourselves thinking, "I have observed hundreds and thousands of times that the same circumstances lead to the same results; I can see perfectly well that the world in which I live is subject to 'causal' order. I won't submit to the folly of believing things are different in any given particular case. But why would that be foolish? This is the very question with which I started, under another guise. I have been trying to support the general procedure of induction by means of a particular inductive inference – like Münchhausen, who tried to pull himself out of the bog by his pigtail. – But I really *do* find it foolish, and furthermore, I'm right. It is quite obvious that I am right in concluding from the fact that the same circumstances have produced the same result in so-and-so many completely different instances that this must be generally true. Thousands of people have made this inference and found it

* [This letter of Winter 1924 was printed in *Erkenntnis*, 3 (1932) as an appendix to Reichenbach's article [1932d] – Ed.]

to be confirmed throughout their lives, so I, too, will . . . ." and now the very pigtail is pulling itself out of the bog by its own little pigtail, and so it goes gracefully along *ad infinitum*.

Thus we really cannot escape the fact *that* we invariably draw inductive inferences, *that* these inferences are extremely useful, *that* the organization of the whole of our lives depends upon inductive reasoning. The millions of times in which experience has confirmed our expectations may certainly lead us to the *resolution* to continue on in the same way, but why it does so is a matter for which we can only offer a tautological explanation.

Perhaps we can put the problem another way. Under what circumstances would we not end up with inductive reasoning and the concept of causality? Surely in cases in which we lack the *material* on the basis of which we *actually* draw the inference (even if we are unable to justify it), i.e., if we lived in a chaotic world in which we did *not* invariably observe similar results following upon similar circumstances. (To be sure, it is hard to believe that rational creatures would develop in such a world.) The retort might be, "But we in fact live in a world which is not chaotic." This would be a repetition of the circular reasoning illustrated above. And yet it may turn out that the idea of causality does have some connection with *realism* after all. Only because we view our world as something real, having an enduring existence, do we come to ascribe to this 'something real' the *characteristics* of causal connectedness. Of course, what lurks behind the idea of 'something real that is relatively continuous' is just the original question: Why can past experience say something about future experience now? Because of the ordered nature of the real, which is to be conceived of as continuous.

(By the way, you will see from my inaugural address [given in 1922; reprinted in *Die Naturwissenschaften* 17, 9 (1929)], which I have taken the liberty of enclosing, that I have no very great faith in this supposed characteristic of order.)

Now for a few comments on your study. I must take exception because of the small degree of cogency with which, in an actual case, the 'governing sequence of functions', let alone the '*complete* governing sequence of functions' can be derived. I myself would say that the sequence is not really given extensionally at all; that is, it is *not even extensional*. It would be extensional if, for example, the *form* of the function were fixed and only certain parameters had to be adapted to it. As you know, which these are can be inferred from given observational data by means of the method of the least square. The discovering of the correct functional form, for which there is absolutely no set method, is, in fact, based upon chains of inference that lie deeply buried in the shadowy realms of instinctive guessing. – But that you know yourself!

What I have called the 'insoluble puzzle of induction' is, of course, included in your 'probability inference'. You say yourself (Section 3) that very little light has been shed on this problem as yet.

But now to the matter I consider most important. You say (Section 4), "Causality exists when the governing sequence exhibits type I."

Given the present state of physical knowledge, I have no doubt that, if applied to an actual case, i.e., to the observation of molar (not molecular) events, the sequence would exhibit type II.

By this I mean two things:

(1) Disregard, for the moment, the so-called fluctuating phenomena. Then, provided that more and more are accumulated or that they are sufficiently dense in the first place, the points of observation will lie, in their two-dimensional aspect, along a surface rather than a *curve*; this is because of observational error. Now these errors are really there; they are not to be willed away. It is clear that if we accumulated observations in a purely mechanical manner by, e.g., observing the ordinate values for excessive abscissa values more and more closely together, and attempting successively the construction of the functions of the governing sequence, these would eventually fluctuate wildly.

Now, you may say, the observer will be clever enough to see what is happening; he will regard deviations from these values as small enough to be ignored and will not construct the function according to them. But what is applied to the ordinate must be applied to the abscissa. Actually, *both* are defective: the 'observing of the excessively high abscissa values' is fictitious, too. The observer will have to admit to himself that once he has attained a certain density of observations he gains nothing through further accumulation of observations for establishing the *form* of the function (including the number and type of the parameters occurring in it), but only for the precision of the parameters. Thus the sequence necessarily conforms to type I after a finite number of steps. Or rather, after this point, the only question that can remain is the much simpler one whether *perceptibly* equivalent ordinates are observed in conjunction with *perceptibly* equivalent abscissa values. And that was the original question (same circumstances, same result?). In this case – the very one that we generally believe to *be* the case – the sequential schema simply cancels itself out, without having provided us with any criterion.

(2) This time, let us *not* disregard the fluctuating phenomena, to which virtually every measurable physical value is subject in principle. Instead, we

shall leave aside for the moment all observational errors, or at least regard them as sufficiently small relative to the real fluctuations; quite enough realizable cases of this kind exist. Under these circumstances, our observer will discern a sequence of type II in a pure state and will, through further inspection of his instruments, be able to convince himself that observational errors are *not* present. Will he, for that reason, deny causality at all for the phenomenon under investigation? He studies the state of affairs, comprehends it, sees that his points of observation group themselves around a certain function *just like* observational errors, and derives a totally precise *law* (e.g. radioactive decay $a = a_0 e^{-\lambda t}$), not only for the middle curve, but even for the fluctuations – this last being, of course, really a statistical law.

Given these considerations, which I don't believe to be open to attack, most reasonable judges would, I think, regard your arguments like this: Reichenbach's sequence criterion is of no use. If we take into account the limited precision of observation (which it would be senseless to disregard), it necessarily leads to type I, except that real fluctuations exist that perceptibly exceed observational error. On the other hand, fluctuations always exist in principle: it is, in principle, always possible to increase the precision of observation so greatly that they become noticeable. In the end, we are really always led to type II. It would appear, then, that the problem of causality cannot be solved on empirical grounds, that the procedures actually used by scientists vindicate instead the philosophers of the *a priori*: causality is an indissoluble constituent of our mode of comprehension which we cannot avoid applying to every object of observation, no matter what its nature.

But this is surely not the only possible position. We could decide to retain your sequence criterion. But in that case, present-day experience would in all probability speak *against* causality, most likely *in all cases*. I find it of great interest that your careful, impartial analysis led to this result, which you surely did not intend. I would say that your sequences of functions are what the mathematicians call semi-convergent. The point of greatest stability determines the physical law, but this law is of a statistical, not a causal nature – as is shown by the later divergence of the sequence. Suppose we accept this second conception – essentially that of Franz Exner – and implement it throughout our vision of nature, as will probably happen within a few decades. I am far from having a clear idea of the structure which our vision of nature will then assume. Of course, we could not get along without a certain *continuity* in the elements making up the vision; otherwise, the world of the succeeding second would be completely independent of the world of the second before. (In the principles of the conservation of energy and momentum, we are probably already in possession of an important component of this continuity.) What I want to say is this: We must not believe that the 'puzzle of induction' will either disappear or be resolved in this vision. Into what corner it will retreat cannot be said with certainty. Armed with the law of large numbers, we will be able through our picture to pursue the justification of the expectation of the same results under the same circumstances right back into the sphere of atomic events; we are even able to do this today. As I see it, we will need laws which indicate, for *sharply* defined circumstances, a whole continuum of possible results – possibly, with certain restrictions regarding continuity, *all* possible results. The 'puzzle' will then have been reduced to the fact that frequent repetition of sharply defined initial conditions produces a quite definite distribution of results over this continuum, e.g., a uniform distribution. – We certainly cannot know whether this view, which is obviously modelled on games of chance, will prove to be useful. In any case, an axiom no less mysterious than causality is bound to slip in somewhere; problems do not solve themselves.

55. CAUSALITY AND PROBABILITY*

[1930g]

[The following is an excerpt from a three-part essay, originally published in *Erkenntnis*. The first part briefly outlines the historical development of the concept of probability beginning with its application to games of chance and concluding with its current use in quantum physics.

The second part discusses the problem of explicating the concept of probability. It contends that the subjective theory, according to which probability refers to the state of our knowledge, is inadequate and must be replaced by the objective theory which explicates probability as the limit of the relative frequency of physical events.

The third part, which follows in translation, is concerned with the justification of probability statements. – M.R.]

Let us now turn to the problem of the justification of probability statements. Why are we entitled to believe that probability statements are true? There are two aspects to this problem. The first concerns the ground for specific metrical assertions, for example, that the probability of the occurrence of a given face of a die is 1/6; the second concerns the justification of more general probability statements, in particular, the rule of induction.

Traditionally, the problem of the justification of probability statements was considered primarily in connection with an analysis of specific probability statements. This procedure has not always been conducive to a solution of the problem; in the case of games of chance, the origin of the probability metric is obscured by the symmetry of physical conditions. It is well known that a consideration of games of chance led to the suggestion that probability metrics always be constructed on the basis of *equiprobable* cases. The origin of this equiprobability was usually connected with a rather questionable principle, the so-called principle of insufficient reason. According to this principle, we must consider the six faces of the die equiprobable because there is no reason to prefer one of them. The untenability of this principle, which manifests itself in the occurrence of all kinds of unfortunate expressions such as 'equipossible cases', has been pointed out frequently in the literature.[1]

Nevertheless, the probability metric for games of chance is rather easily justified. It seems to us that the solution consists in reducing the mechanisms of games of chance to probability functions; this method was first suggested by von Kries and Poincaré, and has been elaborated by the present author for the purpose of finding a general solution for the problems of the probability

* From *Modern Philosophy of Science: Selected Essays*, ed. and tr. by Maria Reichenbach, Routledge & Kegan Paul, London, 1959, pp. 67–78.

of games of chance and the theory of errors [1915b, p. 22]. It can be shown that equiprobability in games of chance is reducible to the assumption of a continuous probability function.[2] Let us take the game of roulette as an example: if we assume that a continuous probability function determines, for repeated spinnings of the wheel, the frequency with which a given angle (measured in multiples of 2π) of rotation of the indicator occurs, then the division of the wheel into red and black sectors is equivalent to a division of the arguments of the probability function into small equal intervals. It can be demonstrated mathematically that the sum of the areas of the odd-numbered sectors is approximately equal to the sum of the areas of the even-numbered sectors, and that for a given probability function this approximation increases as the width of the sectors decreases. In this manner, the *metrical* assumption of the equiprobability of red and black in the game of roulette can be reduced to the *topological* assumption of the existence of a continuous probability function.[3] Corresponding assumptions hold for all other games of chance. The mysterious occurrence of equiprobability in games of chance is thus revealed as the consequence of an assumption which is accessible to epistemological analysis. The existence of continuous probability functions is assumed in other areas of physics as well, especially in the theory of errors. The equiprobability of discrete cases in games of chance is due to the paraphernalia of the games which are constructed in such a way that they lead to a division of the probability functions into alternating equal intervals. Thus equiprobability does not pose a special epistemological problem; rather, the problem is posed by the existence of continuous probability functions.

The analysis of the kinetic theory of gases can be carried through from the same point of view. In this instance, too, it is the assumption of the existence of continuous probability functions which plays the decisive role; the specific form of these functions is of secondary importance. With regard to the theory of gases, the specific form of the probability function is frequently explained by reference to certain physical conditions, such as those described by Liouville's theorem, etc. These conditions, like those underlying games of chance, lead from a continuous probability function to equiprobability. Yet neither in the theory of gases nor in games of chance can equiprobability be presupposed *a priori*; this becomes obvious in recent elaborations of the theory of gases. Boltzmann has derived Maxwell's distribution law for velocities from the equiprobability of all permutations of the molecules in the velocity space. Modern quantum theory has shown that this assumption is not correct, and that it must be replaced by another assumption which does not distinguish between the various permutations of individual particles. This new assumption was mistakenly interpreted as contradicting the *a priori* principles of the calculus of probability; yet the discovery of equiprobable cases has nothing to do with the calculus of probability. Only experience can determine whether any given cases are equiprobable. Equiprobability can either be obtained from a given statistic pertaining to the problem under consideration, or can be derived from certain physical facts in combination with a hypothesis about certain probability functions as in the theory of games of chance. The determination of the probability metric is ultimately not a problem of justification but of discovery. The actual probability metric is a fact which has to be discovered or which can, at best, be inferred from other facts.

The reduction of a probability metric to probability functions constitutes a great advance in the epistemological analysis of the problem of probability. Whereas probability laws used to be considered as representing a special kind of regularity distinct from the causal regularity of nature, it can be shown on the basis of the theory of probability functions that this distinction is only superficial, and that probability laws and causal laws are logical variations of one and the same type of regularity. We shall explain this result briefly. For a detailed presentation, we must refer to earlier writings of the present author [1915b, p. 48].

The characterization of the causal laws of nature as strict laws is justified only for certain schematizations. When all causal factors are known, then an effect can be predicted with certainty; such an idealization would be irrelevant for science without the addition of further assumptions. It is impossible to know all causal factors; we can only select a limited number of relevant factors and use them to predict future events, but must neglect factors of lesser influence. It is usually assumed that the influence of the less important factors is small, and that we can therefore predict the future within certain limits of exactness. This formulation is inadequate, however, and misses a fundamental point in the epistemological situation. Actually, we can only maintain that it is highly probable that future events will lie within certain limits of exactness. For instance, if we calculate the path of a projectile, we cannot be certain that the influence of simultaneous shocks in the interior of the earth upon the direction of the gun-barrel will be small; as a matter of fact, such shocks always occur. At the moment of firing, this influence may have the force of an earthquake, and thus produce a result outside the limits of exactness. Furthermore, no solution is provided even if we interpret the laws of nature as conditional statements because we can only mention a finite number of relevant factors in the antecedent, and the rest remains unspecified. On the other hand, if we were to stipulate that no disturbing influences above a certain strength occur, the resulting conditional would be certain but it would be deprived of its factual character. The statement would then assert the logical triviality that a future event will lie within certain limits of exactness if it does not lie outside them. There is no way out of this dilemma; either the statement is absolutely certain, in which case it is logically true (i.e. analytic), and does not state anything about reality; or it is a descriptive statement with factual content (i.e. synthetic), in which case it can be asserted only with probability.

The assertion of causal laws is warranted only if probability laws are admitted. The hypothesis that events are causally connected must be supplemented by an hypothesis about the probable effect of neglected factors; only the combination of these two hypotheses enables us to make statements about reality. I have called these two hypotheses the principle of connection (causality) and the principle of distribution (probability). Only the combination of these two principles completely accounts for the assumptions of science.

The application of the principle of distribution is most conspicuous in the theory of errors, where the application of probability functions is well known. The fundamental importance of this application failed of recognition because the theory of errors was thought to be a means of ensuring exactness. But the form of a law of errors such as Gauss' should make it obvious that the smallness of the error can be assumed only with high probability, and that arbitrarily large deviations can never be excluded with certainty. The theory of errors clearly demonstrates the supplementation of causal laws by probability statements, but such a supplementation is necessary even when we are not concerned with such narrow limits of exactness since even the highest probability is not identical with certainty. The statement that a street is between 1 and 100 yards long must therefore be regarded as a probability statement of the same sort as the statement by the geometer that the street is between 74.346 yards and 74.348 yards long; that the first statement has a much higher probability than the second constitutes only a difference in degree, not a difference in principle.

This last point again demonstrates the significance of reducing the equiprobability in games of chance to continuous probability functions. When we consider the red and black sectors, equiprobable in roulette, our assumption about the physical world does not differ in kind from that made by the geometers or physicists when they measure physical magnitudes. We assume that as a result of repeated spinnings, the frequency of a physical magnitude – the angle of rotation of the roulette wheel indicator – is determined by a probability function; the physicist makes similar assumptions when he predicts the path of a projectile, and compares his prediction with observational results. Thus, the enigma of equiprobable cases is solved. The assumption of equiprobability is replaced by a more general assumption which is the indispensible supplement to the principle of causality.

From such a point of view, we must revise our conception of the nature of physical laws, abandoning the old idea that while events are strictly determined, the finite mind of man has only an approximate knowledge of them. The old view claimed, in effect, that a limit exists even though we cannot find it since we are always part of the sequence of events which tends towards a limit. But in fact the only empirical phenomenon is the sequence itself, and statements about the limit are admissible only in so far as they can be transformed into statements about the convergence of the sequence toward a limit. This positivistic principle must be maintained in the face of all idealistic philosophies, for it is the basis of the precision and import which belong to modern science in contrast to metaphysical speculation. Therefore we must replace the traditional formulation of determinism by the following more modest formulation: there is a description of nature which enables us to predict the future with probability, and it is possible to bring this probability as close to 1 as we wish by a more precise consideration of the relevant parameters. This statement says considerably less than does the assertion of determinism. No matter how closely the probability approaches 1, we can never speak of strictly determined events, and it is therefore meaningless to use the language of determinism when speaking about the limit itself; such assertions necessarily remain empty.

At the same time, the revised formulation of determinism leads to a generalization of the concept of causality, a generalization which seems to coincide with the conception of causality prevalent in modern physics. It is not necessary to assume that the probability can be brought as close to 1 as we wish; it is possible that there may be a lower limit. Even this lower limit may be unattainable in practice; then it would be the case that for every attained level of exactness, there is a more exact level. The possibility of a limit less than 1 had been suggested by the present author in 1925 [1925d, pp. 133–75], and has in the meantime been substantiated in quantum mechanics. Heisenberg's Principle of Indeterminacy must be regarded as such a generalization of the concept of causality. We should like to elaborate.

Usually the significance of Heisenberg's considerations is said to consist in the fact that the disturbance of the object by the means of observation can no longer be neglected. I do not think that this formulation is very fortunate.

On the one hand, this disturbance is not unique to quantum physics; even in macroscopic physics one has long been accustomed to the fact that the means of observation cause certain changes in the object observed which cannot be neglected. On the other hand, macroscopic physics has found methods of circumventing this difficulty: disturbances by the means of observation are included in a given theory, and all inferences take into account the disturbance of the phenomena. The physicist is used to taking the influence of the thermometer into account when he measures the temperature of a solution in order to calculate its specific heat; in spite of this influence, or rather because he knows about it, he can determine the specific heat. It would be wrong to argue that in macroscopic physics observational disturbances can be eliminated in principle; in the first place, this is not true, and in the second place, it is not necessary. There are theoretical methods which enable us to draw inferences concerning the physical world from the disturbed phenomena by utilizing a theory about the instruments of observation. Applied to quantum mechanics, this means that if an electron is deflected by the means of observation, i.e. by a light ray, this fact does not make observation impossible in principle; the influence of the light ray must simply be incorporated into the theory, and must be taken into consideration in any inference based upon such observations. The sharp demarcation between the objects of observation and the means of observation is an idealization applicable to certain macroscopic phenomena, but it cannot be regarded as a necessary presupposition of the exact sciences.

The significance of Heisenberg's Principle of Indeterminacy seems to us to lie in a different direction. The crucial point is not that the observed event is compounded of object and means of observation, but that it is impossible to uniquely determine the objective state of affairs (which may include the so-called means of observation) on the basis of the observed phenomena. According to Heisenberg, we are able to determine either the position or the velocity of an electron as precisely as we wish, but cannot determine both these parameters simultaneously as precisely as we wish. An increase in the probability of one parameter is always accompanied by a decrease in the probability of the other in such a way that the product of the two probabilities does not increase. Here we have an exact analogue of our generalization of the principle of causality. The probability of a comprehensive description of an objective state of affairs, one which includes all relevant parameters, does not converge toward 1 as the exactness of the observation increases, but converges toward a lower limit.

These remarks should be taken as more or less provisional; so far we do not possess an exact epistemological analysis of quantum physics upon which we can base a final formulation. On the other hand, we can already conclude from what we said above that the present situation need not be regarded as an epistemological crisis. The replacement of causal laws by probability laws has been deemed a failure of scientific method. Indeed, the issue has acquired moral overtones. The physicist has been charged with the duty of searching for more precise methods even though none are at hand at the moment. This interpretation seems to stem from a neglect of the fundamentally probabilistic character of science. The concept of probability is so indispensable for the foundations of science, including classical physics, that it must be given the same epistemological status as the concept of causality. What we are experiencing today is merely the explicit recognition of this epistemological fact; unfortunately, misinterpretations of the concept of probability have hindered the acknowledgement of its fundamental importance. Even classical physics, if it is not to deteriorate into metaphysical speculation, can maintain strict causal determinism only as the assertion that the probability of prediction can be brought as close to 1 as we wish. If this view is accepted, a generalization asserting that the limit of the converging sequence is less than 1 appears feasible. If physics were compelled to accept such an assertion, this acceptance would not be due to any deficiency in its observational methods, but to an objective property of the physical world. Nothing would be a greater mistake than to reproach physics for such a result. On the contrary, the discovery of this fundamental feature of the microcosm and the development of conceptual methods that can do justice to this property of nature must be hailed as achievements of the highest order.

We turn now to the final epistemological question: How do we know that probability laws hold? Why are we justified in asserting probability laws? We should like to pose this question in its most general form. It goes beyond the problem of justifying the choice of a specific probability metric; as explained above, the latter problem raises no special epistemological difficulties since we are able to account for a probability metric in terms of probability functions whose specific form is simply an empirical fact. Our present investigation concerns the general question: how do we know that probability functions determine events; why are we justified in believing in the regularity of repeated events, a regularity postulated in the theory of errors, in games of chance, in the kinetic theory of gases, etc.? These questions touch upon the problem of induction which is at the core of the theory of probability. Why are we justified in inferring that the observed relative frequency in a sequence of events will be preserved in a future continuation of the sequence?

The historical discussion of the problem of induction has clearly revealed that the inductive inference is not logically necessary. It is Hume's merit to have recognized this fact and nothing essential has been added to his discovery. Hume has also shown that it is impossible to justify induction by experience because any such inference presupposes induction on a higher level. This epistemological fact cannot be denied, and philosophical theories which do not accept it cannot be taken seriously. For this reason, we shall discuss only two philosophical treatments of the problem of induction which have been developed in response to Hume's criticism.

The first of these theories is the conventionalist attempt to solve the problem of induction. According to this conception, the principle of induction is not a statement about the physical world, but merely constitutes an ordering principle of science. Therefore, scientific theories are to be constructed by interpreting observational data in accordance with the principle of induction; if exceptions to the principle of induction are observed, the theory under consideration is to be declared false and must be supplemented by additional assumptions which satisfy the principle. It is certainly correct to say that such a flexible procedure is used in constructing theories; yet we do not think that it is possible to base a justification of induction upon this practice. The following consideration will help to clarify this objection. A given observational state comprises a finite number of observational data, and it may be possible to interpret the data by means of theories in such a way that the relative frequency observed within a finite sequence satisfies the principle of induction. However, in this way it is impossible to justify the belief that additional observational data will preserve this frequency. A simple example will illustrate this consideration. When a physicist has plotted a number of observed points on a co-ordinate system, he can always draw a simplest curve which approximately connects these points and which shows the least possible oscillations. But why does the physicist draw the simplest curve through the points? He might just as well draw a different curve, one that oscillates several times between two neighbouring points; such a curve would correspond to a complicated physical hypothesis which is nevertheless compatible with the given observations. The physicist has a good reason for preferring the simplest curve; he believes that the simplest curve is that curve which approximates the results of future measurements. If this is correct, then the conventionalist solution is not tenable. From the conventionalist point of view, one might just as well choose a curve through the measured points that always oscillates twice between any two of them; if the resulting curve is not confirmed by subsequent measurements, it can be replaced for the increased number of observational data by another curve which connects the additional observed points in a similar manner. From the conventionalist point of view, the preference for the simplest curve is unreasonable. There is a definite reason that prompts the physicist to choose the simplest curve: he believes that only the simplest curve will enable him to make predictions. This belief cannot be justified on the basis of conventionalism.

The second theory we shall discuss attempts to justify induction as a practical rule. According to this theory, the justification of the principle of induction is not a problem of epistemology because the principle of induction has nothing to do with the content of science. Allegedly, it is not the task of science to make predictions; science is only supposed to order the given observational data, and it is of no concern to science whether predictions of future observations can be based upon this order. The problem of justifying predictions is a practical one. Only after we have ceased to theorize, and seek to influence the world actively, be it by making experiments, be it by means of technology, be it by performing the simple tasks of daily life, does the principle of induction come into its own. Therefore the belief in the principle of induction is not a problem of science, but of ethics.

It is our opinion that this attempt at solving the problem is also untenable. To regard the principle of induction as extra-scientific is to render unintelligible the preference of science for that order which it actually employs. We can illustrate this point by our previous example; it would be impossible to understand why science uses the simplest curve connecting the measured points instead of a more complicated one if induction were an extra-scientific principle. Such a view would deprive the scientific ordering procedure of its very aim. It is the aim of science to formulate statements which apply to the physical world, but this aim can be achieved only by means of the principle of induction, for it is this principle alone which enables us to distinguish between arbitrary conceptual constructions and empirical theories. Why has science abandoned the phlogiston theory of combustion in favor of the theory of oxidation? There is no single observational fact that is not equally compatible with either theory if the principle of induction is neglected. The theoretical assumptions of the phlogiston theory are more complicated, but the theory itself is not refuted. Although Lavoisier's observation of the increased weight of a burned body is regarded as the *experimentum crucis* in establishing the oxidation theory since a burned body should decrease in weight if a hypothetical heat substance were escaping from it, the increase in weight can still be reconciled with the phlogiston theory if certain assumptions are added; for instance, one might assume that there are certain hypothetical force fields which attract phlogiston, and that it is the upward pull of these forces upon a body containing phlogiston which accounts for the difference in weight between a burned body and an unburned body. The physicist may call such assumptions highly improbable, yet he cannot exclude them without the use of induction.

Now we can recognize the central position which the principle of induction occupies in science: the princple enables us to judge the truth of scientific theories. To eliminate it from science would be to dispense with any decision about the truth or falsity of theories in science; in this case, there would no longer be any basis for distinguishing scientific theories from the fanciful creations of the poet.

We said that the principle of induction provides us with the means of judging the truth of scientific theories. To be more precise, we should say that it serves to decide their degree of probability. The alternatives in science are not truth and falsehood; instead, there is a continuous scale of probability values whose unattainable limits are truth and falsehood. We shall employ this idea in presenting our own conception of the problem of induction.

We admit that probability inferences cannot be justified logically; indeed, we have previously admitted much more: probability statements are not even meaningful unless the principle of induction is presupposed. We have seen that the principle of induction plays a decisive role in the interpretation of probability statements since predictions that observed relative frequencies will be preserved in the future presuppose the principle of induction. The explication of the meaning of probability statements throws new light upon the problem of the justification of induction. Probability statements are not meaningful within a two-valued logic that requires every statement to be either true or false. There is a correspondence between Hume's exposition of the problem of justification and our analysis of the problem of interpretation; it turns out that one cannot justify the assertion of probability laws if two-valued logic is regarded as the only criterion for testing our knowledge of reality. Both the problem of interpretation and the problem of justification remain insoluble if only two-valued logic is presupposed in science. Yet we do not infer from this fact that the justification of probability statements is impossible. We merely infer that the assumption of two-valued logic alone will not help. It is not possible to justify the system of scientific statements simply on the basis of deductive logic together with observational reports; this is our epistemological result.

When we are asked why, under these circumstances, we continue to believe in probability laws, we have but one answer: we cannot help believing in them. The requirement that the ultimate foundations of our knowledge of the physical world must be justified has turned out to be untenable; rather, the task of epistemology is to *discover* the ultimate foundations by means of analysis. If one is not content with this discovery, if one requires that the theory of probability be reduced to logic, one makes an unreasonable demand that seems to stem from a misconception of the epistemological status of logic. The foundations of logic cannot be justified either. The contention that the laws of logic need not be proved because they are empty overlooks the fact that 'empty' means nothing but 'corresponding to the laws of logic alone', and that such a justification would be circular. For our belief in logic there is no justification but the fact that we simply cannot think differently. Analogous considerations hold with respect to probability: we cannot help but believe in probability laws.

These remarks are closely connected with our previous discussion of the interpretation of probability. We found that a probability statement is meaningful only if the principle of induction holds; therefore the statement that probability laws do not hold is itself meaningless unless the principle of induction holds. This important fact results from our analysis of the problem of interpretation. To say that probability laws do not hold is equivalent to predicting that the observed relative frequency of sequences of events will not be preserved in the future, that the regularity implied by the principle of induction does not hold – and this statement is empirically meaningful only if it can be decided inductively, i.e. if the principle of induction holds. The statement that probability laws do not hold is selfcontradictory and makes no sense. We do not maintain that we have justified probability laws by this argument; this is no more a justification of probability than showing that every violation of the laws of logic leads to contradiction is a justification of logic. It is no justification because the occurrence of contradiction can be regarded as proof of inconsistency only if we have already assumed that the laws of logic hold. There is no justification of logic. We can only say that to challenge its foundations is not even possible. Similar considerations apply to probability laws; we cannot justify them, but we cannot imagine them not to hold.

Our answer to the problem of justification is therefore not an answer to Hume's question. Rather, the attempt to give a logical proof of probability statements is an impossible one like squaring the circle. Just as the failure to square the circle did not undermine mathematics, so the failure to justify induction does not impair probability; the problem of squaring the circle was eliminated on the grounds that its formulation was inadmissible, and Hume's problem can be resolved on the grounds that the demand for a justification of probability statements in terms of deductive logic is unreasonable. Instead of concerning ourselves with the pseudo-problem of justifying induction, we would rather analyze the methods of science; such an analysis shows that the concept of probability is an indispensable component of all empirical statements. We are able to construct a probability logic which provides the conceptual framework for all empirical knowledge, but though we can analyze this framework, we cannot justify it. As surely as we believe that statements about the physical world are meaningful, so much are we entitled to trust in the significance of the concept of probability.

NOTES

[1] H. Reichenbach, [1915b] § 31; R. v. Mises, *Wahrscheinlichkeit, Statistik und Wahrheit*, Vienna, 1928, p. 63f.

[2] More precisely: the function is integrable according to Riemann.

[3] More precisely: if one knows that the probability function can be integrated according to Riemann, one can always choose a sufficiently narrow division of sectors so that the deviation from equiprobability will be smaller than a given ϵ. In order to assert that a given division satisfies the conditions of equiprobability within the limits of exactness $\pm \epsilon$, one must know that the probability function does not oscillate too much. It is an empirical fact that the probability function for actual roulette games does not oscillate too much, and that this property disappears when the indicator is spun very lightly.

56. THE PRINCIPLE OF CAUSALITY AND THE POSSIBILITY OF ITS EMPIRICAL CONFIRMATION*

[1932d]

Preliminary remarks by the author:

The present essay, written in 1923 in Stuttgart, could not be published at the time due to adverse circumstances. The manuscript circulated among a small group of philosophers who were interested in precise conceptual analysis of epistemological problems. In the meantime, a development has taken place in quantum mechanics in the course of which the conception of the principle of causality presented has been accepted. In addition, the epistemological positions of apriorism, conventionalism, and probabilism analyzed in this essay concern problems that play an important role in the present epistemological discussion of quantum mechanics. These facts induce me to publish the manuscript in unchanged form, even though some of the suggested ideas have been elaborated in the meantime in other works of mine. Concerning these later publications we refer to the bibliography in *Erkenntnis* **2**, 190, (1931). [See items 1916a, 1920c, 1920e, 1925d, 1929g, 1929l, 1929m, 1930g, 1931i, and 1932f of the Bibliography]

1. CAUSALITY AS A COMPLEX OF PRINCIPLES

The problematic nature of the discussion of causality is due to the fact that the term 'causality' covers a number of principles that are generally not sufficiently distinguished, even though they constitute completely different assertions. One might almost say that, in its most general form, the principle of causality merely designates the scientific method of ordering given observational data in a certain way, and thus leads to the assertion that there is a certain regularity in nature. In this form, which neglects the various ways in which scientific data are ordered, the assertion is not accessible to precise analysis. Therefore, all critics of the principles of causality have tried to formulate it in a more exact manner, with the result that they have always dealt only with *parts* of the complex, and have neglected the fact that their criticism could at most do justice to these *parts*. The present paper likewise intends to analyze only a *part of the complex*, although perhaps the most important one. In order to avoid the mistaken view that these investigations exhaust the total content of the complex, we shall first give a brief survey of the various principles which are subsumed under the concept of causality. This survey will indicate specific differences among these principles, and

* From *Modern Philosophy of Science: Selected Essays*, ed. and tr. by Maria Reichenbach, Routledge & Kegan Paul, London, 1959, pp. 109–134.

the need for treating them separately. We do not claim that the enumeration is exhaustive.

(1) In the first place, the principle of causality asserts that natural events are connected, or at least can be connected, in such a way that from *known* events *unknown* ones can be predicted. The practical significance as well as the unique epistemological position of the principle of causality rests upon this assertion which we shall call the *inductive* principle of causality. The predictability of *unobserved* events by means of *observed* ones is the central problem of causality.

The procedure that is used in this context depends on the existence of *functional relationships.* The core of the principle of causality has therefore been seen in the assertability of functional relationships. This view is correct, but only if one adds that such functional relationships link the *unobserved* state with the observed one. It is obvious that observed events can be brought into a functional relationship since the problem is merely that of connecting a finite number of data by means of an approximate function.[1] Yet the peculiar result is that the *same* function always occurs for the observed cases when the phenomena show certain properties. This re-occurrence would be irrelevant, and would signify nothing but a registration of subjective experiences, were it not for the inference that the same function applies to *all* such cases. What this inference means becomes clearer if we say that the same function applies to all those cases *which we shall experience in the future.* But we need not touch upon the subjective aspect of epistemology; we can interpret the 'all' as referring to objectively existing events never observed by us. This extrapolation transforms the registration of subjective experiences into an objective law of events.

On this basis, the assertion that causality has always held in the past has sometimes been distinguished from the assertion that it will hold in the future. Only the second assertion has been called *inductive.* This distinction is not correct. It can likewise only be inferred inductively that causality has existed in the past. We have merely observed that the same function has been satisfied for the *experienced* cases; but this is not an assertion about *all* past events. The assertion that causality has held in the past means that the *unobserved* cases in the past have satisfied the function. The line of separation does not lie between past and future, but between observed and unobserved events. The statement about observed instances – we shall call it a *reproductive description* in contrast to an *inductive description* (Section 4) – cannot therefore be considered part of the principle of causality since it does not state anything about objective events, but only something about the structure of the data upon which we base the assertion about the events. This is the reason that we do not construe a statement about observed data as logically parallel to the *inductive* principle of causality.

The functional relationship arises from inductive considerations alone. Therefore, we speak of an *inductive principle of causality;* the term 'principle of causality' may express the fact that the induction is performed by means of a *functional relationship* which determines the unobserved events. The inductive principle of causality says that by means of a functional relationship unobserved events can be predicted from observed ones, no matter whether the unobserved events lie in the future, or in the past, or happen at different space points simultaneously with the act of observation. The specific problem of causality originates in this assertion about *unobserved* events. Can the principle of causality be tested empirically, or does the assertion have a different epistemological significance? It is the purpose of the present investigation to answer this question; in order to do so, we must first develop a rigorous formulation of the inductive principle of causality.

(2) In addition, the principle of causality asserts that for the course of events only the existing 'things' and 'forces' are important, not the position of the events in *time* and *space.* This assertion is equivalent to denying the causal relevance of space and time. In other words, the space-time co-ordinates do not occur *explicitly* in the laws of nature. They often occur implicitly, for instance, when the motion of falling bodies is represented as a function of time, but the space-time co-ordinates can always be eliminated by a transformation of the statements into differential laws. This assertion pertains more to the properties of space and time than to those of causality, yet it is regarded by many as inseparable from the *inductive* principle of causality. (Cf. Section 4.)

(3) Frequently, the principle of causality asserts that effects can spread only through continuous propagation in space. An effect which starts at a space point A at the time t will not be observable after a time Δt outside a sphere with the radius r and A as its center; r increases continuously with Δt, and $r = 0$ for $\Delta t = 0$. This law is called the *principle of action by contact.* Occasionally, it is formulated in connection with the inductive principle of causality in the following way: events in a finite volume V can be calculated for all times if the state in the interior of V is given for *one* cross-section, and the state at the surface of V for *all* times.[2] The principle of action by contact is expressed in the assumption that effects from outside can reach V only by way of its surface. This principle is accepted by modern physics; it is closely

connected with the transition to differential laws and the field concept. Yet it would be a mistake to believe that there is a *necessary connection* between this principle and the inductive principle of causality, even though it has been contended that events inside V would be chaotic if all external processes had an immediate effect upon the interior of V. In order to overcome this difficulty, it would be sufficient to assume that every effect decreases greatly with the distance from its origin, and that no finite time interval would be needed for its propagation. In this case, external influences in V would be small compared to internal ones, and would constitute only minor disturbances. Under these circumstances, a knowledge of the surface conditions would no longer suffice for a calculation – this condition expresses the violation of the principle of action by contact – but for any desired exactness of the calculation within the frame of probability statements, the knowledge of a certain volume surrounding V would be sufficient. Newton's theory of gravitation corresponds to such an assumption. Even quite different assumptions could save the inductive character of a determination of the events within V because these events can be calculated, in any case, only with probability.

Since the principle of action by contact can be divorced from the inductive principle of causality, it can be tested empirically if the inductive principle is used for its confirmation. It is obvious, therefore, that the principle of action by contact is not a necessary presupposition of knowledge.

It is erroneous to infer from the principle of action by contact that there must exist a finite upper limit for all velocities of propagation. Although this assertion is made in the theory of relativity, it is not derived from the principle of action by contact, but is based on specific empirical grounds. From the principle of action by contact, one can only infer that every effect spreads with a finite velocity, possibly without a finite *limit*.

(4) Finally, causality asserts that there is a *temporal order* of events: the cause precedes the effect in time. This assertion has caused much confusion due to its imprecise formulation. First, it became associated with Zeno's familiar paradoxes of the continuum, and led to the conclusion that temporal sequence means simultaneity in the infinitesimal; cause and effect were said to happen simultaneously. We refer to Russell's works[3] for an explanation of these paradoxes; the paradoxes have no bearing upon the problem of causality. Secondly, the temporal order of the relation of causality has become the source of metaphysical and anthropomorphic interpretations. Causality has been seen as a mysterious relation of 'being contained in each other' projected into time; the effect is said to be potentially contained in the cause, to emanate from, and to be actually identical with, the cause. In short,

some philosophers became enmeshed in obscure conceptual constructions instead of formulating structures of order. Other philosophers, who rejected these metaphysical trappings, wanted to deny the temporal order of causality altogether, and tried to reduce causality to the assertion of a functional relationship in the sense of the inductive principle of causality. They did not notice that they thereby neglected a part of the principle of causality.

We can formulate the assertion concerning *temporal order* contained in the principle of causality as follows: it is possible to assign *uniform* time indices to *all* natural events in such a way that the cause always precedes the effect in time. *Temporal sequence* is *defined* by the relation of cause and effect; this relation can be discovered on the basis of certain criteria. We arrive at an objective statement by comparing all of the particular time relations obtained in this way, and the significance and the practical value of this temporal order consists in its overall consistency. This is the meaning of the given formulation. It is not a necessary condition, but depends on certain assumptions concerning physical events that can be formulated as axioms. Whether these axioms are actually satisfied can be tested empirically. For a justification of these remarks, I refer to my *Axiomatik der relativistischen Raum-Zeit-Lehre* [1924h]. The *assertion of the temporal order* of causality can thus be distinguished from the *inductive principle of causality*. Whether it holds in reality is a matter of experience.

Now we recognize that assertions 2–4 supplement the inductive principle of causality, but have no independent status. They always presuppose the inductive principle of causality. With respect to the possibility of empirically confirming the principle of causality, which we shall discuss in the following sections, only the inductive principle is important. We shall therefore often speak in the subsequent discussion simply of the principle of causality when we mean its inductive core.

2. THE DISCOVERY OF GOVERNING FUNCTIONS

We shall now proceed to develop a rigorous formulation of the inductive principle of causality. We shall be guided by the procedure that the physicist uses in order to establish an individual law. He begins by formulating a function, which he first maintains as an hypothesis, and then tests by experiments. The hypothesis will be regarded as true or false depending on whether or not the test supplies quantitative confirmation.

We write the function to be tested in the form

$$F_r(p_1 \ldots p_r) \tag{1}$$

where $p_1 \ldots p_r$ are parameters that characterize the process. We can conceive this function as describing a surface in the r-dimensional space of the parameters $p_1 \ldots p_r$; a set of values $p_1 \ldots p_r$ we call a point p of this space. Now we wish to compare the function with the observational data. For this purpose, we take observed sets of correlated values $p^*_1 \ldots p^*_r$, which we shall call *observed points* p^*, and see whether these points p^* lie on the surface F_r, i.e. whether the sets of values $p^*_1 \ldots p^*_r$ satisfy condition (1). The class of the observed points p^* is called the class M.

The gas law $p.v = \frac{R}{m} \cdot T$ which can be written in the form

$$F_r = \frac{p \cdot v \cdot m}{R \cdot T} - 1 = 0$$

supplies an example of a function to be tested. The parameters $p_1 \ldots p_r$ are the five magnitudes p, v, T, m, R. If m and R are regarded as constants, then F_r can be visualized as a surface in the three-dimensional space of p, v, T. Then p, v, and T can be varied experimentally and measured individually. Sets of correlated values of these measurements furnish an *observed point* p^* in this three-dimensional space. The question arises whether the points p^* lie on the surface (1). The fact that certain parameters are constant and cannot be varied creates no difficulty. Supposedly, this fact is only due to a lack of technical means and of scientific knowledge; we may assume that all physical constants can be represented as functions of other constants. The problem is irrelevant in this context, and our following considerations will not be affected if constants occur among the parameters $p_1 \ldots p_r$. Incidentally, in our example, the molecular weight m can be varied if other gases are used.

Another example is furnished by Newton's equation describing the orbit of a planet. Here the space-time co-ordinates occur among the parameters, but this fact does not contradict what was stated above as the second assertion contained in the principle of causality because these co-ordinates occur only implicitly; if the solar system is at a different space point at a different time, the relative co-ordinates have the same numerical values. Therefore, we can treat the relative co-ordinates in the same way as other parameters. Strictly speaking, they are not co-ordinates but space-time *distances*. Their role is the same as that of the parameter *volume* in the gas law; this parameter is likewise a result of measurements of spatial distances. Among the parameters are furthermore: the mass of the sun, initial position and initial velocity of the planet, and the gravitational constant. Usually these magnitudes are called orbital parameters. In addition to the co-ordinates, we can also vary the orbital parameters by taking other planets into consideration. In this example, in contrast to the previous one, there are no *experiments,* there are only *observations*; this means that we cannot establish arbitrary sets of values $p_1 \ldots p_r$. But this fact is irrelevant for the problem of testing the principle of causality.

Let us now imagine that the test has been performed. We cannot expect that all points p^* will lie exactly on the surface F_r; but if the deviation is too great, we shall modify our theory. Let us assume that the deviations are considerable. Then we can avail ourselves of two methods of changing our hypothesis:

(1) We assume that our data are false. The magnitudes $p_1 \ldots p_r$, however, are not directly observed, but are usually inferred by means of theoretical considerations from the indications of the measuring instruments, i.e. they are derived with the help of functions

$$p_i = P_i(q_1 \ldots q_m)$$

from other data $q_1 \ldots q_m$. The mistake therefore could lie in the functions P_i. Yet, if one has made use of functions that are highly confirmed through frequent previous observations, and the test has been performed according to the schema we described, we shall regard the values $p^*_1 \ldots p^*_r$ as correct.

(2) We assume that the observed phenomenon is not adequately described by expression (1) and replace it by

$$F'_{r+s}(p_1 \ldots p_r, p_{r+1} \ldots p_{r+s}) = 0 \qquad (2)$$

We call the magnitudes $p_{r+1} \ldots p_{r+s}$ *additional parameters;* they correspond to *other relevant causes.* They are variables, and are determined for every set of values $p^*_1 \ldots p^*_r$ in such a way that (2) is satisfied with the desired exactness. This result can also be reached without the introduction of additional parameters by a change of the function F_r alone; but it can be *better* accomplished by the use of additional parameters, and later we shall make use of the greater adaptability gained in this fashion. The points p^* of the r-dimensional space are now spread over the $r+s$-dimensional space of the parameters $p_1 \ldots p_{r+s}$; two points p^* that were previously close together may now lie far apart because of the difference in their co-ordinates $p_{r+1} \ldots p_{r+s}$. This is the reason that the choice of a simple function F'_{r+s} is more easily attainable.

For F'_{r+s} we lay down the following conditions:

(*a*) The surface F'_{r+s} should lie in the space of the parameters $p_1 \ldots p_{r+s}$ in such a way that *the observed points* p^* *lie sufficiently close to the surface.* This approximation is required 'on the whole', i.e. greater deviations of individual points are permitted. We may require, for instance, that the sum of the squares of the perpendicular distances of the p^* from the surface divided by the number of the p^* be within a given interval of exactness. In this connection, the additional parameters $p_{r+1} \ldots p_{r+s}$ may be regarded as variables, i.e. they may assume specific suitable values for every set of values $p_1 \ldots p_r$.

(*b*) The function F'_{r+s} should be the *simplest* among all those functions satisfying condition (*a*). It is not possible to define the concept of simplicity precisely, but we shall nevertheless be able to apply it with sufficient assurance. Simplicity refers to mathematical form; for example, a linear function is simpler than a quadratic one, a function with few oscillations is simpler than one with many oscillations. One often draws a simplest curve through an observed sequence of points; this is the concept of simplicity that we mean. Furthermore, a function F'_{r+s} needing a smaller number of additional parameters is simpler than another one.

(*c*) The function F'_{r+s} should be *justified,* i.e. it should be possible to incorporate the additional parameters into other relationships. There should exist functions

$$P_{r+i}(q_1 \ldots q_k) = p_{r+i} \qquad i = 1 \ldots s$$

and the special values p^*_{r+i} which the parameters p_{r+i} assume relative to the $p^*_1 \ldots p^*_r$ should be related by means of the functions P_{r+i} to observational data $q^*_1 \ldots q^*_k$ which have been found experimentally to belong to the respective sets of values $p^*_1 \ldots p^*_r$. The functions P_{r+i} should be confirmed empirically in the manner described here for the function F'_{r+s}.

If the *justification* of the function F'_{r+s} is not completely successful, the function should at least be *plausible,* i.e. some of the requirements (*c*) may be replaced by weaker ones. If it is not possible to determine the p^*_{r+i} quantitatively, they should at least be justified qualitatively.

These three requirements are rather inexactly formulated, but they determine with sufficient precision a function F'_{r+s}. We shall call it the *function governing the class M.*

We can trace the introduction of F'_{r+s} in our examples. For the first example, F'_{r+s} is van der Waals' equation

$$p + \frac{a}{v^2} \ (v - b) = \frac{R}{m} \cdot T.$$

It satisfies conditions (*a*) and (*b*). The additional parameters $p_{r+1} \ldots p_{r+s}$ are van der Waals' constants a and b; the relations P_{r+i} are the equations that connect these constants with the force of attraction and the effective volume of the gas molecules. Moreover, these relations explain the significance of the additional parameters. Thus condition (*c*) is satisfied.

In the second example, F'_{r+s} is Einstein's orbital equation which yields a precession of the perihelion. The additional parameters are the magnitudes $g_{\mu\nu}$, the P_{r+i} are given through the connection of the $g_{\mu\nu}$ with light rays, clocks, and measuring rods. Condition (*c*) is also satisfied by Einstein's principle of equivalence. With respect to condition (*a*) we must remember that F'_{r+s} is to supply the observed precession of the perihelion not only for one planet, but for all of them; it should be applicable to variable distances r from the sun. This condition is not satisfied if Newton's law is corrected only in the exponent, whereas Einstein's law is adequate.

We have now found an improved law F'_{r+s}, but we may ask whether this law is correct. In order to answer this question, we proceed with the experimental test. By observation, we determine further values p^*, i.e. sets of values $p^*_1 \ldots p^*_r$, and compare them with the calculated values. These values $p^*_1 \ldots p^*_r$ should lie within the same value domain as before. In this way, the class M of the observed values p^* is extended to the class M' which comprises the old and the new values. Now the decisive question arises: *is the function* F_{r+s} *also the function governing the new class* M'*, or does* M' *lead to a new governing function* F''_{r+s+t}? No increase in precision is required for condition (*a*); the limits of exactness remain the same. Nevertheless, it is possible that the new points p^* may lie too far away from the surface F'_{r+s} in the $r+s$-dimensional space, and thus may be scattered without any relation to the surface if parameters $p_{r+1} \ldots p_{r+s}$ are co-ordinated to them according to the previous rules. Under these circumstances, the introduction of new additional parameters $p_{r+s+1} \ldots p_{r+s+t}$ and a new function F''_{r+s+t} will be needed.

It is always possible to establish a function according to conditions (*a*)–(*c*) because the class of the measured values p^* is *finite* and the new parameters can be used to fulfil the simplicity condition. The establishment of a function F'_{r+s} is therefore compatible with *every* result of the measurements; experience cannot disconfirm it. However, the answer to the above question depends on experience in so far as the new points in M' may be situated in such a manner that a completely different function is indicated.

With respect to our examples, this question may be formulated as follows: If I make further measurements of a gas, will the newly determined values coincide, within the required exactness, with the constructed *p-v-T*-surface of van der Waals' equation? Or will they supply a completely different surface, for instance, with more points of inflection? Will they satisfy Einstein's formula with the previous exactness if I make further measurements of the perihelia of the planets? Or will they require a new formula?

We can imagine the procedure continued; we extend the class M' of the observed points p^* to a class M'' which comprises new observations in addition to the previous ones. We can then establish the function governing M''. By continuing in this way, we arrive at a *sequence* of governing functions; we shall call it the *governing sequence* of the classes $M, M', M''. \ldots$ Depending on the answer to our question, two types of governing sequences result:

$$\text{I.} \qquad F_r \, F'_{r+s} \, F'_{r+s} \ldots F'_{r+s} \ldots$$

$$\text{II.} \qquad F_r \, F'_{r+s} \, F''_{r+s+t} \ldots F^{(i)}_{r+s+t+\cdots+w} \ldots$$

The core of the inductive principle of causality is obviously represented by the assertion that type I and not type II results. Before we enter into a more detailed discussion of this assertion, we want to correct our considerations in one respect. It may be objected that the form of sequences I and II depends on the accidental separation of past from future observations. Yet this circumstance, which is connected with the person of the experimenter, should not have any influence since it is our purpose to characterize an objective state of affairs by the types of the sequences. We can exclude this influence by restricting our assertion to one observed measured sequence.

We imagine that all the measurements p^* forming the class $M^{(i)}$ have been executed. Only now do we start with the test. If we were to treat the total class $M^{(i)}$ as we treated the first class M, we would pass directly from F_r to $F^{(i)}_{r+s+t+\cdots+w}$ (or to the corresponding function of type I), i.e. we would obtain only two elements of the sequence. Therefore we use another method. We form all subclasses of $M^{(i)}$ – it is a finite number since $M^{(i)}$ is finite – and arrange them in a sequence such that an earlier term of the sequence never contains more elements than a later term; subclasses with an equal number of elements are arranged in arbitrary order. We form the governing function for each subclass, and arrange these functions in the order of the corresponding classes. Thus the governing sequence is defined, independently of the observer, as a function of the number of elements in the class. The resulting sequence may again be of type I or type II. Evidently, these two possible forms constitute an objective characteristic of the numerical data.

This procedure will never supply type I in pure form. The subclasses showing 'singularities' furnish different governing functions that occasionally deviate from the sequence, but the regular functions will be more frequent, so that the type can be clearly recognized. We shall call such a sequence a *complete governing sequence,* in contrast to the *incomplete governing sequence* which results if the choice of the subclasses is left to chance. Even this chance selection need not come about in the manner described above, by means of alternate calculation and observation; it is possible to first obtain all the numerical data, and then to select subclasses $M, M', M''\ldots$ according to laws of chance, for instance, by throwing dice.

Sequences I and II characterize an objective state of affairs. One notices that the requirement of simplicity has a decisive influence. It is due solely to the requirement of simplicity that the subclasses $M, M', M''\ldots$ furnish *different* governing functions from that furnished by the total class $M^{(i)}$. The governing function $F^{(i)}{}_{r+s+t+\ldots+w}$ belonging to $M^{(i)}$ satisfies conditions (*a*) and (*c*) for the subclasses, but occasionally does not satisfy condition (*b*). It is due to this fact that the type of the governing sequence represents an objective characteristic of the physical events under consideration.

3. THE PROBABILITY INFERENCE

Let us now investigate which of the two types of sequences, I or II, is realized in the physical world, or rather what we are able to say about this problem.

Here we encounter a difficulty which we have not mentioned before. The main feature of the two sequences I and II is that they are forms of an *infinite* sequence. It is possible that an initial *finite* section might be the *same* for both types, and that the difference might manifest itself only in the sequence of the infinitely many subsequent elements. The construction of a sequence on the basis of experience, however, will always furnish only a finite number of elements, and therefore we cannot know with certainty which type is represented. If we have ten members of form II, we cannot know whether the following elements will represent form I, and vice versa. Here we encounter a fundamental difficulty in physics which does not exist in mathematics. In mathematics, we can make absolutely certain statements about the character of an infinite sequence because we need not know all the elements in order to do so. Thus, we know that the division of 367,895 by 1233 yields an infinite sequence of decimals which becomes periodic after a certain element, whereas, when we calculate $\sqrt{2}$, we obtain an infinite sequence in which continuously new groupings of integers occur instead of periods. We know this because these sequences are given *intensionally,* i.e., all their elements are determined by the given designation, and the properties of the sequence can therefore be inferred from the nature of the defining operations. For an interpretation of the sequences I and II, we do not have such criteria since they can be given only by an enumeration of their elements, i.e. *extensionally.* Although types I and II of the sequences are well defined *conceptually,* and can be precisely distinguished, we cannot know in an actual case which type is represented. It is, however, possible to come to a decision if probability statements are admitted as meaningful.

Let us imagine that we have established the governing sequence for experiments concerning the gas law. As we know – we want to adopt the naive point of view for the moment – van der Waals' equation will satisfy the experiments within a certain range of the variable with a certain exactness. We shall obtain a relatively large number of elements according to type I. Although we must break off the sequence after a finite number of elements, we shall say, nevertheless, that there exists a high *probability* that the following elements will be similar, and that type I will be realized. In any case, this is the kind of inference that is used in practice.

It is at the same time the kind of inference that is used to establish a *particular* law of nature. If we want to know *which* law describes the experiments performed with the gas, we infer from a finite number of the elements of the sequence that the specific function

$$\left(p + \frac{a}{v^2}\right) \cdot (v - b) = \frac{R}{m} \cdot T$$

is the desired law. We make this inference on the assumption that the *infinite* sequence would furnish this function if it could be constructed. We are again left with a probability inference.

We shall later have to investigate whether the probability inference can be avoided if *a priori* principles are admitted. For the time being, we shall base our considerations on the assumption that this inference is meaningful and admissible, and with the help of this assumption try to analyze the meaning of the principle of causality.

4. FORMULATION OF THE PRINCIPLE OF CAUSALITY

When we connect a set of measured values by a function, we can interpret the result in two ways. First, the function may simply describe the numerical data. If $p^*_1 \ldots p^*_{r-1}$ are observed values of parameters, the function F_r can serve the purpose of calculating the value p^*_r of the last observed parameter belonging to this system. The function must furnish the correct value p^*_r within the desired interval of exactness because it was fitted to all observed points p^*. In this case, the function is nothing but an alternative indication of the last parameter. But since this parameter must first be *observed* in order for us to find the function, nothing is gained by the construction of the function; we would not call it a *law.* If the function connecting the set of observed points is to have this extended significance, something more is required from it. For any *unobserved* value $p_1 \ldots p_{r-1}$ of the parameter, it should furnish an *unobserved* value p_r that one *would* have observed as belonging to it if the parameters $p_1 \ldots p_{r-1}$ *had been* observed. Only such a function expresses a *law;* only such a function describes an *objective state* and is independent of the accidental observational data. This interpretation of *law* is always presupposed when we speak of causality. This interpretation raises the fundamental problem of induction: we cannot know what we *would* have observed, only what we *have* observed. Yet we believe that the above interpretation is meaningful. We can make random tests by first calculating the values of the function, and then making observations and checking whether the observed values correspond to the calculated ones. This procedure is equivalent to constructing a governing sequence. But on the basis of such random tests we can only say whether the *performed* random tests agree, not whether *all possible* random tests will do so. If we believe, nevertheless, that frequent confirmation is sufficient to justify the probability statement that the function will *always* be correct, we are applying the inductive inference which we have characterized above as leading to an assertion concerning the type of the sequence. The assertion that our function is more than a mere *description* of observational data, that it represents an *objective law,* a *description of all possible observations,* depends on the application of a probability inference.

Accordingly, a law is not a *description of what is observed,* but *of what is observable.* We distinguish between the *reproductive* and the *inductive description;* the first is merely a reproduction of what is known through observation, whereas the latter describes something that to a large extent is known only by virtue of the description. The inductive description is also called a *law.*

This extended significance of *law* is the reason that the governing sequence cannot have only a finite number of elements. We regard the governing function as a law of nature only because we assume that it will remain the same when the sequence is extended to any number of further measurements; the proper meaning of the principle of causality can therefore be formulated in the following way: *causality exists when the governing sequence exhibits type I.*

So far we have construed physical laws as functions of physical parameters $p_1 \ldots p_r$. Now we have to investigate whether space-time co-ordinates can also occur among the arguments. Obviously, one cannot know beforehand whether measurements in the same system at different times and at different places will furnish the same results; in general, one will have to expect a function $F(p_1 \ldots p_r, x_1 \ldots x_4)$. It is possible, however to eliminate the space-time co-ordinates by introducing functions

$$\begin{aligned} p_{r+1} &= h_1(x_1 \ldots x_4) \\ &\;\;\vdots \\ p_{r+4} &= h_4(x_1 \ldots x_4) \end{aligned}$$

that characterize, instead of the space-time co-ordinates, certain magnitudes $p_{r+1} \ldots p_{r+4}$ determining physical states. The $p_{r+1} \ldots p_{r+4}$ can then be regarded as a space-time field; any influence upon physical events is ascribed to this field, while the space-time co-ordinates are eliminated from the laws of nature. It is a matter of taste whether one prefers this procedure or whether one wishes to retain the space-time co-ordinates explicitly in the laws. Yet the second property of causality mentioned in Section I can always be obtained if causality exists at all, i.e. if the governing sequence assumes type I. The independence of the laws from space-time co-ordinates does not constitute a separate empirical principle. The procedure described above, by which the number of the parameters p_i is increased until a governing sequence of type I is reached, will at any rate result in the elimination of the co-ordinates; one will automatically proceed in this way until the parameters $p_{r+1} \ldots p_{r+4}$ are also introduced.[4]

5. THE SIGNIFICANCE OF THE REQUIREMENT OF SIMPLICITY

We suggested at the end of Section 2 that the requirement of simplicity is of great importance for the establishment of the governing function; without it the function is not uniquely determined. It is well known that simplicity plays a decisive role in the theory of the laws of nature, and we are now in a position to explicate the requirement of simplicity on the basis of our formulations.

In Section 4 we distinguished between *reproductive* and *inductive descriptions.* The requirement of simplicity can be made even for reproductive descriptions. Among the many possible descriptions, it would single out one and in this way determine the choice. The function obtained in this manner would be the simplest description. We call this simplicity *descriptive simplicity;* it is a characteristic of the *description* alone, not of the *objects which are described.* Any other more complicated function would represent the observed sets of values p^*, i.e. the objects of the description, just as well. We can characterize this simplicity also as *logical economy.* There are cases where physical theories make use of descriptive simplicity; in the theory of space and time, for instance, it determines the choice of the simplest geometry.[5] For our present problem, simplicity has a different significance. We saw that the governing function is supposed to be a law, i.e. an *inductive,* not merely a *reproductive description.* This claim is based on the property of simplicity. When we make the requirement of simplicity for the governing function, we assume that it is the simplest function which will furnish a sequence of type I. This simplicity is called *inductive simplicity;* it has nothing to do with economy, but claims to lead to laws which hold objectively.

Let us illustrate the significance of inductive simplicity. A number of measured points may be given in a plane co-ordinate system. We can connect them by a simple curve or by a complicated one that oscillates several times between two neighbouring measured points. If the latter curve were the correct one, namely the curve on which further measurements will lie, the given measured points would constitute an exceptional selection, which is not characteristic of this curve but which suggests, by chance, a more regular shape. Yet we deem it improbable that the measured points represent such a singularity, and we regard the simplest interpolating curve as the most probable one. In this interpretation, we were guided by a 'hunch' that influences our thinking considerably. But we can formulate this idea also as an *hypothesis;* if there exists a random selection of measured values, it is improbable that it represents an exceptional selection. This statement is certainly an hypothesis, i.e. an assumption about nature, and does not refer to a property of the *description.* In this formulation, we have not mentioned the simplicity of the description, but we have assumed that we had the correct description. In this case, our hypothesis is an assumption that future measured values will lie on the curve. Therefore, it is also an *hypothesis,* and the same hypothesis in a different form, that for a given finite number of measured points, it is the simplest curve upon which future measured values will lie.

The considerations holding for the curve in the two-dimensional co-ordinate system hold correspondinly for the surface (2) in the space of the parameters. First, we need the requirement of simplicity in order to establish a connecting function beyond the observed data because the mere requirement of approximation admits of infinitely many functions. Furthermore, we want to arrive by means of this requirement at that function which corresponds to *future* measurements. We can formulate more precisely what is expected of the requirement of simplicity with respect to future measurements. The fundamental shape of the surface is essentially determined by the requirement of approximation, i.e. by the measured points. We need the requirement of simplicity in order to interpolate and extrapolate: to determine the *intermediate* values of the parameters, and the closely located exterior values of the parameters. The requirement of simplicity thus constitutes the hypothesis that the desired function is the *smoothest one* (the one showing the least oscillation) or that *the smoothest surface is the most probable one.* The probability hypothesis represented by the requirement of simplicity is therefore an hypothesis of continuity. This result is noteworthy because continuity plays an important role in the theory of probability.

The foregoing considerations help us to justify the hypothesis of simplicity in terms of the theory of probability. If we wished to test this hypothesis empirically, we would have to proceed in the following way: we would have to find new observed values p^* lying *between* the old ones, and establish the governing function for this larger class of values. If the hypothesis of simplicity is correct, the previous governing function, obtained with its help, must also hold for the larger class. Thus we test the hypothesis of simplicity by increasing the density of the observed points. Two results are possible: either type I is obtained or not. If it is obtained, the hypothesis of simplicity is confirmed. We can know *only with probability* whether type I will result; but we know with certainty: *if* it results the hypothesis of simplicity is confirmed. If type I does *not* result, there are again two possible cases. The failure may be due to the fact that the hypothesis of causality does not hold; then it would be impossible to find any continuous function approximating type I, even by violating the requirement of simplicity. On the other hand, the failure may be due to the requirement of simplicity, and if one were to abandon this requirement and choose a more complicated (yet continuous) function F, a sequence of type I would be obtained. In the latter case, one should continue with the observations; according to the assumption, the points will satisfy the function F approximately. One should be careful to set the instruments or to choose the observations in such a way that the measured points will be distributed as evenly as possible over the space of the parameters so that the distance between any two succeeding observed points will nowhere be too great; the larger the number of observed points, the smaller we make the average distance between them. When enough points have been plotted, the function F, because of its continuity, will also be the simplest interpolating function between them, and from then on the property of simplicity will again be satisfied by the governing function. The failure of the method due to the requirement of simplicity resulted from the fact that under the supposition of an even distribution of observed points, the class of measured values was too small. Summarizing, we can say: *Either no continuous causal laws exist, or they can be obtained by the requirement of simplicity.* This assertion holds with *certainty*, not only with *probability*.

The above argument justifies the requirement of simplicity. Of course, we must not infer with certainty from this argument that the simplest connecting function will correspond to future observational results because we cannot know whether we had a sufficient number of observed points, or whether, within the given interval of exactness, a continuous function can be constructed only on the basis of a larger class of points. But we do know: if *any* function exists that fulfils the requirement of approximation (*a*) for *all* classes $M^{(k)}$, $M^{(k+1)}$. . . , then the probability that this function for $M^{(k)}$ is also the simplest according to requirement (*b*) converges with increasing k toward 1. The above assumption that the simplest function is most probably the correct one follows from this result. If we prefer the simplest among all the functions that connect the observed points, we do so on the basis of the probability inference that a given element with a finite ordinal number exhibits the type represented by the whole sequence. In the present section, the probability inference itself is accepted without proof.

6. THE APRIORIST CONCEPTION OF CAUSALITY

We have formulated the principle of causality precisely enough to enter into a critical discussion of some philosophical doctrines concerning it. It is essential for our conception of causality to assume that probability statements about an infinite sequence can be made when only a finite number of the elements of the sequence are given. This assumption contains the specific difficulty of the problem of causality. It is the source of the great variety of interpretations of the problem of causality because, at closer examination, it appears so puzzling that new attempts to circumvent this assumption are always being made. We can say that scientific discussion of the problem of causality began with Hume's discovery of this difficulty.[6] However, Hume's treatment of this difficulty was not successful because his interpretation of the inductive inference as a *habit* does not solve the problem. Subsequently, two major attempts to avoid using the inductive inference have been made. We shall analyze these two attempts, the apriorist and the conventionalist on the basis of our precise formulation of the principle of causality.

Apriorist philosophy also takes the infinite governing sequence into account, but rejects the inductive inference. It refuses to infer the nature of the sequence from a finite number of empirically established elements, yet maintains that we can make judgments about the structure of the infinite sequence by other means. The source of these judgments is *a priori intuition.* It is claimed that the principle of causality is required on grounds which pertain to the nature of pure reason, and that it is impossible for reason to fail in this respect. These philosophers consider causality a self-evident principle although they admit that it is not logically necessary, and they appeal to the fact that science is permeated by the belief in causality. This justification is of the same type as Kant's first justification of synthetic *a priori* judgments [1920f, p 106, n. 17]. According to this view, knowledge concerning the nature of the infinite sequence springs from a special source which we shall call *a priori intuition.*

Once the inductive inference has been rejected, the apriorist contention can be maintained in the face of all possible experience. In this case, any empirical refutation of their assertion is impossible because the sequences of types I and II are infinite. If I have discovered that a number of elements of the governing sequence correspond to type II, I can still hope that subsequent elements will correspond to type I. It is always possible to search for causal laws which have not yet been discovered.

Even if we grant, for the moment, that the probability inference may be rejected, we maintain, nevertheless, that the resulting *irrefutability* of the apriorist conception is not equivalent to its *justification.* Such an argument overlooks the fact that the assertion 'such and such a type of sequence holds' is always an empirical statement. If type I does *not* hold in reality, one will *never* find causal laws, however long one searches for them. If causality is regarded as *a priori* necessary, it must be possible to show in some way that type I and not type II of the sequence holds in the physical world; if it is impossible to do so on empirical grounds due to the rejection of the inductive inference, the justification must be given on other grounds. We recognize immediately that such a decision is not possible. The type of the governing sequence depends on what measured values $p^*_1 \ldots p^*_r$ are obtained, and the assumption that type I holds constitutes a restriction on the possible values of the measurements. This restriction would mean that the elements of the governing sequence obey an *intensional* rule; but the sequence can only be given *extensionally*. It is given by the measured values, and nobody can know what measured values will be obtained. Since it is mathematically possible to construct classes $M, M', M'', \ldots$ in such a way that type I is *not* realized for the proper intermediate values – every subsequent class $M^{(i)}$ can be chosen in such a way that it will lead to a different governing function – it is possible for the empirically given classes $M^{(i)}$ to have such irregular nature. The assertion that causality is necessary because of the nature of pure reason means that pure reason has an influence upon the values of the measurements; yet nobody would seriously defend such an assertion.

Let us return now to the rejection of the inductive inference by apriorist philosophers. They could rightly reject this inference only if they could avoid it everywhere. But that is impossible. Even if they avoid it in connection with the *general* principle of causality, they have to admit it in order to establish any *particular* law. If we want to know what law of nature applies in a given case, say that of changes in the pressure of gases, we can only use the procedure described above for the formulation of the governing function. We have to infer from a finite number of elements that the resulting function – say, van der Waals' equation – will probably remain the same for all subsequent elements. In this case, apriorist philosophers must also admit the probability inference – why are they entitled to reject it in establishing the general principle of causality? Those philosophers who hold an apriorist conception of causality frequently use the following argument in defending the different ways in which they treat particular laws and the general principle of causality. They contend that the inductive inference concerning a particular law is justified only because the general principle of causality is absolutely certain. If this were true, the general principle of causality could neither be confirmed nor refuted by an inductive inference since there would be no basis for this inference. Let us examine the above contention.

Suppose we had constructed the governing sequence for a given case and we were trying to discover the corresponding particular law. What use would it be to know with certainty that after a certain element $F^{(k)}$ of the sequence the governing function will remain the same? Even if we had found a number of identical elements, we would not know whether we had reached the element $F^{(k)}$. We might not yet have reached $F^{(k)}$ and the elements might change once more. Our *a priori* knowledge of the existence of such an element does not facilitate the inductive inference in the least. In order to clarify these considerations, let us distinguish three cases:

(1) It is certain that the sequence will *not* be of type I.
(2) It is certain that the sequence will be of type I.
(3) It is uncertain whether the sequence will be of type I or not.

In the first case, it is inadmissible to infer any particular law; however many elements corresponding to type I we may have found, we are certain that ultimately we shall arrive at the infinite sequence of the deviating elements. The second and third cases do not differ from each other in their bearing upon our inference; the second one does not give us any more right to infer the existence of a *special* function than the third one does. We can only assert: if it is to be possible to establish particular laws of nature, it *cannot* be an *a priori* principle that causality does *not* hold. However, the existence of causality may not be asserted as an *a priori* principle; that would be a serious mistake.

Thus the last defense of the *a priori* conception of causality breaks down. This conception involves an assertion that would restrict the possible behavior of nature, and therefore it cannot be maintained. The apriorist conception would not even eliminate the difficulties involved in discovering particular laws of nature by means of inductive inferences since it cannot justify such inferences. Therefore, this conception must be abandoned.

7. THE CONVENTIONALIST CONCEPTION OF CAUSALITY

The conventionalist conception is an elaboration of the apriorist view. However, it is not based on the idea that *a priori* principles hold objectively, but rather on Kant's other contention that they are necessary presuppositions of knowledge.[7] Kant had attempted to justify the necessity of *a priori* principles by maintaining that they are indispensable for all knowledge. Conventionalism has adopted the view that there must be specific principles compatible with the totality of experience because we keep ordering our experiences until they satisfy these principles. Conventionalist philosophers draw the correct conclusion that these principles lose their apodictic character because they do not state anything about reality; if these principles are compatible with *every* experience, they do not say anything about the content of experiences but only something about the manner of describing that content. From the epistemological point of view, these principles are therefore arbitrary and empty.

Yet the conventionalist assertion *itself* is *not* empty; to say of a principle that it is compatible with every experience is to make a synthetic assertion since the continuous application of a principle to certain conceivable experiences may finally lead to contradictions. At any rate, contradictions are possible if more than one principle is used, since these principles might conflict when applied to certain experiences. If one wishes to adopt the conventionalist interpretation of causality, one must first find out whether the principle of causality could contradict itself or other principles with regard to conceivable experiences.

The conventionalist conception promises to be a successful solution for the problem of causality as we formulated it. This problem consists in predicting the structure of an infinite sequence. We showed that such a sequence, in so far as it is given at all, is given *extensionally*, and that therefore we have no means of inferring its structure from its definition. For this reason, we introduced the probability inference which encounters so much resistance. The conventionalist conception offers a special method for avoiding the difficulty connected with the extensionally given sequence.

The sequence is given extensionally only because we have constructed it in this way. We defined the governing function in such a way that it is completely dependent on the empirical numerical values. Are we compelled to give such a definition? Is it not possible to include in the definition of the governing function a prescription guaranteeing that the resulting sequence of governing functions will be of type I? Then this sequence would no longer be determined purely in an *extensional* manner, but in some respects it would be determined *intensionally*, and we could make the desired statement about the infinite sequence with certainty because we would be able to derive it from the intensional definition of the sequence.

It is the fundamental tenet of conventionalism to explain the continued confirmation of certain principles by showing that these principles are deliberately maintained. In order to apply this tenet to the principle of causality, we need only add the following condition to the definition of the governing function:

(*d*) Once the governing function has been established for a class $M^{(i)}$ according to conditions (*a*)–(*c*), this function may not be changed for the subsequent classes $M^{(i+t)}$.

If this condition is satisfied when the governing function is established, it follows with certainty that the infinite sequence will be of type I.

But now we have to ask whether conventionalism can carry through this condition without arriving at contradictions. Can we prove that for any given numerical values requirement (*d*) will be compatible with requirements (*a*)–(*c*)? This is the problem of the conventionalist conception of causality.

Let certain classes $M, M', M'', \ldots M^{(i)} \ldots M^{(i+t)}$ of measured values be given, and let us adopt requirement (*d*), i.e. that the governing function should be the same for the classes $M^{(i)}$ to $M^{(i+t)}$. If the governing function for $M^{(i)}$ has been formed according to requirements (*a*)–(*c*), we are no longer certain with respect to $M^{(i+1)}$ that the same function will still satisfy the requirement of approximation. It is possible that (*a*) and (*d*) will contradict each other. We can avoid the contradiction if we first establish the governing function for $M^{(i+t)}$ according to (*a*)–(*c*), and then go backward and use the same function for the earlier classes. In this case, (*a*) is satisfied because the earlier classes are subclasses of $M^{(i+t)}$ and the governing function of $M^{(i+t)}$ approximates their measured values as well. But then the requirement of simplicity may be violated for the earlier classes. Requirement (*d*) would therefore contradict either (*a*) or (*b*).

The requirement of approximation is certainly the most important one because the resulting function could not be called a law without it; an assertion of causality that contradicts the requirement of approximation is self-contradictory. Yet we cannot dispense with the requirement of simplicity either since it is the basis for assuming that the resulting function will also satisfy the requirement of approximation for the subsequent classes $M^{(i+t+k)}$ which have not yet been given. Thus we cannot make a probability prediction that the function fitted to $M^{(i+t)}$ but violating the requirement of simplicity for $M^{(i)} \ldots M^{(i+t-1)}$ can be retained for future measurements. If we believe that this probability inference is unjustified, we have to test it by forming further classes $M^{(i+t+k)}$. Should the requirement of approximation be violated, we could change the function again, and from there on retain it according to requirement (*d*); but we still do now know whether we shall ever arrive at an unchanging function by this method. The intensional rule (*d*) does not help us because we do not know whether or not it will contradict the requirement of approximation when we make new measurements. Since it cannot guarantee a sequence of type I, it cannot enable us to eliminate the probability inference. If we admit this inference, however, we may be able to infer from the violation of the requirement of simplicity for the subclasses $M^{(i)} \ldots M^{(i+t-1)}$ that future measurements will probably contradict requirement (*a*).

These are the reasons for the failure of the conventionalist attempt to justify causality as a necessary principle of knowledge without using an inductive inference to determine the type of the sequence. The principle of causality constitutes a restrictive statement about the behaviour of physical phenomena, and may therefore encounter contradictions. This *possibility* cannot be ruled out by a rejection of the inductive inference. On the contrary, in order to establish that the principle of causality holds, we are dependent upon the inductive inference.

8. THE PROBABILISTIC CONCEPTION OF CAUSALITY

We want to call the conception of causality which we have developed in the present investigation the probabilistic conception, in contrast to the apriorist and the conventionalist conceptions. According to the probabilistic interpretation, the principle of causality is an objective statement about the physical world, and it is possible to determine whether causality holds by means of a probability inference based on observations.

Empiricist philosophy asserts that this determination is based on experience. This point of view is usually defended in the following manner: we notice that certain laws hold in *particular* instances, and we infer that the laws will hold in *all* instances. This inference is an inductive inference and therefore not certain; exceptions to the principle of causality are possible. We can only infer with high probability, on the basis of all past experience, that causality holds always and everywhere.

Yet these considerations do not solve the problem. The main problem concerning the principle of causality is not the assertion that it holds for *all* phenomena. The inductive problem arises earlier, namely, with regard to each specific determination of causality, for instance, with regard to the determination of the gas law. Whether causality holds in a specific instance can ultimately be decided only by investigating that instance. If causality holds in other cases, the probability that causality holds in the specific case under consideration merely *increases,* but this probability must be determined initially on the basis of the relevant governing function. The increase of the probability due to other cases is not very great because special circumstances may intervene in the specific instance. If the probability that causality holds in a specific case could not be independently established for at least one case, it could not be increased by an appeal to other cases. Such an appeal presupposes that the probability that causality holds has been determined for these other cases.

The correct inference has the following form: since we have observed that the same function governs a finite number of observations, we conclude that it governs *all* observations. The presupposition of this inference is not that causality holds in certain cases, i.e. that an *inductive* description is true, but only that a *reproductive* description exhibits regularity. This inference can be carried through separately for every particular natural phenomenon without considering its result in other cases. By means of this inference alone we come to a decision concerning the given case.

Let us recall the space of the parameters in which the observed points p^*_i have been plotted. When we construct the governing function F, we assert that certain other points p_i, which are determined by F on the basis of the p^*_i, *would* be measured if we were to observe the corresponding states. We must distinguish two statements:

(*a*) These other points are the points p_i.
(*b*) It is possible to determine these other points on the basis of the p^*_i.

Statement (*a*) represents the *assertion of a particular law,* and statement (*b*) the *general assertion of causality for the particular case.* We can formulate the two statements in a different way if we assume that the governing sequence whose last actually determined element is F has been established:

(*a*) F is that element of the sequence after which all elements remain the same.
(*b*) There is an element of the sequence after which all elements remain the same.

Now we can ask for the probability that these statements are true; let us call these probabilities $P(a)$ and $P(b)$, respectively.[8] Then we have

$$P(b) \geqslant P(a) \tag{1}$$

$P(b) < P(a)$ is excluded because (*b*) follows from (*a*); if (*a*) is true, then (*b*) is always true, and therefore the probability of (*b*) is not lower than the probability of (*a*). How are $P(a)$ and $P(b)$ determined? We do not wish to calculate these probabilities precisely; we know, however, that they depend upon how many elements of the governing sequence have been determined, and how many of them are the same. The greater the number of identical elements, the higher $P(a)$ as well as $P(b)$. If we consider only the established elements of the governing sequence for the determination of (*b*), then $P(b) = P(a)$. But we may infer with high probability, from other experiences in which $P(a)$ is high, that the sequence will still be of type I, even though the elements established so far do not exhibit this type, and therefore $P(b) > P(a)$. The assertion that $P(b)$ is higher than $P(a)$

expresses the empiricist idea that we may infer, from the fact that causality holds in other cases, that it holds in a given case; but this assertion exhausts the significance of the inference. $P(b)$ is largely determined by $P(a)$ and when $P(a)$ is low, $P(b)$ is also low. If $P(b)$ is low we say it is probable that causality does not hold in this case.

The apriorist and the conventionalist conceptions determine $P(a)$ in the same way as we do, but they set $P(b) = 1$ by claiming that (b) holds with certainty. However, this claim is incorrect because $P(b)$ is essentially determined by the observed initial elements of the governing sequence. If a probability inference from the observed elements is admitted for $P(a)$, it must also be admitted for $P(b)$. The assertion

$$P(b) < 1 \tag{2}$$

characterizes the probabilistic conception of causality. The fundamental idea, expressed incorrectly by the apriorist conception, that the certainty of the general assertion of causality is a necessary presupposition for the inductive ascertainment of every particular law, is taken over and formulated adequately by the inequality (1).

If we infer in a particular instance where $P(b)$ is small that causality does *not* hold, we have presupposed that causality holds in *other* cases. By regarding the observed points p^* as correct, we have presupposed causality in the relation of the ${p^*}_i$ to the given measured values ${q^*}_k$, i.e. we have made the inductive inference that the functions $P_i(q_1 \ldots q_k)$ hold (cf. Section 2). This assumption does not involve us in a contradiction. It is possible that causality holds for certain phenomena and not for others; the assertion that causality holds for certain cases may lead to the inference that it does *not* hold for others.

We infer $P(b)$ by means of the probability inference characterized in Section 3. For the time being, it seems to be impossible to justify this inference, although we cannot dispense with it. The other conceptions of causality cannot avoid it either, because they need it for the determination of $P(a)$. Furthermore, we pointed out in Section 6 that the probability inference concerning $P(a)$ cannot be eliminated by setting $P(b) = 1$; therefore, the statement about $P(b)$ is as well founded as that about $P(a)$. It seems that the conception of causality developed here is as well grounded as the probability inference. Even though this inference has not yet been clarified sufficiently, it appears to be our best means of making statements about reality. Without it we could hardly formulate a single law of nature.

The inductive inference has greater generality than the principle of causality. This result follows from the fact that we use this inference in order to determine whether causality holds. On the other hand, the inductive inference is also employed for the determination of statistical laws. The two forms of laws, causal and statistical, must be regarded as special forms of a more general assertion dealing with the existence of probability laws in the physical world.[9]

The different forms of probability show that causality does not occupy the central position which philosophy has traditionally ascribed to it. In modern physics, statistical laws have acquired the same significance as causal laws, and their practical application does not lag behind that of causal laws. It is therefore not impossible that physics will some day be confronted by phenomena that compel it to abandon causality. We know that causal laws are becoming more and more complicated as we require greater and greater precision, and it seems that the microcosm is much more intricate than the macrocosm. Some of the causal laws of the macrocosm could be explained as statistical laws of the microcosm, and even though certain causal laws could be taken over in atomic dimensions, it does not seem possible to describe atomic processes by their help alone. Perhaps this result indicates that causality does not hold throughout the microcosm. It would certainly be premature to make such a definite assertion; yet if physics is not able to solve the problem of the quanta in a different fashion, it must not be handicapped by a faith in the necessary existence of causality. In principle, it is possible to determine on the basis of experience whether causality holds.

NOTES

[1] For a finite number of data, an analytic function can always be established that connects the data strictly, not only approximately.

[2] This formulation is given by M. Schlick, *Naturw.* 8, 461f. (1920).

[3] Bertrand Russell, *Our Knowledge of the External World*, Lectures V and VI (London, 1914).

[4] At this place, the original manuscript contained detailed investigations concerning the possible occurrence of co-ordinates in the laws of nature; for the sake of brevity, we are omitting these passages whose results we have summarized in the preceding paragraph.

[5] Compare [1924h]. (In this book, I distinguished between the two kinds of simplicity for the first time.)

[6] Hume's merit consists, first of all, in his having shown that the inductive inference is not *logically necessary*. His discovery that the inductive inference is not an inference of the logical type has since been accepted in philosophy. There is another point in

which the principle of causality does not possess logical necessity. It is not logically necessary for the specific relation F to hold between the parameters; a different relation F' could hold just as well. If we say that the principle of causality is not logically necessay, we are making two assertions: (1) it is not logically necessary that the regularity discovered for certain observations will persist for all observations; (2) a particular law of nature is not logically necessary (for instance, it is not logically necessary that heat expand bodies instead of contracting them). Furthermore, Hume has shown that the inductive inference cannot be justified by experience; if we were to infer its truth from the fact that it has held in the past, we would be making another inductive inference.

[7] I analyzed these two aspects of Kant's philosophy in detail in [1920f] Vol. VII.

[8] According to my recent probability notation ([1932f], p. 568), each of these probability statements would have to be written with two arguments since probability is a relation; but this consideration is irrelevant for the present discussion.

[9] The preceding paragraph is a summary of a longer discussion in the original manuscript which has been omitted for the sake of brevity. Let us add the following clarification: according to the above considerations, the distinction between the two kinds of laws must be interpreted in such a way that the statistical law may apply even if the sequence is of type II, i.e. even if causality does not hold. According to our discussion, the establishment of a function of type I is a method of predicting with high probability the occurrence of each particular element of the sequence. If such a prediction is not possible because the governing sequence is of type II, an alternative method of transforming the existing *low* probability of the occurrence of a certain instance into a *high* probability is available. This method consists in using a large number of cases, in going from the particular case to a multitude of cases by means of statistics. There may be two reasons for the occurrence of statistical laws: either it is not possible to determine a function of type I *for practical reasons,* i.e. it is impossible to establish a high probability for the particular case (as in the classical interpretation of the kinetic theory of gases); or it is impossible *in principle* to determine a function of type I because the governing sequence is of type II (as in quantum mechanics).

57. INDUCTION AND PROBABILITY

Remarks on Karl Popper's *The Logic of Scientific Discovery*

[1935e]

I

In a recently published book,[1] Karl Popper has set himself the task of analyzing the methods of scientific research and finding a solution for those problems that lie at the focal point of all logical questions of science: the problems of induction and probability. An investigation of these problems is certainly a worthy endeavor, if only because it seems to be necessary to attempt every conceivable solution in this complex field before deciding on the definitive answer. Furthermore, Popper deserves credit for relying heavily upon mathematics and physics and making extensive use of the conceptual tools of logical positivism in the development of his ideas. The discussion is thereby elevated to the plane of scientific philosophy, and we need not be bothered with hazy trains of thought that cannot progress beyond metaphor – as is, alas, too often the case with philosophical discussions. But with all due recognition of the sincerity of the intention, I nonetheless unfortunately feel compelled to declare my opposition to the theses presented in Popper's book, for they appear to me to be completely untenable. I see this as related to the fact that the earnest of the methods fails to measure up to the earnest of the intention. It is simply not enough to make use of the conceptual constructs of logic; in the treatment of epistemological problems, it is necessary to think through the applicability of these concepts to their most remote consequences and thus to create a method that transfers to material problems the certainty attaching to formal logic. It can be shown in detail, point for point, that failure to carry out a critique of this kind bears the responsibility for the defects in Popper's book. To analyze here every single one of Popper's ideas would take us too far afield. I will therefore limit myself to picking out the principal points and criticizing them in turn.

For Popper, the problem of induction stands in the foreground. He lists the epistemological difficulties of inductive inference first pointed out by Hume and declares them to be insuperable. As a substitute, he wishes to show that it is possible to construct the methods of science without using the principle of induction (Section 1, pp. 29–30).

The argument he develops runs roughly as follows. In proposing theoretical assumptions, science does not claim them to be true propositions. Rather, all theoretical assumptions in science are proposed provisionally. They are maintained so long as they stand up to scrutiny, rejected as soon as they prove false.

It is not this positive portion of Popper's argument that I wish to attack at present. For the thesis that the propositions of science make no claim to absolute truth and are to be rejected on the appearance of contradictory evidence accords equally with the epistemological theory I have advocated for some time and is most certainly the view that has been held by most scientists for centuries. What distinguishes Popper's view from mine is the negative part of his discussion. Over and above the idea that all knowledge is provisional, some thinkers have suggested that the construction of hypotheses has something to do with probability, that the process of proposing and testing hypotheses is carried out through probability judgments. But Popper rejects this idea, which stands in the forefront of my epistemological works, among others, and he sees the value of his arguments as consisting precisely in their elimination of the concept of probability from the epistemological problems of science.

The idea that scientific processes *exclude* the concept of probability certainly does not seem to me to be new either. On the contrary, widely varying schools of philosophical thought have developed methods of natural science not including the concept of probability. To be sure, closer inspection has invariably revealed that this elimination of the probability concept rests upon a certain schematization. That is, through the equating of a low probability with impossibility and a high probability with certainty, a characterization of knowledge has been achieved that gets along with two alternative truth-values. The elimination of the concept of probability, then, is attributable solely to this schematization, and the disappearance of the problems of probability is merely illusory. They reappear just as soon as the schematization itself is scrutinized more carefully.

Popper, too, has fallen victim to this fallacy. It is possible to identify the precise points in his argument at which, apparently unaware, he makes use of the above-mentioned schematization. Contrary to his claims:

1. The process of falsifying a theory contains the concept of probability.
2. The procedure for constructing a new theory contains the concept of probability.

In the following sections I will demonstrate these points in greater detail.

II

The possibility of strict falsification plays a major role in Popper's thinking. He admits – and here he approaches the view I have long taken – that scientific theories can never be judged true, but he believes that they can be judged false under certain circumstances. He appeals for justification to the logical form of a universal proposition. The universal propositions of natural science relate, indeed, to cases that are only extensionally given. While in mathematics – that is to say, for intensionally given cases – it is possible to decide from the definition whether a universal proposition is true, science is only able to run through and test each case one by one. But this process cannot be carried out for an infinite number of cases, and as the universal propositions of natural science apply to an infinite number of cases, it is impossible to verify them. The situation is different for refutation; a single contrary case is sufficient to refute a universal proposition. Popper's theory is based upon this logical asymmetry; he believes it to have the consequence that scientific theories can at least be falsified, viz., by the discovery of a single case in which the theory fails to apply.

However, it is easy to see that this idea, while correct in itself, is not applicable within the precincts of natural science. It fails because that which is called a fact in natural science can never be asserted to be absolutely certain. Rather, the fact possesses a degree of uncertainty, and we always have the choice of explaining a conflict with a theory as an error in the establishing of the fact. For instance, if we want to test the theoretical assertion that an electrical current produces a magnetic field by demonstrating the deflection of a magnetic needle by the current, a failure of the experiment need not be seen immediately as a refutation of the theory. The failure of the magnetic needle to turn may also stem from the compass needle being clamped somewhat too tightly onto its mooring. And while we can perhaps exclude this possible explanation of the failure by doing another experiment, there are still many other possible explanations for it. Here the physicist simplifies matters by treating certain possibilities possessing a low degree of probability as though they were impossible. He tests his instruments with the utmost precision, and ultimately declares the pertinent theory responsible for the negative outcome of the experiment. Thus Michelson's experiment was conceived at the time as a refutation of the mechanical theory of a luminiferous ether, and yet it is clear that it cannot be regarded as an absolutely certain refutation, for the absence of the interference effect can be explained in many other ways. That there exists no absolute falsification of a theory becomes particularly plain whenever it is less than clear that the probability of another interpretation of the experiment is low. For instance, the latest observations of the Einsteinian deflection of light have revealed an excessive deflection, so that a conflict with Einstein's theory has arisen; nonetheless, no one has yet dared to declare the pertinent formula in the theory false, for it is not clear what other effects have played a role in the origin of the observation.

The elimination of the concept of probability through the idea of falsification can only succeed, then, because a certain low degree of probability is equated with zero. But if we accept a schematization of this kind, we have no need to confine ourselves to falsification, for it can also be used to *verify* theories. Under certain circumstances, that is, the probability of a theory being correct is so high that it may be regarded as virtual certainty. For instance, no one today will seriously contest the idea that light waves are electrical phenomena; we can declare this assertion to be verified with as much certainty as we assert the mechanical ether theory to be falsified. It is therefore untenable to maintain the existence of an asymmetry between verification and falsification within scientific knowledge. Either we resort to a schema by equating low degrees of probability with zero and high degrees with one, in which case we are able to verify as well as falsify scientific theories, or we accede to more precise observation and reject the schematization, in which case we open the door to the concept of probability, for falsification as well as for verification.

III

We turn now to the second point at which Popper erroneously believes himself able to eliminate the concept of probability. This concerns the construction of scientific theories. At the beginning of this article, I indicated, as the common ground between his theory and mine, the idea that scientific propositions (hypotheses, theories) cannot be asserted to be true propositions and may later be discarded under certain circumstances. But the step beyond this, to which I assent but Popper does not, is the claim that hypothetical propositions are established by means of a procedure in which the concept of probability plays a role. It is my contention that past observations make certain propositions more probable than others and that the consequent *ranking of hypotheses* determines the scientific judgment. For instance, the similarities between the specimens of various races together with the relations between fossil and types of rock established in paleontology render probable our descent from a common ancestor, so that we must regard the Darwinian theory as more probable than a theory holding all races to have existed side by side ever since the existence of dry land. The procedure employed here in setting up the hypothesis is called *induction*. This, then, is how I would sum up my theory of scientific knowledge:

(1) In constructing scientific theories, we employ the process of induction; (2) we pronounce hypotheses with probability; (3) the probability of hypotheses is of fundamentally the same nature as the probability of phenomena; and (4) we require, for the logical characterization of this process, a generalization of logic which I have developed under the name of *probability logic*. Popper objects to these ideas, and I wish immediately to address myself to his objections to my theory.

Popper differentiates 'logical probability' from 'numerical probability', asserting that theories are based on logical, not numerical, probability (Section 33, pp. 115–6, and Section 83, p. 269). The concept of logical probability is introduced by Popper in the following way (Section 33–4, pp. 115–9). Let P and R be the classes of testable possibilities of two theories, i.e., the classes of verifiable phenomena asserted by the theories. Now, it is possible that P is a subclass of R, in which case Popper calls the P-asserting theory more 'logically probable' than the R-asserting theory. Popper goes on to assign the designation 'incommensurable' to two theories the testability of which does not stem from one including the other as a subclass. It is obvious that this is a totally inadequate definition of so-called logical probability, for it does not permit any determination of the degree of this probability and allows for comparability only under quite specific conditions. The term, 'logical probability', and the reverse term, 'degree of falsification', also used by Popper, is not adequately defined by him, and he therefore has no right to make use of this concept. He nonetheless does use it constantly, for he has before his mind simply the ordinary concept of probability. It can easily be shown that the ranking relation introduced by Popper for logical probability is immediately fulfilled for ordinary probability. If P is a subclass of R, then R may be replaced by the logical product $P \cdot Q$, where Q is the portion of R that does not contain P. In probability calculus, it is possible to prove the universally applicable inequality[2]:*

* [*W* is used here and throughout because it is the first letter of the German word 'Wahrscheinlichkeit' (probability). From the standpoint of perspicuity, *P* should be used in its place in English, but as *P* already has a different use in various formulas, this could only create ambiguity. I have therefore left the notation just as Reichenbach set it down. – E.H.S]

$$W(O, P \cdot Q) \leqq W(O, P),$$

where O is the common antecedent for the relevant probabilities. Under certain circumstances, then, it is possible to set up a probability inequality on the basis of purely logical relations. But this is far from justifying any talk of a 'logical probability', for such logical relations do not suffice to define a metric. On the contrary, the usual treatment of the concept in such cases shows that we are faced here, too, with nothing other than genuine probability.

Behind Popper's expression, 'logical probability', there presumably lies the idea that, with reference to the degree of probability, we can separate *assertions of order* from *metric assertions*, as is indeed possible for, e.g., geometrical measurements along a straight line. Such assertions of order are expressed through inequalities like the one given above. Yet we are not to believe that, say, we have here any sort of independent concept of probability; it would be totally erroneous to assume that we can get along simply with ordering assertions in judging the probability of theories.

In order to refute my view that the probability of hypotheses is identical with numerical probability, Popper introduces yet another argument. In my theory, probability is attributed, not to propositions, but to sequences of propositions, and Popper undertakes to deny (Section 80, p. 257) that hypotheses are sequences of propositions in the way that I have described, asserting (pp. 257–9) that hypotheses do not have the form, 'For every value k, it is the case that such and such occurs at the place k.' No doubt it is true that hypotheses do not possess this form, yet they do have the form of a general implication: 'For all k, it is the case that if the attribute O is present at a place k, then the attribute P is also present there.' If we add to this the fact that, when we are dealing with laws of nature, this implication is never a general implication, but invariably a probability implication, it turns out that hypotheses possess precisely the same form that I developed for probability implication and have transcribed in the form:

$$(k)(x_k \in O \Rrightarrow y_k \in P).$$

(Cf. my *Theory of Probability* [1935h], Section 9). Both here (Section 71) and in my earlier work on the logic of probability ['Wahrscheinlichkeitslogik', (1934c)], I showed that a transition can be made from this implicative conception of a probability proposition to a predictive conception by cancelling out in the sequences all those pairs of members for which $x_k \in O$ is *not* true. I am thus correct in asserting that hypotheses may be conceived as propositional sequences.

Popper considers it a nonsensical consequence of this theory that the probability $\frac{1}{2}$ is attributed to an hypothesis if every second member of the propositional sequence contradicts it (p. 257). But I can see nothing nonsensical about it. Let us take a sequence of throws of a die and consider the hypothesis, 'When the die is thrown, side six will show.' With respect to the sequence of throws, this hypothesis is to be assigned the probability 1/6. We can, in turn, inquire into the probability of the hypothesis, 'The probability that side six will show is 1/6.' The probability of this hypothesis has, of course, a quite different value, yet it, too, can be statistically defined, namely, by the enumeration of a sequence the elements of which are sequences of throws of the die. This probability is given through the total number of sequences displaying the frequency 1/6 relative to the total number of sequences. Probability of hypotheses, then, is simply a probability of a higher order, the logical-mathematical form for which I presented in my *Theory of Probability* (Section 8). Thus there is no difficulty whatsoever in interpreting the probability of an hypothesis as genuine probability.

IV

In this connection, I wish to insert a few remarks concerning the probability of theories that will serve to complement my previous very brief comments on this subject and also, perhaps, to clear up a certain confusion that still hangs over it. We can follow either one of two routes in defining the probability of a theory. To begin with, we can enumerate the totality of experimentally testable propositions belonging to the theory and calculate the relative frequency of the pertinent propositions. This relative frequency can be viewed as a measure of the probability of the theory. Let us call this *probability of the first form*, or *first-form probability*. But we can also treat the theory as an ideological construct and order it into a class along with other, similar ideological constructs – that is, with other theories proposed by scientists – and then calculate the relative frequency within this class. This we will designate *probability of the second form*, or *second-form probability*. Which of these two probabilities should be described as the probability of the theory?

It would be a mistake to insist upon one or other of the two. Both have a meaning, and it is simply a matter of clarifying this meaning. Let us take quantum theory as an example. This theory embraces a whole series of groups of propositions, which may be symbolized as follows:

$$\varphi_1(x_{11}) \ldots . \varphi_1(x_{1n})$$
$$\varphi_2(x_{21}) \ldots . \varphi_2(x_{2m})$$
$$\varphi_r(x_{r1}) \ldots . \varphi_r(x_{rs}).$$

For instance, $\varphi_1(x_{11})$ might stand for, 'Photograph x_{11} shows a spectral line at 589 $\mu\mu$'; $\varphi_1(x_{12})$ for, 'Photograph x_{12} shows a spectral line at 669 $\mu\mu$', and so forth; $\varphi_2(x_{21})$, for 'The Geiger counter is deflected at moment x_{21}'; $\varphi_2(x_{22})$ for, 'The Geiger counter is deflected at moment x_{22}', and so on. Each of the horizontal sequences will have a certain number of positive outcomes, which are to be conceived of as the probability of the applicability of the assertion $\varphi_k(x_{ki})$ or as the probability of the propositional sequence $\varphi_k(x_{ki})$ (where i is consecutive), as the case may be.[3] The quantum theory ψ then becomes the logical product of the propositional sequence φ_1 through φ_r:

$$\psi \equiv \varphi_1 . \varphi_2 \ldots \varphi_r,$$

and its probability, if we also presuppose the independence of the factors in the product, becomes:

$$W(\psi) = W(\varphi_1) . W(\varphi_2) \ldots W(\varphi_r).$$

This is the first-form probability for quantum theory. We note that it is not equal to one, but is somewhat smaller, for in the report of $W(\varphi_k)$ there are also included the 'misses' that occur whenever a theory is subjected to experimental tests and that are taken into account, e.g., in calculating the margin of error. Only when we round off the numbers, setting the individual $W(\varphi_k)$ equal to one, does $W(\psi)$ become equal to one. This is the schematization to which I referred earlier.

Let the value of the given product equal q; then the first-form probability for quantum theory is:

$$W(\psi) = q.$$

We can now make a transition to the higher-level probability that $W(\psi) = q$, i.e., to the probability that:

$$W[W(\psi) = q].$$

Rather than asking whether the probability in question is precisely equal to q, we will find it more to the purpose to indicate an interval $q \pm \delta$, within which the probability lies. We will ask, that is, after the probability:

$$W[q - \delta \leqq W(\psi) \leqq q + \delta] = w,$$

where, for a large q, δ is so chosen that $q + \delta \geqq 1$; w then becomes the probability that the probability of the quantum theory lies near one within a certain interval $\eta = \delta + 1 - q$.

In order to determine this probability w, we will have to include the proposition:

$$q - \delta \leqq W(\psi) \leqq q + \delta$$

in a class containing other, similar propositions, and here we find the possibility of making the transition to the class of scientific theories of such and such a type.

Carrying out an enumeration of this class is relatively complicated. Conceptually, the probability sought for corresponds to Bayesian probability. In my *Theory of Probability*, I present this idea in connection with the theory of inductive inference (Section 62), treating it subsequently in connection with the correctional procedure (Section 89). What is in question is the probability that a sequence, the frequency of which, after n members, lies within the interval $q \pm \delta$, will, upon further extension, approach a limit lying within this interval. This probability could be calculated by considering other scientific theories which indicate an initial probability within $q \pm \delta$ and then counting among these the narrower class of those that have continued to show this probability, i.e., have maintained it up to the present. That, in making this calculation, we do not know the limit of the frequency for certain, but simply posit it, is a peculiarity not only of this problem but also of all other probability calculations (Section 89).

In this way the second-form probability for quantum mechanics arises. Thus it is not to be conceived of as the probability of quantum theory but as the probability that the probability of the quantum theory lies near one within the interval η, or, more briefly, that the quantum theory is correct within a probability interval η.

Thus the probabilities of the first and second form are probabilities of different levels, comparable to the two levels in the above example, 'Side six shows when the die is thrown' and 'The probability that side six will show when the die is thrown is 1/6.' The reason these two levels are not usually kept separate in considering the probability of theories is that theories have not been treated as having a probabilistic nature but have been regarded as being either true or false. From this viewpoint, the group of propositions in quantum theory corresponds to what we have called the first level, while the proposition, 'The quantum theory is true', corresponds to the second level. It is well known that two-valued logic generally makes no distinction between the proposition *a* and the proposition '*a* is true'. In that context, equating these two propositions is perfectly safe, for if *a* is true, the proposition '*a* is true' is true. In probability logic, however, this distinction becomes essential, for the two probabilities in question, *q* and *w* in our example, are independent of one another and may be quite different. (Cf. my *Theory of Probability*, Section 60, p. 324).

The probability of theories fits naturally into probability logic, and it is quite erroneous to suppose that it constitutes a different type of probability. Perhaps the idea that a theory means a single case and thus creates difficulties in the treatment of probability plays a role in Popper's objection. But the question arises in just the same way for phenomena and can be handled with my concepts, *posit* and *appraisal* (*Theory of Probability*, Section 73). The probability *w* appearing above for the validity of quantum mechanics is accordingly to be conceived of in the sense of an appraisal. Furthermore, the arbitrary element in the choice of the 'scientific theories of such and such a type' raises no new difficulty that pertains only to theories. The very same difficulty arises for individual cases, e.g., in considering the question whether the probability that a particular case of tuberculosis will prove fatal should be calculated in accordance with the death rate from tuberculosis in the general population or in accordance with the death rate among persons having tuberculosis with certain specified X-ray findings. In fact, we choose the narrowest class for which reliable statistics can be compiled (*Theory of Probability*, Section 72, p. 374). These are practical difficulties which arise in the determination of any probability but which can have no effect upon the fundamental possibility of the statistical comprehension of probabilities. It is the possibility in principle that matters here, and consequently the fact that the available historical material does not suffice for developing a usable statistical theory for confirming theories does not constitute an objection to my views.

V

In this connection I must correct yet another of Popper's opinions, this one belonging more to the mathematical theory of probability, but nonetheless playing a role in his discussions. He believes (Section 64, p. 185) that he can eliminate from probability calculus the concept of a limit by using a certain interpretation of the concept of probability. I have already made note of this interpretation in my article 'Causality and Probability' ([1930g], p. 165) and employ for this concept the term *partial limit* in *The Theory of Probability*

(Section 64). The frequency sequence coordinated with the probability sequence has a partial limit near p if the frequency invariably approaches the value p without restriction, i.e., if p is a point of accumulation in this sequence. It can easily be shown that the partial limit becomes a genuine limit if p is the only place for which a partial limit exists. On the other hand, it can also be shown that if two values p_1 and p_2 are *partial* limits, all the intermediate values are also *partial* limits. This follows from simple considerations concerning the progression of the frequency sequence.

Popper wishes to use the concept of the partial limit in his interpretation of probability. This portion of his discussion is extraordinarily unclear. He fails altogether to understand the logical order required for the construction of the probability calculus, so that it is difficult even to correct his arguments. I must confine myself to sketching briefly the proper way in which to construct such conceptions.

Probability calculus can be developed axiomatically, as a formal discipline, without appeal to any particular interpretation of the concept of probability. The axioms then take the form of implicit definitions. All the propositions of the probability calculus will be derivable from them; but these propositions will, of course, always contain the concept of probability as a term without interpretation. On the other hand, the concept of probability can be given interpretation in terms of its contents, e.g., by means of the limit. It must then be shown that this interpretation conforms to the axioms, i.e., that the axioms are fulfilled for it. This is the case, for instance, for the interpretation in terms of the limit. Once it is shown to conform to the axioms, the validity on this interpretation of the propositions deduced from it requires no further demonstration.

We could make the experiment of interpreting probability by substituting for the limit the partial limit, which will result in a weaker interpretation. The fulfilment of the axioms must then be demonstrated separately, requiring more careful formulation. For instance, in the case of the theorem of addition, it is not permissible to say: If P has a partial limit at p and Q has a partial limit at q, then there exists for *P or Q* a *partial* limit at $p + q$. It must instead be reformulated as: If the *partial* limits for P lie entirely between p_1 and p_2 and, similarly, the *partial* limits for Q entirely between q_1 and q_2, then the *partial* limits for *P or Q* lie between $p_1 + q_1$ and $p_2 + q_2$ (without necessarily filling the entire interval). Carrying out this interpretation becomes extremely complicated, as account must be taken of the intervals for the *partial* limits. But if the conception is carried out with appropriate care, all the propositions of the probability calculus will be valid on this interpretation as well. It is therefore impossible to comprehend the special stress Popper lays on the fact that Bernoulli's theorem is valid on his interpretation through the *partial* limit, which he calls 'middle frequency' (Section 64, pp. 186–8). That it is valid follows as a matter of course, once the axioms are shown to be fulfilled on this interpretation and the applicability of the special theorem of multiplication is assured by means of certain stipulations as to the structure of the sequence. Popper, however, omits the initial investigation, and as a result his presentation is incorrect in that it does not include information about the intervals for the partial limits.[4] Yet it is quite wrong to see any special mathematical significance in the fact that Bernoulli's theorem can be inferred if the limit interpretation is rejected. Bernoulli's theorem as such has no more connection with the limit interpretation than does every other proposition in the probability calculus and can be derived even in formal probability calculus.

The view might be taken that an interpretation through the *partial* limit possesses certain advantages over interpretation through the limit. I myself was formerly of this opinion, but I gave it up when I showed, in my article, 'Causality and Probability' (p. 165), the possibility of inferring the interpretation by means of the true limit from the interpretation through the *partial* limit in conjunction with the axioms of the formal probability calculus (cf. my *Theory of Probability*, Section 64). It is demonstrable that interpretation through the *partial* limit possesses no other logical advantages, either. Popper perceives as a logical advantage of this interpretation the fact that the existence of a *partial* limit is mathematically necessary for an infinite sequence of frequencies. He believes that it avoids the difficulties arising for the assertion of the limit from the fact that the value of the limit and the points of convergence for an extensionally given sequence cannot be given. But he neglects to note that the same difficulties arise for the partial limit as soon as we try to *determine* it. For the proposition, '*p* is a *partial* limit', contains the same logical difficulties as the proposition, '*p* is a true limit.'

For these reasons, the mathematical portion of Popper's argument is also completely untenable.

VI

In the preceding sections, I was able to do no more than select for comment the most important details in Popper's presentation. There remain some points of which I shall not offer a thorough refutation here, e.g., his discussion of quantum mechanics, which has been subjected to criticism from another direction (Weizsäcker, C. F. von, *Naturwissenschaften* 22, 808 (1934)). I only wish now to comment on the positive portion of Popper's discussion, in which he attempts to replace the principle of induction with a different conception. His argument is related to the idea, described in Section I above, that we do not need the concept of probability in order to construct new theories. Popper disputes that we are working according to a system at all in such cases. "We do not know," he says, "we guess" (Section 85, p. 278). He evidently sees the substitution of guessing for methodical investigation as resolving the problem of induction.

Now I can see no advantage whatsoever in tossing aside the systematic endeavors to rationalize the process of constructing scientific hypotheses with the remark that the process is not rational. Ever since Hume, eminent philosophers and scientists have been of the opinion that the principle of induction is used in this process, and careful investigation has, in fact, shown that the construction of scientific theories is based upon this principle. To be sure, no one had as yet come up with a justification for its use, which made an unpleasant situation for epistemology. Yet I do not believe we can escape from this situation by contesting the use of the principle of induction as such. Certainly, it is difficult to establish which principles a person has used in his thinking; all we can show is that his actions are logically ordered according to a certain principle. But if we succeed in this latter proof, we will not put much trust in his word, even if he constantly assures us that he does not employ the aforesaid principle in his thinking. I shall avail myself of a somewhat dramatic analogy. Fruit-vendors on the street have a habit of placing the good apples at the front of their cart – that is, on the side visible to the public – while leaving the bad apples at the back; then, when they come to fill the customers' bags, they invariably take the apples from the back of the heap. If we take up this matter with a fruit-vendor, he will deny emphatically that he employs any such principle in selling his fruit, describing his choice of apples for the customers as independent of such considerations. I put as little trust in those who claim to form their predictions about the future without using induction as I do in the fruit-vendor. For time after time it turns out that they believe precisely those assertions about the future that accord with the principle of induction – that they, for instance, expect a train to depart at the time indicated in the railway schedule and press the bell when they want to ring. If anyone retorts, "We don't know; we guess", I can only point out that this guessing follows a course which accords quite strikingly with the principle of induction – just as I had to indicate, in the case of the fruit vendor, the agreement of his actions with a principle he denies using. Popper admits that "our guessing is guided by the unscientific, metaphysical belief (which can be explained biologically) that there exist regularities which we can unveil, discover" (Section 85, p. 278). If he goes this far, he ought rather to admit that this belief is nothing other than the principle of induction, for it is, in fact, this principle that determines our opinions about the future.

As long as Popper goes on speaking of a metaphysical belief in this connection, I certainly do not share his opinion. We have no right to a metaphysical belief, and contenting ourselves with this Popperian solution would mean the end of all scientific philosophy. I am therefore unable to understand why Popper believes his investigation to constitute even the smallest step forward in resolving the problem of induction. Hume had already reached this point when he spoke of an unshakable belief, but he was honest enough to admit that he knew of no justification for this belief. Whatever good can it do us today to be offered a theory of epistemology that allegedly makes no use of the principle of induction and substitutes for it a metaphysical belief? How such a philosophy can even be described as a scientific conception of the universe is incomprehensible to me.

Thus it seems to me to lie in the interests of any advocate of a scientific world-view to take up my probability theory of induction, which Popper attacks without success. This theory shows that there is only *one* concept of probability, the same for phenomena and for scientific theories, and, further, that the process of constructing scientific theories fits into a probabilistic method that I have called the method of correction. That scientific theories are never treated as absolutely true by this method, but as merely provisional – a point which Popper appears to regard as the most important discovery in his book (Section 85, pp. 280–1) – is assumed as a matter of course for the probability theory of knowledge and is not, by the way, a discovery that first came to light in recent times. But the successful establishment of a procedure conforming to probability logic for predicting propositions and constructing hypotheses, leading ultimately to a rational justification of the rule of induction, seems to me a result that should induce us to favor the probability theory of knowledge over all other epistemological theories.

* * *

Postscript. Carnap has published in this journal (*Erkenntnis* **5**, 290–4 (1935)) a discussion of Popper's book in which he makes two attempts to defend Popper's arguments against my criticisms. I wish to show briefly that his attempted defense of Popper is untenable.

Carnap first tries to defend Popper's theory of falsification by replacing the *single* observed contradictory case with a finite sequence of contradictory cases. (This is what Carnap's comments on p. 290 amount to.) However, this alters nothing. For we then have the choice, as we did before for a single case, of asserting either that the theory correctly predicts the future or that an induction based upon the finite sequence of cases does so – a genuine probability problem in the sense of my method of correction. Thus a falsification, like a verification, can only be pronounced with probability – which is just what I said, in contrast to Popper. Carnap says, "A theory is said to be falsified . . . if it is incompatible with an empirical hypothesis that has been confirmed." But this is identical to my assertion at the end of Section II above that a theory can only be falsified through the employment of a schema in which high probabilities are equated with one and low probabilities with zero: that is, by regarding Carnap's 'empirical hypothesis' as true. And then the asymmetry between falsification and verification disappears.

Next, Carnap believes that he can salvage Popper's mathematical theory by appealing to his requirement of freedom from after-effects (p. 291). This portion of Carnap's defense is answered by note 4 of my article, with which Carnap was unacquainted when he wrote his review. Let me emphasize, by the way, that Popper's mathematical error was not the decisive factor in my rejection of his idea. Even if his theory were mathematically viable, it would not hold the slightest advantage for the problem of probability. For it is possible to raise against Popper's definition of probability as a *partial* limit free of after-effects all the logical and epistemological objections that have been brought against the simple limit, so that Popper's mathematical theory simply introduces a superfluous complication into the total complex of problems.

Thus my objections to Popper remain intact. I do not wish at this time to go into Carnap's attempt to give his own interpretation of the problem of the probability of hypotheses. I will say simply that I do not regard this question as the problem of a 'decision'. Decisions and conventions do play a certain part in science, but the question of how we are able to attain the best prediction of the future – which is the aim of scientific theories – is not to be answered by an arbitrary decision.

NOTES

[1] Popper, *Logik der Forschung*, (Springer, Vienna, 1935) [translated with new appendices and notes as *The Logic of Scientific Discovery*, Hutchinson, (London, 1959); section numbers are identical].

[2] Cf. Reichenbach, *Wahrscheinlichkeitslehre*, [1935h], p. 97. [The citation given appears in the English translation in Section 20, p. 85.]

[3] Here, of course, we can only count out the frequency for the finite number of observed cases and regard this value as being a limit in the sense of a posit for the frequency of a sequence. I shall not at present go into this problem, which arises for *every* probability determination, not just, say, for the probability of theories; I discuss the matter thoroughly in my *Theory of Probability* [1935h].

[4] Popper does not render such an investigation superfluous by adding to the probability sequences another requirement, over and above the existence of the partial limit, for which he uses my expression *freedom from after-effects* (Reichenbach, 'Axiomatik der Wahrscheinlichkeitsrechnung' [1932f], p. 600). It is incumbant upon Popper to prove the applicability of the addition theorem to the partial limit determined through this characteristic. However, I think it unlikely that this is possible. In any case, Popper's theory cannot be discussed in mathematical terms until such a proof is proffered.

Incidentally, Popper appears to have overlooked that sequences without after-effects cannot be regarded as a complete substitute for 'random' sequences because the special theorem of multiplication does not necessarily apply to them in enumerations by sections. (It does so only in enumerations by over-lapping segments; cf. my *Theory of Probability*, Sections 28–9.) If, for instance, we make, in a sequence without after-effects, a line after every fourth member and regard the resulting four-member sections as elements in a new system of counting, the special multiplication theorem will not necessarily apply to the probability of these combinations. This is the reason that I have introduced the normal sequence as a type alongside the sequence without after-effects.

58. THE SEMANTIC AND THE OBJECT CONCEPTIONS OF PROBABILITY EXPRESSIONS

[1939b]

1. INTRODUCTORY REMARKS

My probability theory has been criticized from a number of different quarters on the grounds that it leaves unanswered the question whether probability is to be conceived of as a syntactic, a semantic, or an object relation. Some of my critics have even attempted to draw the conclusion that my probability logic is untenable. It is my purpose, in this article, to investigate this question more closely and to show that considerations of this nature pose no difficulties for my theory[1].

First, a few comments about my use of terminology. I call a relation between things an *object relation,* as does Carnap[2]; a relation between signs a *syntactic relation;* and a relation between signs and things a *semantic relation,* as does Tarski[3]. I regard states of affairs as belonging in the category of things; sentences, in turn, as belonging to the category of signs. An example of an object relation is the attraction between a magnet and a piece of iron; of a syntactic relation, the relation of logical derivability of one sentence from another. A semantic relation is exemplified by the truth of a sentence, for it represents a relation between the sentence (that is, a group of signs), and objects (that is, states of affairs).

There are also signs for signs, that is, second-level signs; these occur in discussions of signs. Like Carnap, I propose to derive these second-level signs from first-level signs by setting the first-level signs in quotation marks. If, for instance, I am discussing the word 'magnet', this word, *including* the quotation marks, signifies a second-level sign. Example: The word 'magnet' is spelled with six letters. Signs of the first level could likewise be constructed from objects by laying the objects themselves on the paper and surrounding them by quotation marks. Instead of writing: A magnet attracts iron, we could replace the word 'magnet' on the paper with a magnet and put quotation marks to the right and to the left. – While such a procedure would be highly unsuitable for signs of the first level, it would be appropriate for second-level signs.

If I say: a magnet exercises a force of attraction upon a piece of iron, the relation of attraction exists between two objects, viz., between the magnet and the iron; the sign of relation, viz., the word 'attraction' stands between object signs, viz., between the words 'magnet' and 'piece of iron'. The objects, magnet and iron, are the argument signs of the relation; the argument signs are the words, 'magnet' and 'iron'[4].

I wish to add one more remark concerning the elementary sentential connectives, such as the or-relation and the and-relation. These relations are to be conceived as object relations; for instance, the sign 'or' stands between sentences, not between signs for sentences. Accordingly, the elementary sentential connectives are relations of the same logical level as, e.g., the magnetic force of attraction.

Yet there is a remarkable difference here. It is immediately clear that the logical sentential connectives are relations between states of affairs, not between things (such as the magnetic force of attraction). Thus the object relations are to be classified into thing relations and state-of-affairs relations. Moreover, the sentential connectives present state-of-affairs relations of a special sort.

Let us show this by going from the sentence,

'*a* or *b*' (1*a*)

to the sentence

'"*a*" is true or "*b*" is true.' (1*b*)

We call the second 'or' the *semantic correlate* of the first 'or', for it stands between sentences of higher levels of language. Instead of saying, 'The or-relation pertains to states of affairs', we may instead say, 'The semantic correlate of the or-relation pertains to sentences'; these are equivalent expressions. The second expression may be abbreviated as, 'The or-relation connects sentences.' The word 'connect', then, offers us a short-cut in place of the circuitous method of characterizing a relation by means of an assertion about its semantic correlate.

The semantic 'or' may also be conceived as the logically primary sign, reducing the object 'or' to it by construing the sentence (1*b*) as the definition of the sentence (1*a*). The peculiarity of the sentential connectives, as distinct from other state-of-affairs relations, finds expression here: they are determined by their semantic correlates, and their presence therefore depends solely upon the truth value of the sentences in question. In symbolic logic, this peculiarity is ordinarily expressed by designating the sentential connectives as *truth functions*, an expression that, strictly speaking, applies only to the semantic correlates of the sentential connectives. If we speak of the sentential connectives as state-of-affairs relations, we must say, accordingly: The sentential connectives are specifically those state-of-affairs relations that depend only upon the *existence* of the pertinent states of affairs. We will therefore call them *existential relations.*

2. PROBABILITY IMPLICATION

I turn now to the question of probability implication. I have introduced this relation[5] in the form

$$(i)\ \ (x_i \epsilon O \underset{p}{\ni} y_i \epsilon P) \tag{2a}$$

which is equivalent to the form[6],

$$(i)\ \ (\varphi x_i \underset{p}{\ni} \psi y_i). \tag{2b}$$

The sign '$\underset{p}{\ni}$', then, is used between sentences and is therefore of the same logical level as the sign 'or', the implication sign, and so forth. Probability implication is, accordingly, an object relation – specifically, an existential relation – just like the elementary sentential connectives.

In order to show this, let us take a closer look at probability implication. I have described probability implication as a generalization of logical implication[7]. This description is justified by its factual meaning for the pertinent states of affairs, for logical implication signifies a relation between states of affairs, which may be conceived as a special case of probability implication: A case in which y_i is P whenever x_i is O is an instance of logical implication. Probability implication, on the other hand, is exemplified by the cases in which, if x_i is O, y_i is only sometimes P (the only condition here being that the relative frequency tend toward a limit). We can formulate the same point as a semantic sentence by using the word 'connect', introduced above; if, whenever '$x_i \epsilon O$' is true, '$y_i \epsilon P$' is also true, these two sentences are connected by logical implication, whereas a connection through probability implication is present also in the case in which, if '$x_i \epsilon O$' is true, '$y_i \epsilon P$' is only sometimes true.

I have expressed this logical state of affairs in Axiom II, 1:

$$[(i)\ \ (x_i \epsilon O \supset y_i \epsilon P)] \supset [(i)\ \ (x_i \epsilon O \underset{p}{\ni} y_i \epsilon P) \cdot (p = 1)]$$

This axiom has been mistakenly attacked by Nagel[8]; logical implication and probability implication are relations occupying the same logical level, viz., existential relations. To put it another way: their semantic correlates are truth functions.

Despite the analogy presented here, there exists in another respect a fundamental difference between logical implication and probability implication. If two series of events are related by logical implication, this signifies something not only about the series of events as a whole, but also about each individual pair of events: namely, that whenever an event of type O occurs, an event of type P also occurs (semantic correlate: If the sentence φx_i is true, then the sentence ψy_i is also true). But there exists no corresponding meaning for the single case in the realm of probability implication. The sign '$\ni\!\!\!\rightarrow$', then, is meaningful only within expression such as (2*a*) or (2*b*); in the absence of the universal quantifier at the beginning of the expression, it possesses no meaning. There is no analogue here to individual implication[9]. Only advocates of the interpretation of the concept of probability in terms of the individual case will endow an individual probability implication with a meaning. For advocates of the frequency interpretation, the sign '$\ni\!\!\!\rightarrow$', has a merely fictitious meaning when applied to the single case: i.e., every individual application of this sign must be capable of being converted into a general application[10].

3. THE W-NOTATION*

I have introduced yet another notation[11]: in place of (1), I write

$$W(O, P) = p \tag{3}$$

or, using sentence functions, I write in place of (2)[12]

$$W(\varphi x_i, \psi y_i) = p \tag{4}$$

I will present here a more detailed analysis of this notation than was carried out in *The Theory of Probability* [1935h].

In this second notation, probability implication is divided into two operations. The first is distinguished by the appearance of the comma between the argument signs 'O' and 'P' in (3) and between 'φx_i' and 'ψy_i' in (4), and is to be called the *operation of selection.* According to the frequency interpretation, the operation of selection signifies the construction of a new sequence of events from the two original sequences. This is achieved by omitting from the second sequence of events all those members the corresponding member of which in the first sequence does not belong to

* [See translator's note on p. 376 above – Ed.]

class O: through the construction, that is, of a subsequence from the second sequence of events, selected by means of the first sequence of events. A corresponding procedure may be carried out at higher language levels for the sentence sequences corresponding to the sequences of events, if we take note of the semantic correlate of the operation of selection. This operation signifies the construction of a new sentential sequence from the two original sentential sequences, accomplished by omitting from the second sentential sequence all those members the corresponding member of which in the first sequence is false; through the construction, that is, of a partial sequence from the second sentential sequence, selected by means of the first sentential sequence. We see here that the operation of selection is similar to the elementary sentential connectives, except that it connects entire series of sentences and not individual sentences. Just as the or-operation connects two sentences to a new sentence, the operation of selection connects two sentential sequences to a new sentential sequence.

For the new sentential sequence, let us introduce the abbreviation '(ψz_k)' using the definition

$$(\psi z_k) =_{Df} (\varphi x_i, \psi y_i) \tag{5}$$

(where z_k represents a subsequence taken from y_i by a new, complete enumeration). Then (4) takes on the form

$$W(\psi z_k) = p\,. \tag{6}$$

In this case, only *one* sequence of events comes into question; the selective operation disappears in this conception.

The second constituent in (3) or (4), the only one remaining in form (6), is the sign '$W(\ )$'. On the frequency interpretation, this means we are to determine the limit of the frequency of events of type P (or, as the case may be, type ψz_k) in the new sequence of events. Thus the probability appears here as a characteristic of *one* sequence of events and no longer as a relation between *two* sequences of events. In my *Theory of Probability*, (p. 396) [1935h], I make the corresponding distinction between the *predicative* conception and the *implicative* conception of probability. For purposes of of producing the predicative conception, the character of the relation was first put into the distinctive form of a special logical operation, the operation of selection.

Let me just note that we can, in the same way, also divide the sign '$\overset{p}{\supset}$' into the sign '$\supset$' [*sic* – Ed.], which corresponds to the comma, and the degree of probability 'p'. I have used the sign '$\supset$' in this sense – that is, as equivalent to the comma – in the truth table in *Theory of Probability*, pp. 400–401. But I must add the qualification that the sign '$\ni$' is not strictly identical with the indefinite probability implication that I introduced on p. 52 of *Theory of Probability*, which asserts somewhat more; it includes the additional assertion that the new subsequence possesses a limit of the frequency.

4. SEMANTIC CONCEPTION OF THE W-NOTATION

We have interpreted the operation '$W(\ )$' as the determination of a frequency in a sequence of *events*. It now becomes obvious that we can likewise apply the frequency in the corresponding sequence of *sentences*. If our symbol $W(\ )$ is to have this meaning, we must write, in place of (4),

$$W(\text{`}(\varphi x_i, \psi y_i)\text{'}) = p \tag{7}$$

and, in place of (6),

$$W(\text{`}(\psi z_k)\text{'}) = p \tag{8}$$

Thus the argument of the logical function '$W(\ )$' is no longer the new sequence of *events*, constructed by means of the operation of selection, but is the new sequence of *sentences* coordinated to it. Hence (7) and (8) must contain as an *argument sign* an expression that contains the new sentence sequence in quotation marks. The selective operation will not be altered, but will retain the character of an object relation, analogous to the elementary sentential connectives[13].

In notations (3), (4), and (6), probability is defined as a frequency of events; in (7) and (8), as a frequency of sentences. Thus the first conception is an *object* theory of probability, while the second is a *semantic* theory of probability. The first conception corresponds to the *mathematical* theory of probability, the second to the *logical* theory of probability[14].

It is now apparent that both theories are isomorphic, for the result will invariably be the same whether we count events or the corresponding sentences. Thus the difference between (4) and (7) on the one hand and (6) and (8) on the other is of no practical significance. If, in the course of a precise logical investigation, we wish nonetheless to carry through this distinction, the result is as follows. For all mathematical purposes, the notations (3), (4), and (6) are correct; only when the transition is made to a conception of probability theory as a probability logic do the notations (7) and (8) become correct. I should therefore, strictly speaking, have used notations (7) and (8) throughout the last chapter of my *Theory of Probability*; wherever I neglected to do so, it is to be interpreted as a stipulation that the parentheses in the symbol *W*( ) are to take over the function of the quotation marks, i.e., that they are to transform sentential sequences into signs for sentential sequences. In addition, there comes into play the stipulation enforced in my *Theory of Probability* that the parentheses in the *W*( ) symbol take over the function of the parentheses of the sentential sequence symbol at the same time. The notation with the quotation marks seems to me impracticable, but anyone whose logical conscience will not let him rest is expressly justified in inserting quotation marks.

I believe this disposes of the objections according to which I erred in presenting my concept of probability as a generalization of the concept of truth[15]. Just as truth is a relation between sentences and things, probability in its semantic conception is a relation between sequences of sentences and of events. Incidentally, I have already indicated, in *Theory of Probability*, p. 396, the semantic character of probability assertions in their logical form by describing assertions of probability as parallel to assertions of the form '*a* is true'.

On the other hand, it must be clearly recognized that, in view of the complete isomorphism, no deeper epistemological meaning is to be attributed to the difference between the semantic and the object conceptions of the concept of probability. On the contrary, the logical and the mathematical concepts of probability can for most purposes be treated as identical; it is only within the framework of certain logical investigations of a specific kind that the difference becomes significant. Thus we do not, in general, differentiate between, e.g., the object relation 'or' and the semantic relation 'or'; this distinction plays a part only in highly specialized studies. The structural equivalence of the concepts expressed in the isomorphism may for most purposes be interpreted as an identity.

Differentiation of high-level probabilities is of the same kind (*Theory of Probability*, Chap. 8). The probability of expressions that are themselves probability expressions is in question here, so that these expressions contain probability concepts of yet higher levels. We must, then, distinguish not just two kinds of probability concepts, but must construct a whole hierarchy of such concepts; the first step will consist of the probability concept of the object conception, the second step, of the first semantic probability concept, represented in (7) and (8). Because of the existing isomorphism, however, we are justified of speaking instead of a *single* concept of probability that is repeated at various logical levels. This is how I have expressed the point in my *Theory of Probability*.

5. DUALITY OF THE SEMANTIC AND THE OBJECT CONCEPTIONS

I will now investigate the question of how it happens that the concept of probability admits of the duality of a semantic and an object conception, whereas the concept of truth may only be conceived semantically. The reason is that the concept of probability has a more complicated structure: while the concept of truth is merely qualitative, the concept of probability also contains quantitative information. An object relation corresponds to the quantitative component in the concept of probability.

In order to understand this, we must study in greater detail a peculiar difference between two-valued and many-valued logics. In writing down a two-valued sentence,

$$\varphi x_1 \tag{9}$$

we may have one of two purposes in mind: we may be setting down the sentence in order to *consider* it, or we may be *asserting* it. The latter function of (9) can be replaced by a second, semantic sentence concerning the truth of 'φx_1', which (including, for once, the quotation marks) shall be written as follows:

$$W(\text{`}\varphi x_1\text{'}) = 1. \tag{10}$$

This is again a two-valued sentence. If we consider (9) and assert (10), the result will be the same as when we assert (9).

This is the reason that, for practical purposes, we do not need to use semantic sentences in two-valued logic. We stipulate instead that, in being *written down,* a complete sentence is also being *asserted;* false complete sentences are simply not written down. The truth value remains indeterminate only for partial sentences with a complete sentence. For instance,

$$a \supset b$$

leaves open the question whether 'a' or 'b' is true; on the other hand, the truth of the complete sentence is asserted.

But this does not work in multi-valued logic. If we write down the sentential sequence

$$(\varphi x_i) \tag{11}$$

we can only consider it; since there are more than two degrees of truth possible, we cannot simply substitute for it the alternative, "Write it down or do not write it down." Consequently, the semantic sentence

$$W(`(\varphi x_i)') = p \tag{12}$$

that is coordinated to it is not superfluous. As semantic sentences in probability logic – at least in its simplest form – are two valued, (12) is a two-valued sentence, and therefore the act of writing it down may be interpreted as an assertion of it.

Only when they are taken together do (11) and (12) make an assertion. Expression (11) alone asserts nothing; it would read, for example,

$$\begin{aligned} &x_1 \text{ is a throw of six,}\\ &x_2 \text{ is a throw of six,}\\ &x_3 \text{ is a throw of six, } \ldots . \end{aligned} \tag{13}$$

This says nothing as to which of the individual sentences in this sequence are true and which are false. Only in (12) are we given information on this point[16].

However, (11) and (12) can be replaced by a two-valued object sentence about the frequency of events:

$$W(\varphi x_i) = p\,. \tag{14}$$

Assertion of this sentence is equivalent to a combination of considering (11) and asserting (12). The object correlate of the semantic sentence (12) is (14). A correlate exists because a two-valued presentation of the event is also possible alongside the multi-valued presentation through (11) and (12).

That this distinction vanishes in two-valued logic becomes clear if we consider the transition to sentential sequences of value 1, through which probability logic is converted into two-valued logic. In this process, (12) is converted into (10), while (14) becomes sentence (9) if p equals 1. The two latter sentences then assume the form, 'x_1 is a single event with the characteristic φ.' Thus the object correlate of the semantic sentence here becomes identical with the object sentence.

With respect to the higher levels of language, however, there exists a perfect parallel between the concept of truth and the concept of probability. In observing the truth of semantic sentences of the form "'φx_i' is true", we come to truth concepts of higher levels, corresponding to the probability concepts of higher levels.

6. INDIVIDUAL CONCEPTION OF THE OPERATION OF SELECTION

A remark about the operation of selection is in order here. We have called the operation of selection an analogue of the elementary sentential connectives, but with this difference: the operation of selection connects sentential sequences as a whole, while the elementary sentential connectives connect individual sentences. In my *Theory of Probability*, I show how it is possible for the latter to be transformed into operations connecting sentential sequences as a whole. This is accomplished through definitions[17]. For illustrative purposes, I shall present here only the definition concerning the or-relation:

$$(\varphi x_i) \vee (\psi y_i) =_{Df} (\varphi x_i \vee \psi y_i)\,. \tag{15}$$

We can now pose the question whether a similar relation may be laid down for the operation of selection, connecting an individual operation with the general operation.

The following must be borne in mind. Our notation of the operation of selection in (4), etc., is so chosen that it corresponds to the right side of (15); i.e., we set the comma between the individual sentences. (Only the expression '(φx_i)' in parentheses signifies a sentential sequence.) Here too, of course, we may introduce a notation corresponding to the left side of (15), writing

$$(\varphi x_i), (\psi y_i) =_{Df} (\varphi x_i, \psi y_i)\,. \tag{16}$$

If we now take a look at our definition of the contents of the right side of (16), as presented in Section 3, we find that it applies only if the case of whole sequences of sentences, not to individual sentences. There we defined the general operation of selection, and this is more correctly symbolized by the left side of (16) than by the right side.

However, we can proceed in the opposite direction and assign an intuitive meaning to the right side of (16) by defining an individual operation of selection, repeated application of which to individual sentences leads to the general operation. This will happen if we conceive of the individual operation of selection as a limiting case of an operation between sentence sequences which occurs when the value of the sentence sequence equals 1. From the aforementioned definition of the contents of the operation of selection, the following immediately becomes evident: If 'φx_1' is true, then the pair

$$`\varphi x_1, \psi y_1\text{'} \tag{17}$$

simply means the sentence 'ψy_1', but if 'φx_1' is false, then the pair (17) has no meaning at all. The corresponding idea is expressed in the truth table IIc[18], which I derived here expressly for the operation of selection. Omitting the quotation marks, we obtain the following results for sentence sequences with the value 1:

$$\text{If } W(\varphi x_1) = 1, \text{ then } W(\varphi x_1, \psi y_1) = W(\psi y_1);$$
$$\text{If } W(\varphi x_1) = 0, \text{ then } W(\varphi x_1, \psi y_1) = \text{indeterminate.} \qquad (18)$$

The operation, then, has an individual meaning, but it is defective; i.e., its table of values contains indeterminacies.

The transition to the probability values for sentential sequences from the truth values of the individual operation is accomplished here analogously to that for elementary sentential connectives. The frequency interpretation requires that the probability of the sentential sequence is determined through enumeration of the truth values assumed by the relevant operation in the individual members of the sentential sequence; this applies likewise to the operation of selection. In this process the members with indeterminate truth values are omitted[19].

A further difference between the operation of selection and the elementary sentential connectives lies in the fact that the operation of selection is not commutative and not associative[20]. The pointlessness of considering expressions of the form

$$(\varphi x_i, \psi y_i) \vee (u_k)$$

is connected with this fact; such expressions are not capable of further dissolution. It is preferable to introduce the sign (ψz_k) into the new sentence sequence in accordance with (5) and transform it in accordance with (15).

7. A DIFFERENT INTERPRETATION OF THE OPERATION OF SELECTION

We have interpreted the operation of selection as an operation between sentential *sequences.* But we can give it another interpretation, according to which it represents an operation between sentential *functions.*

To recognize this, we must consider a peculiar difference between the predicative and the implicative notation. In the predicative notation

$$(\psi y_i) \qquad (19)$$

y_i is not a free variable but a sequence of individual signs, i.e., signs that are coordinated to certain individual events. The advantage of the implicative notation lies in its replacing the individual signs with a description carried out with the help of a free variable; that is, the sequence of events (x_i) is described with the help of another sentential function in which there then occurs a free variable. Thus the implicative notation

$$(\varphi x_i, \psi y_i) \tag{20}$$

may be conceived in the following way: The characteristic ψ is seen as relating to all those y_i belonging to x_i that make φx_i true. The sentential function φx_i, then, assumes the definition of those elements that are to be filled in as arguments in the sentential function ψ. In order to obtain an ordered series from these elements, we must add a prescription about order (for instance, that the y_i are to be taken in the temporal order in which they occur); the addition of such a prescription shall be indicated by the parenthesis in (20). In *Theory of Probability*, p. 396, I call the series of events (y_i) the *fundament* of the sentential sequence (ψy_i), the sentential sequence itself then being defined as the combination of a sentential function with a fundament. Using the given interpretation of (20), we may also say: A sentential sequence is equivalent to a pair of sentential functions plus an order prescription.

The resulting difference in the interpretation of the operation of selection is certainly not very profound. If we regard this operation sometimes as an operation between sentential *sequences,* sometimes as an operation between sentential *functions,* the sole difference involved is whether we introduce the order prescription *before* or *after* applying it. For we must make use of such a prescription in the second case, too, if we wish to set up the new sentential sequence.

In some respects, the implicative notation may appear more correct than the predicative notation, but we must bear in mind that the same may be said of individual assertions. In the sentence

$$\psi y_1 \tag{21}$$

'y_1' is a individual sign; we may instead characterize the argument by a description 'φx_i', writing, as before,

$$(\varphi x_i, \psi y_i). \tag{22}$$

This is an individual sentence if 'φx_i' is so chosen that there is only *one* x_i that makes 'φx_i' true. In that case the single sentence is also to be represented as a pair of sentential functions with the help of a free variable; then, of course, there is no order prescription. Then (22) has the significance of the individual operation of selection.

As an example of (21), let us take the sentence, 'Wolfgang Bolyai was an Hungarian.' Then (22) would have a form along the following lines:

y_i	"The man
x_i	whose son
φ	discovered non-Euclidean geometry in 1823
ψ	was an Hungarian."

Just as we may use either of the two notations for individual sentences, it will also be permissible to use either the implicative or the predicative notation for sentential sequences.

8. EXTENSIONALITY

I turn now to the accusation that my probability logic is not *extensional*[21]. The objection is based upon a definition of extensionality according to which it is present only if the truth value of a sentential connective is determined by the truth values of the two individual sentences. This characteristic is certainly obliterated in probability logic, for, as I have shown, the probability of a connection between sentential sequences depends upon yet a third parameter, viz., upon the probability of the sentential sequence determined by the operation of selection, which I also refer to as the 'degree of coupling'.

However, I do not believe it is proper to regard this peculiarity as a violation of the principle of extensionality. It appears far more to the point to expand the definition of extensionality so as to take account of the relations in question. For probability logic is also not *intensional*, as probability is obviously not dependent upon the *content* of the sentential sequences. What is accomplished in probability logic is a natural extension of those relations found in two-valued logic; instead of being determined by *two* parameters, the probability values of the connection requires *three* parameters. The additional parameter is of the same logical type as the other two; viz., it is also a probability value. It should be noted, too, that in place of the probability of the sentential sequence determined by the operation of selection, the probability of another connection may be used as the third independent parameter, for instance, the probability of the logical product; this determines instead the probability of the connective of selection.

In calling my probability logic extensional, I was, of course, thinking of extensionality in the broader sense. We are invariably compelled to expand concepts when we make the transition to logical systems of a more general nature. Simply consider the expansion of the concept of numerical equality necessitated by infinite sets or the expansion of the concept of a sum required for vector calculus. The circumstances that have promoted the requirement of extensionality to an important epistemological principle – especially the independence of extensional logic from the content of the sentences, the possibility of constructing a theory of meaning upon the concept of verifiability, and the like – are as strongly in force in the case of the more general concept of extensionality. And on the other hand, this discussion appears to me to illustrate how dangerous it is to restrict the development of theories by postulates that are too narrowly conceived. A tenacious insistence upon the narrow conception of extensionality would exclude precisely those of the multi-valued logics that may be conceived of as a logical interpretation of the concept of probability.

9. PROBABILITY LOGIC AS A GENERALIZATION OF TWO-VALUED LOGIC

My probability logic has been further criticized for failing to constitute a genuine generalization of logic in that, by employing the frequency interpretation, it reduces probability to an enumeration of two-valued sentences or assertions[22]. I gave a brief answer to this objection[23] by indicating the necessity of differentiating between the formal system of probability logic as defined through the truth tables and the interpretation through frequency. The latter results in a mapping of probability logic on two-valued logic, comparable to the mapping of a non-Euclidean geometry on Euclidean geometry. The formal system itself is untouched and is structurally different from two-valued logic. In the same place I indicated that a corresponding mapping on two-valued logic would also be possible for the logical systems of Lukasiewicz and Tarski, and that Post, for example, subjected his logic to such a frequency interpretation.

I have presented a more thorough investigation of this problem elsewhere[24], showing that a reduction to two-valued logic can be strictly executed only for *mathematical* probability calculus; in *applied* probability calculus it can only be approximated.

For this reason the physical concept of truth must, strictly speaking, be regarded merely as probability of a high degree. For the so-called *assertions* of physics are, as I indicated earlier, really *posits* to which a *weight* is attributed (I now use the word 'weight' in place of the word 'appraisal', which I used in my *Theory of Probability*); the weight is to be conceived as a probability applied to a single case. The logic of posits, or weight logic, replaces the two-valued logic of assertions; this weight logic is a probability logic made up not of sentential sequences but of posits analogous to individual assertions. Here, then, we are dealing with an interpretation of probability logic that does not represent a mapping upon two-valued logic.

J. Hosiasson[25] has now raised the more sweeping objection that not even the formal system of probability logic may be regarded as a generalization of two-valued logic. She bases this idea upon the fact that all tautologies of two-valued logic retain their tautological character in probability logic, that is, that the set T_o of all tautologies in two-valued logic is a subset of set T of all tautologies in probability logic. This being so, says Hosiasson, T may not be designated a generalization of T_o.

But I never asserted anything of the kind; so far as I can see, T actually becomes identical with T_o. In calling probability logic a generalization of two-valued logic, I meant by 'logic' more than the set of tautologies. The exhaustive characterization of a logic must include, in addition to the set of tautologies, the syntactic and semantic rules of logic, and it is here that we find the differences between probability logic and two-valued logic. This can be seen clearly from, e.g., the truth tables, which are to be construed as semantic rules. In my *Theory of Probability*, I showed that the truth tables of two-valued logic are included as a special case in the truth tables of probability logic. To further illustrate the point, let us take the following example, in which 'a' is to represent the constituent of the pertinent logic: a sentence in two-valued logic, a sentential sequence or a posit in probability logic (quotation marks omitted).

Two-valued logic

Object sentence $\quad a \vee \bar{a} \quad$ (23)

Semantic sentences $\quad W(a \vee \bar{a}) = 1 \quad$ (24)

$[W(a) = 1] \vee [W(\bar{a}) = 1] \quad$ (25)

Probability logic

Object sentence $\quad a \vee \bar{a} \quad$ (26)

Semantic sentences $W(a \vee \bar{a}) = 1$ (27)

$$W(a) + W(\bar{a}) = 1 \qquad (28)$$

Here (23) is a sentence that may be considered or asserted, while (26) may only be considered, in accordance with our previous discussion. Sentences (27) and (28) are subject to the rules of two-valued logic, as are, of course, (24) and (25). It is obvious that (25) would be *false,* taken as a semantic sentence of probability logic; it is replaced by (28). Sentence (28) is a generalization of (25); that is, (28) coincides with (25) in the special case in which the truth value can assume only the value 1 or the value 0.

At the same time, it is clearly impossible to consider the meaning of a formula independent of the attendant syntax. Formulas (23) and (26) are written in just the same way, but they do not have the same meaning. This is seen from the fact that, at higher language levels, formula (25) may be derived from the one, formula (28) from the other. Thus it will not do to define as 'logic' only the system of formulas of the first level, without regard to the attendant syntax.

Thus probability logic may clearly be correctly regarded as a generalization of two-valued logic. That the tautologies of two-valued logic are not disturbed by this arises from the fact that my system is adapted to the probability concept used in physics. It is only physical sentences, not tautologies, that physics wishes to regard as merely probable. Certainly, it is possible to think up other generalizations of logic in which not only the semantic rules but also the tautologies of two-valued logic are altered. But the construction of such systems will take on practical significance only if these systems are applicable to the concept of knowledge in physics.

NOTES

[1] I refer here to the presentation of my theory given in my book, *Wahrscheinlichkeitslehre* [1935h; revised by the author for English translation as [1949f]; page references to the English edition indicated, where appropriate, – Ed.] A more recent summary of my probability theory, including the distinction between the semantic and the object conception, appeared under the title, 'Les fondements logiques du calcul des probabilités' [1937b].

[2] Carnap, R., *Logische Syntax der Sprache,* Vienna, 1934 [trans. as *The Logical Syntax of Language,* London, 1937].

[3] Tarski, A., *Actes du Congres international de philosophie scientifique*, Paris, 1936, III, 1.

[4] Some authors use the terms 'relation' and 'argument' in the sense in which I here use 'relation sign' and 'argument sign', but I prefer this latter terminology.
[5] *Wahrscheinlichkeitslehre*, p. 57 [p. 46].
[6] *Wahrscheinlichkeitslehre*, p. 374 [p. 395].
[7] *Wahrscheinlichkeitslehre*, pp. 65 and 68 [pp. 54 and 57].
[8] Nagel, E., *Mind* **45**, 503 (1936).
[9] I have developed this idea in 'Axiomatik der Wahrscheinlichkeitsrechnung', [1932f] p. 572.
[10] Incidentally, such an individual probability implication would be intensional; general probability implication, on the other hand, is extensional, since – according to the frequency interpretation – it depends only upon the truth values of the individual sentences.
[11] *Wahrscheinlichkeitslehre*, p. 61 [p. 51].
[12] *Wahrscheinlichkeitslehre*, p. 374 [p. 395].
[13] Note that in place of (7) we could also introduce the notation

$$W(`(\varphi x_i)', `(\psi y_i)') .$$

Here the operation of selection is replaced by its semantic correlate; i.e., the comma signifies the semantic correlate of the operation of selection. Incidentally, this notation amounts to the same as (7). We prefer form (7) because we are able through it to treat the operation designated by the comma as an analogue of the elementary sentential connectives.
[14] This distinction stems from Boole (*An Investigation of the Laws of Thought*, (London, 1854), pp. 247–8), who also emphasized the equivalence of the two conceptions.
[15] Nagel, *op. cit.*, p. 510. Hosiasson, J., *Actes du Congrès international de philosophie scientifique* 4, 63 (1936).
[16] A precise enumeration replacing (13) and showing whether each constituent is true or false would offer yet more precise information than (12); but the resulting enumeration is not the sentence sequence '(φx_i)' but a different construct, to which the probability 1 is to be attributed.
[17] *Wahrscheinlichkeitslehre*, p. 379 [p. 398].
[18] *Wahrscheinlichkeitslehre*, p. 381 [p. 400].
[19] The peculiar characteristic of my probability theory (pointed out by C. Hempel), according to which the operation of selection leads, in the case of finite sentential sequences, to sentential sequences for which the number of truth values may be smaller than for the original sentential sequences, results from this feature, i.e., from the defective character of the operation.
[20] At least, if the first sequence is not 'dense', for then the presence of the parentheses signifies the transition to a new enumeration; cf. (5).
[21] Tarski, *Erkenntnis* 5, 174 (1935); Nagel, *loc. cit.*
[22] Tarski, *loc. cit.;* Nagel, *op. cit.*, 509–10.
[23] Tarski, *op. cit.*, p. 177.
[24] In *Experience and Prediction* [1938c], Section 36. Cf. also my article 'On Probability and Induction' in [1938a].
[25] Hosiasson, *op. cit.*, p. 58.

59. A LETTER TO BERTRAND RUSSELL

[1978–59]

456 Puerto del Mar
Pacific Palisades, California
March 28, 1949

Professor Bertrand Russell
Trinity College
Cambridge, England

Dear Mr. Russell:

I have read with great pleasure your book on Human Knowledge.* I like the book, in particular, because it presents philosophical analysis in a form accessible to non-experts and yet precise enough to constitute a basis for discussion among experts. May I congratulate you upon this successful presentation of your views!

I should like to write to you today principally about your views on probability and induction and your criticism of my theory. I like it that your criticism is always to the point and witty – it is so different from the antagonism springing from neurotic arrogance, typical of a certain group of modern logicians. In fact, I am reading your book at present with my graduate students in a seminar; we study your objections against my theory and have great fun in going through the various arguments.

The more I read your book, the more I think that a discussion of an hour or two between you and me would settle the dispute and reveal that you merely misunderstand my theory in some essential points. You would then see that your abandonment of empiricism is unnecessary and that you need not resort to an "extra-logical principle not based on experience." But I will rather not talk about this general problem and go into more details [instead].

There is first the question of interpreting my theory as referring to finite sequences. I have said repeatedly that this is the ultimate meaning of my theory and that the infinite sequence is used merely for mathematical convenience

* B. Russell, *Human Knowledge: Its Scope and Limits* (Simon and Schuster, New York, and Allen & Unwin, London, 1948).

(*Experience and Prediction* [1938c], pp. 360–362; *Wahrscheinlichkeitslehre* [1935h], p. 419; *Zeitschr. f. Physik* **93** [1935b], p. 792). I have always emphasized that my axioms are satisfied strictly even before going to the limit (*Wahrscheinlichkeitslehre*, p. 88). But it is also clear that such a 'finitization' does not excuse us from studying the order of the probability sequences, which is indispensable; in fact, my theory of probability includes a chapter on the theory of the order of probability sequences. The view that probability refers to classes, not to series, is one of the errors of the line of development expounded by Keynes.

My distinction between the frequency up to an element of the series and the limit, criticized by you on p. 364, remains valid, however, since it is only the last value of the frequency in a finite sequence which is decisive. Probability is always used as a property of the sequence as a whole, not as a property varying from element to element.

I will now answer the objection of the infinite regress raised by you. You use it in two ways: first, you say that going from a probability statement to the statement that it is probable, i.e., to probabilities of higher levels, there arises an infinity of levels. Second, you say that going conversely from a probability sequence to the analysis of the statements about elements, which are merely probable, there is [also] an infinity of levels.

As to the first form, there is a mathematical error in your presentation on p. 416. Combining the probabilities of different levels into one probability is permissible only if special conditions are satisfied; but even then, this combination cannot be done by mere multiplication. Let '*a*' be the statement: 'the probability of the event is $\frac{3}{4}$', and let '*b*' be the statement 'the probability of "*a*" is $\frac{1}{2}$'. In order now to find out what is the probability of the event, you have to know what is the probability of the event if '*a*' is false. This probability might be higher than $\frac{3}{4}$. And the final probability of the event is then found by a formula called by me the *rule of elimination*. It is a mean value, which may be greater than $\frac{3}{4}$. These values do not go to 0 (see *Wahrscheinlichkeitslehre*, pp. 316–317). The product which you calculate is the probability, not of the event, but of the total conjunction of the infinite number of propositions on all the levels, which is of course = 0.

I have said that we get away from the regress because we cut it off, on a certain level, by a blind posit. You say (pp. 415–416) "I cannot see what ground he can have for making one posit rather than another except that he thinks it more likely to be true." And on p. 413: "A blind posit is a decision to treat some proposition as true although we have no good ground for doing so." This is the decisive point in which you misunderstand my theory of induction. I have shown that blind posits are justified as a means to an end, and that no kind of belief in their truth is required. This I regard as the essential merit of my theory: I have shown that there are other reasons to make assertions than reasons based on belief. So I do have a good ground for dealing with the posit as if [it were] true; but the ground is derived neither from a probability, nor from a belief.

I therefore do not need such a thing as your 'credibility'. In fact, I think, whoever assumes that there is a credibility other than a probability interpretable as a frequency has committed himself to an error which makes an empiricist solution of the problem of knowledge impossible. The idea that there is such a thing as a 'rational belief' is the root of all evil in the theory of knowledge and is nothing but a remnant from rationalistic philosophies. I was so glad to see that you have made a great step toward the frequency interpretation of probability and recognize its merits – why do you still employ a concept of 'credibility,' which is unjustifiable and redundant?

For these reasons, my theory does not lead to an infinite regress. The scale of probabilities is cut off at some level by a posit asserted for other grounds than probabilities.

I will now discuss your objection that there is a similar regress in the converse direction (p. 368). It is true that the elements of statistics, which we regard as true and use as the basis of an inductive inference, are strictly speaking only probable. It is permissible to neglect this probability character for many purposes. But of course you can apply the frequency interpretation to these probabilities.

Going down the ladder in this direction, you arrive after a finite number of steps at those elements which you call 'sense data' and I call 'immediate things'. So your objection can be formulated: why do we base all statements about the world on sense data, or immediate things? Must we not first prove that statements about sense data are credible?

Here again I answer that the acceptance of the data is not a matter of credibility, but that they are selected as the basis of knowledge for other reasons. We choose them because they are the things that interest us – after all, my toothache and my sensations of pleasure are the things that matter to me. So once more, it is not credibility that determines the choice, but another reason (*Experience and Prediction*, pp. 289–290).

You might perhaps object: But I want to be sure that what I select as a basic datum is something of the kind that interests me; in other words, that it is a perception and not an invented thing. Asking that it be physically existing would of course be too much, but it should at least have subjective existence;

a *Protokoll* sentence should not be a lie (something that Neurath never understood). So you would insist that there is some credibility involved. I would answer as follows. If I say "This is a true *Protokoll* sentence of mine", such a statement is merely probable, and this probability can be translated into a frequency. In doing so, I would have to introduce other *Protokoll* sentences that are not yet checked. So I must always rely on some *Protokoll* sentences accepted without a test. These 'ultimate *Protokoll* sentences' are posited for the simple reason that I have nothing else to start with, that I simply cannot do otherwise. There is no ultimate *Protokoll* sentence in the sense that it cannot be tested; in fact, every *Protokoll* sentence can be tested. But there are always a number of last *Protokoll* sentences for a given context of inquiry. They are accepted because I cannot help doing it. I cannot even *ask* whether they are reliable – the question is meaningless. Such a question has meaning only if other *Protokoll* sentences have already been accepted, since only then is it verifiable; it is so by the use of inductive methods. (In a terminology to be explained presently, it is meaningful only in advanced knowledge.) The answer is similar to what must be said about analytic self-evidence: we cannot help accepting it, and we cannot even ask whether it is reliable unless we use analytic self-evidence for other sentences.

This argument cannot be extended to induction, because the inductive principle is not self-evident; in fact, while using it, we can very well imagine that it leads to false conclusions. This is the reason why I have introduced a very different argument for the use of induction. I shall now answer your criticism of my theory of induction.

First, you say (p. 413) that the inference that m/n goes to a limit is "only apparently more general" than the inference for $m = n$. It is true that to some extent the general inference is reducible to the special one, although a precise formulation would lead to complications (the improvement of the convergence for larger n is then cumbersome to formulate). But that does not make the inference less general. Riemannean geometry can be mapped on Euclidean geometry, but still is more general. Secondly, you would then have to formulate the special inference so as to state that the relative frequency will remain $= 1$ within an interval of exactness δ, which formulation includes, essentially, all the features of the general inference, because it admits frequencies different from 1.

On p. 417 you say that hypothetical induction does not differ from induction by enumeration because we can regard the class of observed consequences that conform to the hypothesis as a class B, and then infer 'all A are B' by simple enumeration. But this trivial way of transforming hypothetical induction into induction by enumeration does not supply an equivalent inference. The schema of the inference by hypothetical induction, which is given by the rule of Bayes and covered by my general theory of probability, is much more powerful than the derived inference by enumeration. If this schema 'collapses' into induction by enumeration, the degree of probability attained for the conclusion is much lower. The thesis that all induction is reducible to induction by enumeration, which has often been maintained, cannot be proved this way. It was proved for the first time in my axiomatic construction of the calculus of probability, which shows that the axioms are derivable from the frequency interpretation and that therefore the application of the calculus to physical reality is ultimately reducible to induction by enumeration.

On p. 369 you say that the inductive principle has no empirical content. You come to this view because you assume that the principle says: the inductive conclusion is probable. So what you discuss is not the meaning of induction, but that of a probability statement. But if probabilities are interpreted as frequencies in finite sequences, the probability statement is verifiable.

However, the inductive conclusion can be called probable only when many other inductions have already been made, which tell us something about the second level probabilities. I speak here of a state of *advanced knowledge*. In *primitive knowledge*, i.e., before any inductions were made, the inductive conclusion is not probable. To make this clear, let us assume that the event B has happened m times among n events; the inductive conclusion is now

$$\text{the probability of B in the sequence is } = \frac{m}{n} \pm \delta$$

This statement is not made probable by the inference. It is maintained as a blind posit, i.e., as a step within a procedure which on continuation will furnish the probability of B if such a probability exists. The rule of induction, in primitive knowledge, is an asymptotic rule, not supported by probabilities nor conferring probabilities on its conclusions.

Strangely enough, you say on p. 413 that the inductive posit can be shown to be false. You construct an example where an individual inductive conclusion is false. But thousands of such examples have been known for thousands of years! Who has ever said that the inductive conclusion *must* be true? Induction merely supplies an asymptotic rule which eventually must come true if there is a limit of the frequency. And this is true for your example, too. Suppose you define your class B as the class of all things except the elements of the sequence beyond n. Then all elements up to n are in B, and the inductive conclusion would be: the next element is in B. Further observation would show this conclusion to be false. Continuing the procedure, you would soon find that the relative number of elements in B gets smaller and smaller, and the inductive conclusion would furnish for an element B a probability that converges to 0. So the asymptotic rule works correctly. Of course, in your example no one would apply the rule because you have a deductive proof that the element $n + 1$, or all elements beyond n, are not in B. Your inference violates the rule: use the narrowest reference class available. You will find a similar example, but of a more general form, in the forthcoming English edition of *Wahrscheinlichkeitslehre* [1949f], § 87.

Induction does not require an intensional logic, as you say on p. 415. And the inductive rule is not too general. That our actual inductions are directed by many other considerations, such as [those] concerning likely connections or suitable variations of instances, that we are even able to tell when a number of observations is large enough to warrant an induction, is a consequence of the fact that all such inductions are made in advanced knowledge. The current literature on induction is abundant with errors resulting from not distinguishing between these two states of knowledge. Once many inductions have been made, the inductive inference becomes much better through a concatenation of inductions. But these 'better' properties cannot be required for primitive knowledge, nor can they be proved *a priori*.

One more remark: I was surprised to find myself hyphenated to von Mises (p. 362) – as much surprised, presumably, as he. You even call my theory a development of that of von Mises. I do not think this is a correct statement. My first publication on probability [1915b], which is earlier than Mises' publications, has already a frequency interpretation and a criticism of the principle of indifference, although I later abandoned the Kantian frame of this paper (see *Wahrscheinlichkeitslehre* p. 342, footnote). Mises' merit is to have shown that the strict-limit interpretation does not lead to contradictions and, further, to have provided a means for the characterization of random sequences. I then could show that my earlier frequency interpretation (which was weaker than a strict-limit interpretation) in combination with Bernoulli's theorem leads to the limit interpretation and thus took over this interpretation. But my mathematical theory is more comprehensive than Mises' theory, since it is not restricted to random sequences; furthermore, Mises does not connect his theory with the logical symbolism. And Mises has never had a theory of induction or of application of his theory to physical reality.

I have answered those of your criticisms which appeared to me the important ones. I wish I could send you the English edition of my book *Wahrscheinlichkeitslehre*, which is now ready in the proofs. I think I have improved the presentation of my views, so that in fact most of your objections are answered in it, although I wrote this edition long before your book came out. When the book appears (it is scheduled for May) I will send you a copy.

Incidentally, when my book on symbolic logic [1947c] came out, about a year ago, I asked my publisher to send you a copy. Did you get it? If not, I will gladly send you one.

Recently I got an invitation from Argentina for their congress, and I saw that you, too, were invited. I hoped to meet you there. But then we came to the decision not to attend the congress in order to demonstrate against the dictatorship in that country. What did you decide?

It is a great pity that we cannot talk about induction together! The distances are so large. I had plans to go to Europe last summer in combination with the congress at Amsterdam, but finally had to renounce them because of the expenses of the trip.

Please give my best regards to Mrs. Russell and to Conrad! We all miss you very much here and would be so happy once again to have you among us!

Yours very cordially,

Hans Reichenbach

BIBLIOGRAPHY OF WRITINGS OF HANS REICHENBACH

1912: (a) 'Die Studienreise englischer Studenten nach Deutschland', *Münchner Akademischer Rundschau* **6**, no. 2: 27 (Oct. 24, 1912)

(b) 'Die Gleichberechtigung der Reifezeugnisse höherer Lehransalten', *Münchner Akademische Rundschau* **6**, no. 3: 46 (Nov. 9, 1912)

(c) 'Studentenschaft und Katholizismus', *Das Monistische Jahrhundert*, W. Ostwald, ed., No. 16: 533–538 (Nov., 1912) (For English trans. see [1978 a-2])

(d) 'Die Neuorganisierung der Münchner Freistudentenschaft', *Dresdner Studentische Blätter*, no. 6: 1–4 (Nov. 20, 1912)

(e) 'Student und Schulproblem', *Münchner Akademische Rundschau* **6**, no. 6: 94–97 (Dec. 16, 1912)

(f) 'Der Student', *Münchner Studentisches Taschenbuch* (Munich: Max Steinebach, Wintersemester, 1912–13) pp. 42–44 (For English trans. see [1978 a-1])

1913: (a) 'Der Student und die pädagogischen Bestrebungen der Gegenwart', *Münchner Akademische Rundschau* **6**, no. 10: 178–179 (March 5, 1913)

(b) 'Die Militarisierung der deutschen Jugend. I. Der Tatbestund', *Die Freie Schulgemeinde* **3**, no. 4: 97–110 (July 1913)

(c) (*with Carl Landauer*) 'Die freistudentische Idee. Ihr Inhalt als Einheit', *Freistudententum: Versuch einer Synthese der freistudentischen Ideen*, by Hermann Kranold, Carl Landauer and Hans Reichenbach (Max Steinebach, Munich, 1913) pp. 25–40 (For English trans. see [1978 a-3])

(d) 'Der Wandervogel und die Juden', Beilage des *Berliner Börsen-Courier* no. 539 (Nov. 16, 1913)

(e) 'Tagebuch', *Der Student* [Neue Folge der *Berliner Freistudentischen Blätter*] **6**, no. 6: 80–81 (Nov. 28, 1913)

(f) 'Freischar oder Freistudentschaft', *Der Student* [Neue Folge der *Berliner Freistudentischen Blätter*] **6**, no. 7: 88–89 (Dec. 5, 1913)

(g) 'Warum treiben wir Körperkultur?, *Berliner Freistudentischen Blätter*, no. 3: 1–4 (1913) (For English trans. see [1978 a-4])

(h) 'Verhandlungen' (Zweiter Deutscher Kongress für Jugendbildung und Jugendkunde, Munich, Oct. 3–5, 1912) *Arbeiten des Bundes für Schulreform* no. 6 (B. G. Teubner, Leipzig and Berlin, 1913) pp. 165–166

1914: (a) 'Militarismus und Jugend', *Die Tat* 5, no. 12: 1234–1238 (March 1914)

(b) 'Von der Georgia Augusta. Die Hochschule', *Göttinger Akademische Wochenschau* **10**: 21 (1914)

(c) 'Der Sinn der Hochschulreform', *Studentenschaft und Jugendbewegung* (Max Steinebach, Munich, 1914) pp. 7–11 (For English trans. see [1978 a-5])

(d) 'Die Jungdeutschlandbewegung. Die Jugendbewegung der Gegenwart und

ihre Bedeutung für die Hochschule', *Studentenschaft und Jugendbewegung* (Max Steinebach, Munich, 1914) pp. 12–33
(e) 'Zum Lietzschen Vortragsabend', *Göttinger Akademische Wochenschau* **10** no. 5: 38 (June 12, 1914)
(f) 'Jugendbewegung und Freie Studentenschaft', *Münchner Akademische Rundschau*, no. 13: 224 (Nov. 13, 1914)
(g) 'Zwei Fichtefeiern', *Münchner Akademische Rundschau*, no. 4: 29 (Dec. 21, 1914)

1915: (a) 'Hans Wegener, *Wir jungen Männer*', *Die Tat* **6**, no. 2: 218–220 (May 1915)
(b) *Der Begriff der Wahrscheinlichkeit für die mathematische Darstellung der Wirklichkeit*. Inaugural Dissertation, University of Erlangen (Barth, Leipzig, 1915) 79 pp.

1916: (a) 'Der Begriff der Wahrscheinlichkeit für die mathematische Darstellung der Wirklichkeit', *Zeitschrift für Philosophie und Philosophische Kritik* **161**, 210–239 (1916); **162**, 9–112, 223–253 (1916). Reprint of Reichenbach's dissertation, [1915b].

1918: (a) 'Die Sozialisierung der Hochschule', unpublished manuscript, 1918 (For English trans. see [1978 a-7])
(b) 'Programm der sozialistischen Studentenpartei', (orig. source and place of publication have not been established) (For English trans. sec [1978 a-6])
(c) 'Bericht der sozialistischen Studentenpartei Berlin' and 'Erläuterungen zum Programm', Winter semester 1918/19 (orig. source and place of publication have not been established; published in Staatsarchiv München, Staatsanwaltschaft München I, no. 2973, vol. 2 (For English trans. see [1978 a-8])

1919: (a) 'Der Begriff der Wahrscheinlichkeit für die mathematische Darstellung der Wirklichkeit', Autoreferat. *Die Naturwissenschaften* **7**, no. 27: 482–483 (1919). Reichenbach's summary of his dissertation, [1915b].
(b) 'Student und Sozialismus', *Der Aufbau*, no. 5 [Flugblätter der Jugend, Berlin]
(c) 'Sterzinger, *Zur Psychologie und Naturphilosophie der Geschicklichkeitsspiele*' (Review), *Naturwiss.* **7**, 644 (1919).

1920: (a) 'Die Einsteinsche Raumlehre', *Die Umschau* **24**, no. 25: 402–405 (1920)
(b) Über die physikalischen Voraussetzungen der Wahrscheinlichkeitsrechnung', *Zeitschrift für Physik* **2**, no. 2: 150–171 (1920). See also [1921a]
(c) 'Die physikalischen Voraussetzungen der Wahrscheinlichkeitsrechnung', *Die Naturwissenschaften* **8**, no. 3: 46–55 (1920) (For English trans. see [1978 b–52a])
(d) 'Nachtrag', *Die Naturwissenschaften* **8**, no. 19: 349 (1920) (For English trans. see [1978 b–52b])
(e) 'Philosophische Kritik der Wahrscheinlichkeitsrechnung', *Die Naturwissenschaften* **8**, no. 8: 146–153 (1920) (For English trans. see [1978 b-53])
(f) *Relativitätstheorie und Erkenntnis Apriori* (Berlin: Springer, 1920) (For English trans. see [1965a])

1921: (a) 'Nachtrag' (Suppl. to [1920b]), *Zeitschrift für Physik* **4**, no. 3: 448–450 (1921)
(b) 'Fürth, Reinhold, *Schwankungserscheinungen in der Physik*' (review), *Die Naturwissenschaften* **9**, no. 7: 111 (Feb. 18, 1921)

(c) 'Erwiderung auf H. Dinglers Kritik an der Relativitätstheorie', *Physikalische Zeitschrift* **22**, 379–380 (1921)

(d) 'Bericht über eine Axiomatik der Einsteinschen Raum-Zeit Lehre' (Address at the German Congress of Physics, Jena 1921), *Physikalische Zeitschrift* **22**, 683–687 (1921)

(e) 'Erwiderung auf Herrn Theodor Wulfs Einwände gegen die allgemeine Relativitätstheorie', *Astronomische Nachrichten* **213**, 307–310 (1921)

(f) 'Die Einsteinsche Bewegungslehre', *Die Umschau* (Frankfurt) **25**, no. 35: 501–505 (1921)

(g) 'Entgegnung' (answer to Oskar Kraus), *Die Umschau* (Frankfurt) **25**, no. 46: 684–685 (1921)

1922: (a) 'Der Nobelpreis für Einstein', *Neue Zürcher Zeitung* (Nov. 22, 1922) (For English trans. see [1978 a-9])

(b) 'Die Relativitätstheorie in der Streichholzschachtel', *Neue Zürcher Zeitung* (April 29, 1922) (For English trans. see [1978 a-10])

(c) 'La signification philosophique de la théorie de la relativité', *Revue Philosophique de la France et de l'Étranger*, **94**, 5–61 (1922)

(d) 'Relativitätstheorie und absolute Transportzeit', *Zeitschrift für Physik* **9**, nos. 1–2: 111–117 (1922)

(e) 'Erwiderung auf Herrn Andersons Einwände gegen die allgemeine Relativitätstheorie', *Astronomische Nachrichten* **213**, 373–376 (1922)

(f) 'Der gegenwärtige Stand der Relativitätsdiskussion', *Logos* **10**, no. 3: 316–378 (1922) (For English trans. see [1959a]; reprinted in [1978 b-44])

1923: (a) 'Müller, Aloys, *Die Philosophischen Probleme der Einsteinschen Relativatätstheorie*' (Review), *Die Naturwissenschaften* **11**, no. 2: 30–31 (Jan. 12, 1923)

1924: (a) 'Eine neue Erfindung in der astronomischen Messtechnik', *Neue Zürcher Zeitung* (Aug. 6, 1924)

(b) 'Radiotechnik und Kultur', *Radio–Umschau* (Frankfurt), no. 14 (May 18, 1924)

(c) *Was Ist Radio*?, vol. 1 of the series *Die Radio-Reihe* (Berlin: Richard C. Schmidt, and Stuttgart: Verlag die Zeit-Max Kahn (paperbound), 1924; 2nd Schmidt ed. 1929)

(d) 'Die Bewegungslehre bei Newton, Leibniz und Huyghens', *Kantstudien* **29**, 416–438 (1924) (For English trans. see [1959a]; reprinted in [1978 b-45])

(e) 'Entgegnung' (answer to O. Brühlmann), *Annalen der Philosophie* **4**, nos. 4–5: 195–198 (1924)

(f) 'Die relativistische Zeitlehre', *Scientia*, 361–374 (Dec. 1924) (For English trans. see [1978 b-46])

(g) Articles in *Physikalisches Handwörterbuch*, Berliner-Scheel, eds. (Springer, Berlin, 1924):
Bernouillisches Theorem;
Diskret;
Erwartung, mathematische;
Ergodenhypothese;
Fehlertheorie;

Gesetz der grossen Zahlen;
Intramolekularbewegung;
Liouvillescher Satz;
Loschmidtsche Zahl pro Mol:
Maxwellsche Geschwindigkeitsverteilung;
Mittel;
Molare Unordnung;
Statistisches Gleichgewicht;
Stirlingsche Formel;
Stosszahlansatz;
Umkehreinwand;
Unabhängige Ereignisse;
Wahrscheinlichkeit;
Wahrscheinlichkeitsnachwirkung;
Wahrscheinlichkeitsrechnung;
Zufall.

(h) *Axiomatik der relativistischen Raum-Zeit-Lehre* [Die Wissenschaft, vol. 72] (Vieweg, Braunschweig, 1924; reprinted Braunschweig, 1965) (For English trans. see [1969a])

1925: (a) 'Metaphysik und Naturwissenschaft', *Symposion* **1**, no. 2: 158–176 (1925) (For English trans. see [1978 a-31])

(b) 'Planetenuhr und Einsteinsche Gleichzeitigkeit', *Zeitschrift für Physik* **33**, no. 8: 628–634 (1925)

(c) 'Über die physikalischen Konsequenzen der relativistischen Axiomatik', *Zeitschrift für Physik* **34**, no. 1: 32–48 (1925) Supplement to [1924-h]

(d) 'Die Kausalstruktur der Welt und der Unterschied zwischen Vergangenheit und Zukunft', *Sitzungsberichte*, *Bayerische Akademie der Wissenschaft* (Nov. 1925) pp. 133–175 (For English trans. see [1978 b-47])

(e) 'Wahrscheinlichkeitsgesetze und Kausalgesetze', *Die Umschau* (Frankfurt) **29**, no. 40: 789–792 (1925)

1926: (a) 'Tycho Brahes Sextanten', *Hamburger Fremdenblatt* (Dec. 18, 1926) (For English trans. see [1978 a-11])

(b) 'Ist die Relativitätstheorie widerlegt?', *Die Umschau* (Frankfurt) **30**, no. 17 (April 24, 1926)

(c) 'Die Probleme der modernen Physik', *Die Neue Rundschau*, 414–425 (April 1926)

(d) 'Die Auswirkung der Einsteinschen Lehre', *Kunstwart*, 35–39 (Oct. 1926) (For English trans. see [1978 a-12])

(e) 'Ein offener Brief an die Funkstunde A.G., Berlin', *Radio-Umschau* (Frankfurt) **3**, no. 4 (Jan. 24, 1926) (For English trans. see [1978 a-13])

(f) 'Erwiderung auf eine Veröffentlichung von Herrn Hj. Mellin', *Zeitschrift für Physik* **39**, nos. 2–3: 106–112 (1926)

1927: (a) 'Grundsteinlegung für das Haus der Chemie. Marcellin Berthelots Werk', *Berliner Tageblatt* (Oct. 22, 1927) (For English trans. see [1978 a-14])

(b) 'Erinnerungen an Svante Arrhenius', *Berliner Tageblatt* (Oct. 5, 1927) (For English trans. see [1978 a-15])

(c) 'Ein neues Atommodell', *Die Umschau* (Frankfurt) **31**, no. 15 (April 9, 1927) (For English trans. see [1978 a-16])

(d) 'Die Umgestaltung des naturwissenschaftlichen Weltbildes: Atomtheorie-Relativitätstheorie', *Exakte Naturwissenschaften* [Sammelband der Lessinghochschule] **2**, no. 17: 247–272 (1927)

(e) 'Lichtgeschwindigkeit und Gleichzeitigkeit', *Annalen der Philosophie* **6**, no. 4: 128–144 (1927)

(f) *Von Kopernikus bis Einstein* (Ullstein, Berlin, 1927) (Czech trans. Prague, 1928; for English trans. see [1942a])

1928: (a) 'Zum Tode von H. A. Lorentz', *Berliner Tageblatt* (Feb. 6, 1928) (For English trans. see [1978 a-17])

(b) 'Denker der Zeit-Bertrand Russell', *Vossische Zeitung* (Feb. 12, 1928) (Reichenbach's contribution to a series of articles entitled 'Great Thinkers of our Time'; for English trans. see [1967a])

(c) 'Philosophie der Naturwissenschaften', *Vossische Zeitung* (Jan. 3, 1928) (For English trans. see [1978 a-18])

(d) 'Raum und Zeit: Von Kant zu Einstein', *Vossische Zeitung* (Mar. 4, 1928) (For English trans. see [1978 a-19])

(e) 'Kausalität oder Wahrscheinlichkeit?', *Vossische Zeitung* (July 18, 1928) (For English trans. see [1978 a-20])

(f) 'Die Weltanschauung der exakten Wissenschaften', *Die Böttcherstrasse* (Bremen), 44–46 (Nov. 1928) (For English trans. see [1978 a-21])

(g) 'Wandlungen im physikalischen Welbild', *Zeitschrift für angewandte Chemie* **41**, no. 14: 347–352 (1928)

(h) *Philosophie der Raum-Zeit-Lehre* (Walter de Gruyter, Berlin and Leipzig, 1928) 330 pp. (For English trans. see [1958a]; Italian trans., Milan, 1977)

1929: (a) 'Einstein's neue Theorie', *Vossische Zeitung* (Jan. 25, 1929) (For English trans. see [1978 a-26])

(b) 'Neue Wege der Wissenschaft. Physikalische Forschung', *Vossische Zeitung* (March 31, 1929) (For English trans. see [1978 a-22])

(c) 'Neue Wege der Wissenschaft. Philosophische Forschung', *Vossische Zeitung* (June 16, 1929) (For English trans. see [1978 a-23])

(d) 'Neue Wege der Wissenschaft. Mathematische Forschung', *Vossische Zeitung* (Aug. 18, 1929) (For English trans. see [1978 a-24])

(e) 'Die neue Naturphilosophie', *Deutsche Allgemeine Zeitung* (Oct. 13, 1929) (For English trans. see [1978 a-25])

(f) 'Crise de la Causalité', *Documents* (Paris), 105–108 (May 1929)

(g) 'Ziele und Wege der physikalischen Erkenntnis', *Handbuch der Physik*, vol. 4: *Allgemeine Grundlagen der Physik* (Springer, Berlin, 1929) pp. 1–80 (For English trans. see [1978 b-48])

(h) 'Neuere Forschungsergebnisse in der Naturphilosophie', *Forschungen und Fortschritte* [Nachrichtenblatt der deutschen Wissenschaft und Technik] vol. 5, no. 16: 185 (1929)

(i) 'Bertrand Russell', *Obelisk Almanach* (Drei-Masken Verlag, Berlin and Munich, 1929) pp. 82–92 (For English trans. see [1978 a-32])

(j) 'Zur Einordnung des neuen Einsteinschen Ansatzes über Gravitation und Elektrizität', *Zeitschrift für Physik* **59**, nos. 9–10: 683–689 (1929)

(k) 'Die neue Theorie Einsteins über die Verschmelzung von Gravitation und Elektrizität', *Zeitschrift für angewandte Chemie* **42**, no. 5: 121–123 (1929)

(l) 'Stetige Wahrscheinlichkeitsfolgen', *Zeitschrift für Physik* **53**, nos. 3–4: 274–307 (1929)

(m) 'Das Kausalproblem in der gegenwärtigen Physik', *Zeitschrift für angewandte Chemie* **42**, no. 19: 457–459 (1929)

1930: (a) 'Johannes Kepler. Zur dreihundertsten Wiederkehr seines Todestages', *Die Woche*, 1329–1330 (Nov. 1930) (For English trans. see [1978 a-27])

(b) 'Der heutige Stand der Wissenschaften. Die exakten Naturwissenschaften', *Die Literarische Welt*, no. 38: 3 (1930) (For English trans. see [1978 a-28])

(c) 'Probleme und Denkweisen der gegenwärtigen Physik', *Deutsche Rundschau*, 37–44; 131–141 (July–Aug. 1930)

(d) 'Zur Einführung', Introduction to *Erkenntnis* **1**, no. 1: 1–3 (Felix Meiner, Leipzig, 1930)

(e) 'Die philosophische Bedeutung der modernen Physik', *Erkenntnis* **1**, no. 1: 49–71 (1930) (For English trans. see [1978 a-33])

(f) 'Tagung für Erkenntnislehre der exakten Wissenschaften in Königsberg', *Die Naturwissenschaften* **18**, no. 50: 1093–1094 (1930) (For English trans. see [1978 a-34])

(g) 'Kausalität und Wahrscheinlichkeit', *Erkenntnis* **1**, nos. 2–4: 158–188 (1930) (For English trans. of Part III see [1959a]; reprinted in [1978 b-55])

(h) *Atom und Kosmos. Das Physikalische Weltbild der Gegenwart* (Berlin: Deutsche Buch-Gemeinschaft, 1930) (For English trans. see [1932h]; Spanish trans. Madrid, 1931; French trans. Paris, 1934; Hungarian trans. Budapest, 1937)

1931: (a) 'Hundert gegen Einstein', *Vossische Zeitung* (Feb. 24, 1931) (For English trans. see [1978 a-29])

(b) 'Naturwissenschaft und Philosophie', *Frankfurter Zeitung* (April 29, 1931)

(c) 'Schlussbemerkung (zur Diskussion v. Aster-Vogel-Dingler)', *Erkenntnis* **2**, 39–41 (1931)

(d) 'Heinrich Scholz, *Geschichte der Logik*', (Review) *Erkenntnis* **2**, 471–472 (1931) (For English trans. see [1978 a-37])

(e) 'Bernhard Bavink, *Ergebnisse und Probleme der Naturwissenschaften*' (Review), *Erkenntnis* **2**, 468–471 (1931)

(f) 'Zum Anschaulichkeitsproblem der Geometrie', *Erkenntnis* **2**, no. 1: 61–72 (1931)

(g) 'Der physikalische Wahrheitsbegriff', *Erkenntnis* **2**, nos. 2–3: 156–171 (1931) (For English trans. see [1978 a-36])

(h) 'Bemerkungen zum Wahrscheinlichkeitsproblem', *Erkenntnis* **2**, nos 5–6: 365–368 (1931)

(i) 'Das Kausalproblem in der Physik', *Die Naturwissenschaften* **19**, no. 34: 713–722 (1931) (For English trans. see [1978 a-35])

(j) *Ziele und Wege der heutigen Naturphilosophie* (Leipzig: Felix Meiner, 1931) 64 pp. (For English trans. see [1959a], reprinted in [1978 a-38]; French trans. Hermann, Paris, 1932)

1932: (a) 'Ist der menschliche Geist wandelbar?', *Die Woche* (Berlin) **34**, no. 2: 39–40 (Jan. 9, 1932) (For English trans. see [1978 a-30])
(b) 'Der endliche Weltenraum', *Die Woche* (Berlin) **34**, no. 38 (Sept. 17, 1932)
(c) 'Bemerkung (Karl Popper, 'Ein Kriterium des empirischen Charakters theoretischer Systeme')', *Erkenntnis* **3**, 427–428 (1932/33)
(d) 'Die Kausalbehauptung und die Möglichkeit ihrer empirischen Nachprüfung', *Erkenntnis* **3**, no. 1: 32–64 (1932) (For English trans. see [1959a], reprinted in [1978 b-56])
'Schlussbemerkung', pp. 71–72 (included in [1978 b-56] reprinting above)
(e) 'Kausalität und Wahrscheinlichkeit in der Biologie', *Klinische Wochenschrift* **2**, no. 6: 251–253 (1932)
(f) 'Axiomatik der Wahrscheinlichkeitsrechnung', *Mathematische Zeitschrift* **34**, no. 4: 568–619 (1932)
(g) 'Wahrscheinlichkeitslogik', *Sitzungsberichte, Preussische Akademie der Wissenschaften, Phys.-Math. Klasse* **29**, 476–490 (1932)
(h) *Atom and Cosmos. The World of Modern Physics*, revised and updated by H.R., trans. by Edward S. Allen (George Allen & Unwin, London, 1932; MacMillan & Co., New York, 1933; Braziller, New York, 1957) (English trans. of [1930h]; Chap. 18 reprinted in *Basic Problems in Philosophy. Selected Readings*, D. J. Bronstein, Y. H. Krikorian, and P. P. Wiener, eds., Prentice-Hall, New York, 1947, pp. 343–353; 2nd ed. 1955, pp. 270–276)
(i) 'Kant und die moderne Naturwissenschaft. Naturwissenschaftliche Berichte', *Frankfurter Zeitung* **77**, nos. 626–627: 2–3 (Aug. 23, 1932)

1933: (a) 'Vom Bau der Welt', *Die Neue Rundschau*, 39–60; 235–250 (July–August 1933)
(b) 'Kant und die Naturwissenschaft', *Die Naturwissenschaften* **21**, no. 33: 601–606; no. 34: 624–626 (1933) (For English trans. see [1978 a-39])
(c) 'Rudolf Carnap, *Der logische Aufbau der Welt*', (Review) *Kantstudien* **38**, 199–201 (1933) (For English trans. see [1978 a-40])
(d) 'Kausalität und Wahrscheinlichkeit in der gegenwärtigen Physik', *Unterrichtsblätter für Mathematik und Naturwissenschaften* **39**, no. 3: 65–69 (1933)
(e) 'Die logischen Grundlagen des Wahrscheinlichkeitsbegriffs', *Erkenntnis* **3**, nos. 4–6: 401–425 (1933) (For English trans. see [1949e])

1934: (a) 'Friedrich Schilling, *Projektive und nichteuklidische Geometrie*' (Review), *Erkenntnis* **4**, 378 (1934)
(b) 'In eigener Sache', *Erkenntnis* **4**, 75–78 (1934)
(c) 'Wahrscheinlichkeitslogik', *Erkenntnis* **5**, nos. 1–3: 37–43 (1934)
(d) 'Sur les fondements logiques de la probabilité', *Recherches philosophiques* **4**, 361–370 (1934/35)

1935: (a) 'Metaphysik bei Jordan?', *Erkenntnis* **5**, 178–179 (1935)
(b) 'Bemerkung zu H. Blumes finiter Wahrscheinlichkeitsrechnung', *Zeitschrift für Physik* **93**, 792–794 (1935)
(c) 'Zur Induktionsmaschine', *Erkenntnis* **5**, 172–173 (1935)
(d) 'Bemerkungen zu Carl Hempels Versuch einer finitistischen Deutung des Wahrscheinlichkeitsbegriffs', *Erkenntnis* **5**, no. 4: 261–266 (1935)

(e) 'Über Induktion und Wahrscheinlichkeit. Bemerkungen zu Karl Poppers *Logik der Forschung*', *Erkenntnis* **5**, no. 4: 267–284 (1935) (For English trans. see [1978 b-57])
(f) 'Bemerkungen zu Karl Marbes statistischen Untersuchungen zur Wahrscheinlichkeitsrechnung', *Erkenntnis* **5**, no. 5: 305–322 (1935)
(g) 'Wahrscheinlichkeitslogik und Alternativlogik', *Einheit der Wissenschaft* (Prager Vorkonferenz der Internationalen Kongresse für Einheit der Wissenschaft, Leipzig, 1935)
(h) *Wahrscheinlichkeitslehre. Eine Untersuchung über die Logischen und Mathematischen Grundlagen der Wahrscheinlichkeitsrechnung* (A. W. Sijthoff's Uitgeversmaatschappij, Leyden, 1935) 451 pp. (For English trans. see [1949f]; French trans. of 8 sections by H. Savonnet, Hermann, Paris, (1939) For 2nd German ed., see [1977a].

1936: (a) 'Ansprache bei der Begrüssungssitzung des Pariser Kongresses', *Actes du Congrès International de Philosophie Scientifique*, Paris 1935, vol. 1 (Hermann, Paris, 1936) pp. 16–18
(b) 'L'Empirisme logistique et la désagrégation de l'*a priori*', *Actes du Congrès International de Philosophie Scientifique*, Paris 1935, vol. 1 (Hermann, Paris, 1936) pp. 28–35
(c) 'Die Induktion als Methode der wissenschaftlichen Erkenntnis', *Actes du Congrès International de Philosophie Scientifique*, Paris 1935, vol. 4 (Hermann, Paris, 1936) pp. 1–7
(d) 'Wahrscheinlichkeitslogik als Form des wissenschaftlichen Denkens', *Actes du Congrès International de Philosophie Scientifique*, Paris 1935, vol. 4 (Hermann, Paris, 1936) pp. 24–30
(e) 'Logistic Empiricism in Germany and the Present State of its Problems', *The Journal of Philosophy* **33**, no. 6: 141–160 (March 12, 1936)
(f) 'Moritz Schlick†', *Erkenntnis* **6**, 141–142 (1936)
(g) 'Clemens Thaer, *Die Elemente von Euklid*', (Review) *Erkenntnis* **6**, 71–72 (1936)
(h) 'Die Bedeutung des Wahrscheinlichkeitsbegriffes für die Erkenntnis', *Actes du huitième Congrès International de Philosophie*, Prague, Sept. 2–7, 1934 (1936) pp. 163–169
(i) 'Warum ist die Anwendung der Induktionsregel für uns notwendige Bedingung von Voraussagen?', *Erkenntnis* **6**, no. 1: 32–40 (1936)
(j) 'Induction and Probability' (Discussion of H. Feigl), *Philosophy of Science* **3**, 124–126 (1936)

1937: (a) 'La Philosophie scientifique: une esquisse de ses traits principaux', *Travaux du IX Congrès International de Philosophie*, Paris 1937, vol. 4 (Paris, 1937) pp. 86–91
(b) 'Les fondements logiques du calcul des probabilités, *Annales de l'Institut Henri Poincaré* (Paris) vol. 7, part 5: 267–348 (1937)
(c) 'Causalité et induction', *Bulletin de la Société française de Philosophie* **37**, no. 4: 127–159 (1937)

1938: (a) 'On Probability and Induction', *Philosophy of Science* **5**, no. 1: 21–45 (1938)
(b) 'Reply to Everett J. Nelson's Criticism', *The Journal of Philosophy* **35**, no. 5: 127–130 (1938)

(c) *Experience and Prediction. An Analysis of the Foundations and the Structure of Knowledge* (Univ. of Chicago Press, Chicago, 1938; 1st Phoenix paper edition, 1961, Midway paper reprint, 1976) 410 pp. For German trans. see [1977a], vol. 4

1939: (a) 'Dewey's Theory of Science', in *The Philosophy of John Dewey*, P. A. Schilpp, ed. (The Library of Living Philosophers, Evanston, Ill., vol. 1, 1939) pp. 159–192.

(b) 'Über die semantische und die Objektauffassung von Wahrscheinlichkeitsausdrücken', *The Journal of Unified Science* (*Erkenntnis*) **8**, 50–68 (1939) (For English trans. see [1978 b-58])

(c) 'Bemerkungen zur Hypothesenwahrscheinlichkeit', *The Journal of Unified Science* (*Erkenntnis*) **8**, no. 4: 256–260 (1939)

1940: (a) 'On the Justification of Induction', *The Journal of Philosophy* **37**, no. 4: 97–103 (1939); reprinted in *Readings in Philosophical Analysis*, H. Feigl and W. Sellars, eds. (Appleton-Century-Crofts, New York, 1949) pp. 324–329.

(b) 'On Meaning', *The Journal of Unified Science* (*Erkenntnis*) **9**, 134–135 (1940)

1941: (a) 'Note on Probability Implication', *Bulletin of the American Mathematical Society* **47**, no. 4: 265–267 (1941)

1942: (a) *From Copernicus to Einstein*, tr. by R. B. Winn (Philosophical Library, New York, 1942) 123 pp. (English trans. of [1927f] with paperback ed., 1957)

1944: (a) 'Bertrand Russell's Logic', in *The Philosophy of Bertrand Russell*, P. A. Schilpp, ed. (The Library of Living Philosophers, Evanston, Ill., vol. 5, 1944) pp. 21–54

(b) *Philosophic Foundations of Quantum Mechanics* (Univ. of California Press, Berkeley and Los Angeles, 1944; paper edition, 1965) 182 pp.; German trans. Basel, 1949 and [1977a], vol. 5; Italian trans. Torino, 1954. §§ 29–37 reprinted as 'Three-valued Logic and the Interpretation of Quantum Mechanics' in *The Logico-Algebraic Approach to Quantum Mechanics*, vol. I: *Historical Evolution*, ed. C. A. Hooker, (Reidel, Dordrecht and Boston, 1975) pp. 53–97.

1945: (a) 'Reply to Donald C. Williams' Criticism of the Frequency Theory of Probability', *Philosophy and Phenomenological Research* **5**, no. 4: 508–512 (1945)

1946: (a) 'Reply to V. F. Lenzen's Critique', *Philosophy and Phenomenological Research* **6**, no. 3: 487–492 (1946)

(b) 'Reply to Ernest Nagel's Criticism of my Views on Quantum Mechanics', *The Journal of Philosophy* **43**, no. 9: 239–247 (1946)

1947: (a) 'Philosophy: Speculation or Science?', *The Nation* **164**, no. 1: 19–22 (Jan. 4, 1947); reprinted under the title 'The Nature of a Question' in *The Language of Wisdom and Folly*, I. I. Lee, ed. (Harper and Brothers, New York, 1949)

(b) 'The Scientist and Society' (review of Philipp Frank, *Einstein*: *His Life and Times*), *The Nation* **164**, 306–307 (March 15, 1947)

(c) *Elements of Symbolic Logic* (Macmillan Co., New York, 1947). For German trans. see [1977a], vol. 6, 444 pp.

1948: (a) 'Rationalism and Empiricism: An Inquiry into the Roots of Philosophical Error' [Presidential address, Pacific Division meeting, APA, Dec. 30, 1947], *The Philosophical Review* **57**, no. 4: 330–346 (1948); reprinted in [1959a] For German tr., see [1977a]

(b) 'Philosophy and Physics' [Faculty Research Lecture delivered March 25, 1946] (Univ. of California Press, Berkeley and Los Angeles, 1948) 13 pp.

(c) 'A Reply to a Review', *The Journal of Philosophy* **45**, no. 17: 464–467 (1948)

(d) 'The Principle of Anomaly in Quantum Mechanics', *Dialectica* **2**, nos. 3–4: 337–350 (1948); reprinted in *Readings in the Philosophy of Science*, H. Feigl and M. Brodbeck, eds. (Appleton-Century-Crofts, New York, 1953) pp. 509–520.

(e) 'Theory of Series and Gödel's Theorems' (unpublished mimeographed manuscript; first publication in [1978 a-41])

1949: (a) 'The Philosophical Analysis of Quantum Mechanics', *Library of the 10th International Congress of Philosophy*, August 11–18, 1948, Amsterdam, vol. 1 (North-Holland, Amsterdam, 1949) pp. 921–922

(b) 'The Philosophical Significance of the Theory of Relativity', in *Albert Einstein: Philosopher-Scientist*, P. A. Schilpp, ed. (The Library of Living Philosophers, Evanston, Ill., vol. 7, 1949) pp. 287–311; reprinted in *Readings in the Philosophy of Science*, H. Feigl and M. Brodbeck, eds. (Appleton-Century-Crofts, New York, 1953) pp. 195–211, and *Readings in Philosophy of Science*, P. P. Wiener, ed. (Scribners, New York, 1953) pp. 59–76; German trans. Stuttgart, 1955.

(c) 'Philosophical Foundations of Probability', *Proceedings of the Berkeley Symposium on Mathematical Statistics and Probability* (Univ. of California Press, Berkeley and Los Angeles, 1949) pp. 1–20

(d) 'A Conversation between Bertrand Russell and David Hume', *The Journal of Philosophy* **46**, no. 17: 545–549 (1949). For German trans., see [1977a], vol. 5.

(e) 'The Logical Foundations of the Concept of Probability', tr. by M. Reichenbach in *Readings in Philosophical Analysis*, H. Feigl and W. Sellars, eds. (Appleton-Century-Crofts, New York, 1949) pp. 305–323; reprinted in *Readings in the Philosophy of Science*, H. Feigl and M. Brodbeck, eds. (Appleton-Century-Crofts, New York, 1953) pp. 456–474. (English trans. with added footnotes of [1933e])

(f) *The Theory of Probability. An Inquiry into the Logical and Mathematical Foundations of the Calculus of Probability*, tr. by E. H. Hutten and M. Reichenbach (Univ. of California Press, Berkeley and Los Angeles, 1949; California Library Reprint Series No. 23, 1971) (English version and 2nd edition of [1935h])

1950: (a) 'On the Theory of Probability', *Felix Kaufmann: A Memorial* [a special issue of *12th Street*, vol. 3, no. 2, 1950], pp. 11–12

1951: (a) *The Rise of Scientific Philosophy* (Univ. of California Press, Berkeley and Los Angeles, 1951) 333 pp.; German trans. Berlin, 1953, Braunschweig 1968 and [1977a], vol. 1; French trans. Paris, 1955; Spanish trans. Buenos Aires and Mexico City, 1953; Swedish trans. Stockholm, 1957; Italian trans. Bologna, 1958; Japanese trans. Tokyo, 1958; Polish trans.

Ksiazka i Wiedza, Warsaw, 1960; Yugoslav trans. Belgrade, 1964; Korean trans., Seoul, 1960; Marathi Trans., Bombay, 1974. Chapter 14 reprinted in *Contemporary Philosophy, A Book of Readings*, J. L. Jarrett and S. M. McMurrin, eds. (Holt, New York, 1954) pp. 366–376.

(b) 'Why I wrote *The Rise of Scientific Philosophy*', *Book Find News*, no. 102 (Braziller, New York, 1951)

(c) 'The Verifiability Theory of Meaning', *Proceedings of the American Academy of Arts and Sciences* **80**, no. 1: 46–60 (1951); reprinted in *Readings in the Philosophy of Science*, H. Feigl and M. Brodbeck, eds. (Appleton-Century-Crofts, New York, 1953) pp. 93–102

(d) 'On Observing and Perceiving', *Philosophical Studies* **2**, no. 6 (1951) pp. 29–39

(e) 'Foreword', Catalog No. 130, Zeitlin and Ver Brugge [Booksellers], Los Angeles, California (May 1951)

(f) 'The Value of Old Books', *Antiquarian Bookman* (July 7, 1951) p. 2

(g) 'Über die erkenntnistheoretische Problemlage und den Gebrauch einer dreiwertigen Logik in der Quantenmechanik', *Zeitschrift für Naturforschung* (Tübingen) 6a, no. 11: 569–575 (1951) (For English trans. see [1978 b-49]

(h) 'Probability Methods in Social Science', in *The Policy Sciences. Recent Developments in Scope and Method*, D. Lerner and H. D. Lasswell, eds. (Stanford Univ. Press, Stanford, 1951) pp. 121–128

(i) 'The Freedom of the Will' (unpublished manuscript; first publication in [1959 a-7])

1952: (a) 'Are Phenomenal Reports absolutely Certain?', *The Philosophical Review* **61**, no. 2: 147–159 (1952)

(b) 'The Syllogism Revised', *Philosophy of Science* **19**, no. 1: 1–16 (1952)

(c) 'Logical Empiricism. Philosophy: Summaries of a Series of Meetings', *The Humanist*, 7–10 (1952)

(d) 'Les fondements logiques de la mécanique des quanta', *Annales de l'Institut Henri Poincaré* **13**, part 2: 109–158 (1952/53) [Four lectures given at the Institut Henri Poincaré, June 4–7, 1952] (For English trans. see [1978 b-50])

(e) 'On the Explication of Ethical Utterances' (unpublished manuscript; first publication in [1959 a-8])

1953: (a) 'La signification philosophique du dualisme ondes-corpuscules', tr. by O. Costa de Beauregard, in *Louis de Broglie, Physicien et Penseur* (Albin Michel, Paris, 1953) pp. 117–134 (For original English text see [1978 b-51])

1954: (a) 'The Emotive Significance of Time', *Idea and Experiment* **4**, no. 1: 3–9 (1954)

(b) 'Les fondements logiques de la théorie des quanta. Utilisation d'une logique à trois valeurs', *Collection de logique mathématique*, Série A, vol. 5, *Applications scientifiques de la logique mathématique* [Actes du 2e Colloque International de Logique Mathématique, Paris, Aug. 25–30, 1952] (Institut Henri Poincaré, Paris, 1954) pp. 103–114, incl. discussion

(c) (Discussion of) 'Jean-Louis Destouches, La logique et les théories physiques',

Collection de logique mathématique, Série A, vol. 5, *Applications scientifiques de la logique mathématique* [Actes de 2e Colloque International de Logique Mathématique, Paris, Aug. 25–30, 1952] (Institut Henri Poincaré, Paris, 1954) p. 126.

(d) 'Exposé introductif: Remarques sur l'application de la méthode inductive dans la physique', *Collection de logique mathématique*, Série A, vol. 5, *Applications scientifiques de la logique mathématique* [Actes du 2e Colloque International de Logique Mathématique, Paris, Aug. 25–30, 1952] (Institut Henri Poincaré, Paris, 1954) pp. 163–172, incl. discussion.

(e) *Nomological Statements and Admissible Operations* (North-Holland, Amsterdam, 1954) 140 pp. Reissued with new foreword as *Laws, Modalities and Counterfactuals*, see [1976a]

1956: (a) 'Can Operators reach through Quotes?', *Philosophical Studies* 7, no. 3: 33–36 (1956)

(b) *The Direction of Time*, M. Reichenbach, ed. (Univ. of California Press, Berkeley and Los Angeles, 1956; Paper edition, 1971) 280 pp; Spanish trans. Mexico City, 1958; German trans., [1977a], vol. 8

1958: (a) *The Philosophy of Space and Time*, tr. by M. Reichenbach and J. Freund (Dover, New York, 1958) 295 pp. (English trans. of [1928h])

1959: (a) *Modern Philosophy of Science*: *Selected Essays*, ed. and tr. by M. Reichenbach (Routledge & Kegan Paul, London, 1959) 214 pp. Italian trans. Milan, 1966 (1st. 4 articles); Spanish trans. Madrid, 1965. Among the contents are:

1. 'The Present State of the Discussion on Relativity', pp. 1–45, (English trans. of [1922f]; reprinted in [1978 b-44])
2. 'The Theory of Motion According to Newton, Leibniz, and Huyghens', pp. 46–66 (English trans. of [1924d]; reprinted in [1978 b-45])
3. 'Causality and Probability', pp. 67–78 (English trans. of Part III of [1930g]; reprinted in [1978 b-55])
4. 'Aims and Methods of Modern Philosophy of Nature', pp. 79–108 (English trans. of [1931j]; reprinted in [1978 a-38])
5. 'The Principle of Causality and the Possibility of its Empirical Confirmation', pp. 109–134 (English trans. of [1932d]; reprinted in [1978 b-56])
6. 'Rationalism and Empiricism', pp. 135–150 (reprint of [1948a])
7. 'The Freedom of the Will', pp. 151–192 (first publication; reprinted in [1978 a-42])
8. 'On the Explication of Ethical Utterance', pp. 193–198 (first publication; reprinted in [1978 a-43])

1965: (a) *The Theory of Relativity and A Priori Knowledge*, tr. with an introduction by M. Reichenbach (Berkeley and Los Angeles: Univ. of California Press, 1965) 116 pp. (English trans. of [1920f])

1967: (a) 'An Early Appreciation', tr. by M. Reichenbach in *Bertrand Russell – Philosopher of the Century* (George Allen & Unwin, London, 1967) pp. 129–133 (English trans. of [1928b])

1969: (a) *Axiomatization of the Theory of Relativity*, tr. by M. Reichenbach with an introduction by W. C. Salmon (Univ. of California Press, Berkeley and Los Angeles, 1969) 208 pp. (English trans. of [1924h])

1976: (a) *Laws, Modalities, and Counterfactuals*, with a foreword by W. C. Salmon (Univ. of California Press, Berkeley and Los Angeles, 1976) xliii + 140 pp. (expanded reissue of [1954])

1977: (a) Reichenbach's Collected Works are under publication in German, edited by Andreas Kamlah and Maria Reichenbach (Vieweg, Braunschweig, subsequently Wiesbaden, 1977ff.:

Vol. 1 *Der Aufstieg der Wissenschaftlichen Philosophie* [German trans. of *The Rise of Scientific Philosophy* (1951a)], with an introductory essay to the *Gesamtausgabe* by W. C. Salmon and the essay 'Rationalism and Empiricism' [1948a]. (1977)

Vol. 2 *Philosophie der Raum-Zeit-Lehre*, with a brief foreword by R. Carnap to the 1st English trans. [1928h], and the essay 'Zur Einordnung des neuen Einsteinschen Ansatzes über Gravitation und Elektrizität' [1929j]. (1977)

Vol. 3 *Die philosophische Bedeutung der Relativitätstheorie*.
1. *Relativitätstheorie und Erkenntnis Apriori* [1920f]
2. *Axiomatik der relativistischen Raum-Zeit-Lehre* [1924h]
3. 'Die relativistiche Zeitlehre' [1924f]
4. 'Über die physikalischen Konsequenzen der relativistischen Axiomatik' [1925c]
5. 'Planetenuhr und Einsteinsche Gleichzeitigkeit' [1925b]
6. 'Die philosophische Bedeutung der Relativitätstheorie' [German trans. of 1949b]
7. 'Der gegenwärtige Stand der Relativitätsdiskussion' [1922f]
8. 'Die Bewegungslehre bei Newton, Leibniz und Huygens' [1924d]

Vol. 4 *Erfahrung und Prognose* [German trans. of *Experience and Prediction* (1938c) by M. Reichenbach]

Vol. 5 *Philosophische Grundlagen der Quantenmechanik und Wahrscheinlichkeit*
1. *Philosophische Grundlagen der Quantenmechanik* [German trans. of *Philosophic Foundations of Quantum Mechanics* (1944b)]
2. 'Über die erkenntnistheoretische Problemlage und den Gebrauch einer dreiwertigen Logik in der Quantenmechanik' [1951g]
3. 'Der Begriff der Wahrscheinlichkeit für die mathematische Darstellung der Wirklichkeit' [1915b]
4. 'Die physikalischen Voraussetzungen der Wahrscheinlichkeitstheorie' [1920c]
5. 'Philosophische Kritik der Wahrscheinlichkeitsrechnung' [1920e]
6. 'Die logischen Grundlagen des Wahrscheinlichkeitsbegriffs' [1933e]
7. 'Eine Unterhaltung zwischen B. Russell und D. Hume' [German trans. of 'A Conversation between Bertrand Russell and David Hume' (1949d)]

Vol. 6 *Grundzüge der symbolischen Logik* [German trans. of *Elements of Symbolic Logic* (1947c)]

Vol. 7 *Wahrscheinlichkeitslehre* (*Theory of Probability*) [1935h; as revised in 1949f]

Vol. 8 *Kausalität und Zeitrichtung*

1. *Die Richtung der Zeit* [German trans. of *The Direction of Time* (1957b)]
2. 'Die Kausalstruktur der Welt und der Unterschied zwischen Vergangenheit und Zukunft' [1925d]
3. 'Die Kausalbehauptung und die Möglichkeit ihrer empirischen Nachprüfung' [1932d]
4. 'Das Kausalproblem in der Physik' [1931i]
5. 'Kausalität und Wahrscheinlichkeit', Part III [1930g]

Vol. 9 *Wissenschaft und logischer Empirismus*

1. *Gesetzesaussagen und sinnvolle logische Verknüpfungen* [German trans. of *Nomological Statements and Admissible Operations* (1954e)]
2. 'Eine erneute Untersuchung des Syllogismus' [German trans. of 'The Syllogism Revised' (1952b)]
3. 'Bertrand Russell's Logik' [German trans. of 1944a]
4. 'Deweys Wissenschaftstheorie' [German trans. of 'Dewey's Theory of Science' (1939a)]
5. 'Die Verifizierbarkeitstheorie der Bedeutung' [German trans. of 'The Verifiability Theory of Meaning' (1951c)]
6. 'Sind Wahrnehmungsberichte absolut sicher?' [German trans. of 'Are Phenomenal Reports Absolutely Certain?' (1952a)]
7. 'Die Willensfreiheit' [German trans. of 'The Freedom of the Will' (1951i)]
8. 'Über die Explikation ethischer Äusserungen' [German trans. of 'On the Explication of Ethical Utterances' (1952e)]
9. 'Kant und die [moderne] Naturwissenschaft' [1932i]
10. 'Die philosophische Bedeutung der modernen Physik' [1930e]
11. *Ziele und Wege der heutigen Naturphilosophie* [1931j]
12. 'Metaphysik und Naturwissenschaft' [1925a]
13. 'Der physikalische Wahreitsbegriff' [1931g]
14. 'Logistic Empiricism in Germany and the Present State of its Problems' [1936e]
15. Complete Bibliography of the Works of Reichenbach
16. Table of Contents for the Full Collected Works

1978: *Hans Reichenbach, Selected Essays: 1909–1953*, Maria Reichenbach and Robert S. Cohen, eds., principal translations by Elizabeth Hughes Schneewind, further translations by Laurent Beauregard, Sheldon Gilman, Maria Reichenbach, and Gisela Lincoln.

(a): Vol. I

(1) 'The Student', pp. 102–103; English trans. of [1912f]
(2) 'The Student Body and Catholicism', pp. 104–107; English trans. of. [1912c]
(3) 'The Free Student Idea: Its Unified Contents', pp. 108–123; English trans. of [1913c]
(4) 'Why do we Advocate Physical Culture?', pp. 124–128; English trans. of [1913g]
(5) 'The Meaning of University Reform', pp. 129–131; English trans. of [1914c]
(6) 'Platform of the Socialist Students' Party', pp. 132–135; English trans. of [1918b]
(7) 'Socializing the University', pp. 136–180; English trans. of [1918a]
(8) 'Report of the Socialist Student Party, Berlin and Notes on the Program', pp. 181–185; English trans. of [1918c]
(9) 'The Nobel Prize for Einstein', pp. 189–191; English trans. of [1922a]
(10) 'Relativity Theory in a Matchbox: A Philosophical Dialogue', pp. 192–195; English trans. of [1922b]
(11) 'Tycho Brahe's Sextants', pp. 196–200; English trans. of [1926a]
(12) 'The Effects of Einstein's Theory', pp. 201–206; English trans. of [1926d]
(13) 'An Open Letter to the Berlin Funkstunde Corporation', pp. 207–211; English trans. of [1926e]
(14) 'Laying the Foundations of Chemistry: The Work of Marcellin Berthelot', pp. 212–215; English trans. of [1927a]
(15) 'Memories of Svante Arrhenius', pp. 216–218; English trans. of [1927b]
(16) 'A New Model of the Atom', pp. 219–225; English trans. of [1927c]
(17) 'On the Death of H. A. Lorentz', pp. 226–227; English trans. of [1928a]
(18) 'Philosophy of the Natural Sciences', pp. 228–231; English trans. of [1928c]
(19) 'Space and Time: From Kant to Einstein', pp. 232–235; English trans. of [1928d]
(20) 'Causality or Probability?, pp. 236–240; English trans. of [1928e]
(21) 'The World View of the Exact Sciences', pp. 241–244; English trans. of [1928f]
(22) 'New Approaches in Science: Physical Research', pp. 245–248; English trans. of [1929b]
(23) 'New Approaches in Science: Philosophical Research', pp. 249–253; English trans. of [1929c]
(24) 'New Approaches in Science: Mathematical Research', pp. 254–257; English trans. of [1929d]
(25) 'The New Philosophy of Science', pp. 258–260; English trans. of [1929e]
(26) 'Einstein's New Theory', pp. 261–262; English trans. of [1929a]
(27) 'Johannes Kepler', pp. 263–269; English trans. of [1930a]
(28) 'The Present State of the Sciences: The Exact Natural Sciences', pp. 270–272; English trans. of [1930b]
(29) 'One Hundred Against Einstein', pp. 273–274; English trans. of [1931a]
(30) 'Is the Human Mind Capable of Change?, pp. 275–282; English trans. of [1932a]
(31) 'Metaphysics and Natural Science', pp. 283–297; English trans. of [1925a]
(32) 'Bertrand Russell', pp. 298–303; English trans. of [1929i]

(33) 'The Philosophical Significance of Modern Physics', pp. 304–323; English trans. of [1930e]
(34) 'The Königsberg Conference on the Epistemology of the Exact Sciences', pp. 324–325; English trans. of [1930f]
(35) 'The Problem of Causality in Physics', pp. 326–342; English trans. of [1931i]
(36) 'The Physical Concept of Truth', pp. 343–356; English trans. of [1931g]
(37) 'Heinrich Scholz' *History of Logic*', pp. 356–358; English trans. of [1931d]
(38) *Aims and Methods of Modern Philosophy of Nature*, pp. 359–388; reprinted from [1959a]
(39) 'Kant and Natural Science', pp. 389–404; English trans. of [1933b]
(40) 'Carnap's *Logical Structure of the World*', pp. 405–408; English trans. of [1933c]
(41) 'Theory of Series and Gödel's Theorems' (Sections 17–22), pp. 409–428; first publication of the mimeographed text
(42) 'The Freedom of the Will', pp. 431–473; reprinted from [1959a]
(43) 'On the Explication of Ethical Utterances', pp. 474–479; reprinted from [1959a]

(b): vol. II

(44) 'The Present State of the Discussion on Relativity', pp. 3–47; reprinted from [1959a]
(45) 'The Theory of Motion According to Newton, Leibniz and Huyghens', pp. 48–68; reprinted from [1959a]
(46) 'The Relativistic Theory of Time', pp. 69–80; English trans. of [1924f]
(47) 'The Causal Structure of the World and the Difference between Past and Future', pp. 81–119; English trans. of [1925d].
(48) 'The Aims and Methods of Physical Knowledge', pp. 120–225; English trans. of [1929g]
(49) 'Current Epistemological Problems and the Use of a Three-Valued Logic in Quantum Mechanics', pp. 226–236; English trans. of [1951g]
(50) 'The Logical Foundations of Quantum Mechanics', pp. 237–278; English trans. of [1952d]
(51) 'The Philosophical Significance of Wave-Particle Dualism', pp. 279–289; original English text of [1953a]
(52a) 'The Physical Presuppositions of Probability Calculus', pp. 293–309; English trans. of [1920c]
(52b) '*Appendix*: A Letter to the Editor', pp. 310–311; English trans. of [1920d]
(53) 'A Philosophical Critique of the Probability Calculus', pp. 312–327; English trans. of [1920e]
(54) 'Notes on the Problem of Causality' [A Letter of January 25,1924 from Erwin Schrödinger to Hans Reichenbach], pp. 328–332; English trans. of a letter printed in *Erkenntnis* **3**, 65–70 (1932).
(55) 'Causality and Probability', pp. 333–344; reprinted from [1959a]
(56) 'The Principle of Causality and the Possibility of Its Empirical Confirmation', pp. 345–371; reprinted from [1959a] with the brief 'Schlussbemerkung' from [1932d]
(57) 'Induction and Probability: Remarks on Karl Popper's *The Logic of Scientific Discovery*', pp. 372–387; English trans. of [1935e]

(58) 'The Semantic and the Object Conceptions of Probability Expressions', pp. 388–404; English trans. of [1939b]
(59) 'A Letter of March 28, 1949 from Hans Reichenbach to Bertrand Russell', pp. 405–411.

INDEX OF NAMES TO VOLUMES ONE AND TWO*

The separate volumes are indicated by Roman I and Roman II

* 'n' refers to note

VIENNA CIRCLE COLLECTION

1. OTTO NEURATH, *Empiricism and Sociology.* Edited by Marie Neurath and Robert S. Cohen. With a Section of Biographical and Autobiographical Sketches. Translations by Paul Foulkes and Marie Neurath. 1973, xvi + 473 pp., with illustrations. ISBN 90-277-0258-6 (cloth), ISBN 90-277-0259-4 (paper).

2. JOSEF SCHÄCHTER, *Prolegomena to a Critical Grammar.* With a Foreword by J. F. Staal and the Introduction to the original German edition by M. Schlick. Translated by Paul Foulkes. 1973, xxi + 161 pp. ISBN 90-277-0296-9 (cloth), ISBN 90-277-0301-9 (paper).

3. ERNST MACH, *Knowledge and Error. Sketches on the Psychology of Enquiry.* Translated by Paul Foulkes. 1976, xxxviii + 393 pp. ISBN 90-277-0281-0 (cloth), ISBN 90-277-0282-9 (paper).

4. MARIA REICHENBACH and ROBERT S. COHEN, *Hans Reichenbach: Selected Writings, 1909-1953 (Volume One).* 1978, in press. ISBN 90-277-0291-8 (cloth), ISBN 90-277-0292-6 (paper). *Hans Reichenbach: Selected Writings, 1909-1953 (Volume Two).* 1978, in press. ISBN 90-277-0909-2 (cloth), ISBN 90-277-0910-6 (paper). Sets: ISBN 90-277-0892-4 (cloth), ISBN 90-277-0893-2 (paper).

5. LUDWIG BOLTZMANN, *Theoretical Physics and Philosophical Problems. Selected Writings.* With a Foreword by S. R. de Groot. Edited by Brian McGuinness. Translated by Paul Foulkes. 1974, xvi + 280 pp. ISBN 90-277-0249-7 (cloth), ISBN 90-277-0250-0 (paper).

6. KARL MENGER, *Morality, Decision, and Social Organization. Toward a Logic of Ethics.* With a Postscript to the English Edition by the Author. Based on a translation by E. van der Schalie. 1974, xvi + 115 pp. ISBN 90-277-0318-3 (cloth), ISBN 90-277-0319-1 (paper).

7. BÉLA JUHOS, *Selected Papers on Epistemology and Physics.* Edited and with an Introduction by Gerhard Frey. Translated by Paul Foulkes. 1976, xxi + 350 pp. ISBN 90-277-0686-7 (cloth), ISBN 90-277-0687-5 (paper).

8. FRIEDRICH WAISMANN, *Philosophical Papers.* Edited by Brian McGuinness with an Introduction by Anthony Quinton. Translated by Hans Kaal (Chapters I, II, III, V, VI and VIII and by Arnold Burms and Philippe van Parys. 1977, xxii + 190 pp. ISBN 90-277-0712-X (cloth), ISBN 90-277-0713-8 (paper).

VIENNA CIRCLE COLLECTION

9. FELIX KAUFMANN, *The Infinite in Mathematics. Logico-mathematical writings.* Edited by Brian McGuinness, with an Introduction by Ernest Nagel. Translated from the German by Paul Foulkes. 1978, xviii + 236 pp. ISBN 90-277-0847-9 (cloth), ISBN 90-277-0848-7 (paper).

10. KARL MENGER, *Selected Papers in Logic and Foundations, Didactics, Economics.* 1978, in press. ISBN 90-277-0320-5 (cloth), ISBN 90-277-0321-3 (paper).

11. HENK L. MULDER and BARBARA F. B. VAN DE VELDE-SCHLICK, *Moritz Schlick: Philosophical Papers Volume I (1909-1922)*. Translated by Peter Heath. 1978, xxxviii + 370 pp. ISBN 90-277-0314-0 (cloth), ISBN 90-277-0315-9 (paper).

12. EINO SAKARI KAILA, *Reality and Experience. Four Philosophical Essays.* Edited by Robert S. Cohen. 1978, in press. ISBN 90-277-0915-7 (cloth), ISBN 90-277-0919-X (paper).